T0145366

Advances in Information Security

Volume 102

The purpose of the *Advances in Information Security* book series is to establish the state of the art and set the course for future research in information security. The scope of this series includes not only all aspects of computer, network security, and cryptography, but related areas, such as fault tolerance and software assurance. The series serves as a central source of reference for information security research and developments. The series aims to publish thorough and cohesive overviews on specific topics in Information Security, as well as works that are larger in scope than survey articles and that will contain more detailed background information. The series also provides a single point of coverage of advanced and timely topics and a forum for topics that may not have reached a level of maturity to warrant a comprehensive textbook.

Yassine Maleh • Mamoun Alazab • Imed Romdhani
Editors

Blockchain for Cybersecurity in Cyber-Physical Systems

Editors
Yassine Maleh
Sultan Moulay Slimane University
Sultan Moulay, Morocco

Mamoun Alazab
Charles Darwin University
Canberra, ACT, Australia

Imed Romdhani
Edinburgh Napier University
Edinburgh, UK

ISSN 1568-2633 ISSN 2512-2193 (electronic)
Advances in Information Security
ISBN 978-3-031-25508-3 ISBN 978-3-031-25506-9 (eBook)
https://doi.org/10.1007/978-3-031-25506-9

This Springer imprint is published by the registered company Springer Nature Switzerland AG
The registered company address is: Gewerbestrasse 11, 6330 Cham, Switzerland

Preface

The growth of bitcoin over the past decade has helped spread the word about blockchain BC, but the technology's applications go well beyond bitcoin alone. IoT, CPS, Manufacturing, Supply Chain, etc. are just some other places it finds use.

The inherent properties of blockchain are fueling its rapid ascent in popularity and the development of several industrial applications. Since it is distributed and decentralized, the ledger cannot be altered and is open to scrutiny. A blockchain network can operate without a trusted authority. Peers on the network verify the information recorded in the blockchain ledger. Ultimately, it is safe and reliable, with few vulnerabilities that malicious actors can exploit. These characteristics make blockchain technology a game changer in information technology in the twenty-first century.

Due to its operating principle, BC technology meets the challenges mentioned above by CPS. In these systems, protection against data manipulation, cyber attacks, and data integrity are significant areas in which BC technology has significant advantages over other solutions. Blockchain presents a new, practical way to address this issue with data storage and exchange. The tremendous success of cryptocurrencies has indeed helped bring blockchain to the forefront. However, the same technology that enables the anonymity and security of cryptocurrencies like Bitcoin and Ethereum might be used to protect against cybercrime, identity theft, and other forms of financial fraud.

This book brings together the latest research results on blockchain technology and its application to cybersecurity in CPS. It will address important issues in the field and provide a sample of recent advances and insights into the research progress and practical use of blockchain technology to address cybersecurity and cyber threat challenges and issues. In this regard, this book provides readers with a good foundation for the fundamental concepts and principles of blockchain-based cybersecurity of cyber-physical systems, guiding them through the core ideas with expert ease.

As far as we know, this will be the first book reference covering the application of blockchain technology for cybersecurity in cyber-physical systems (CPS). This book will assist decision-makers, managers, professionals, and researchers develop a new CPS paradigm by presenting many new models, practical solutions, and technical breakthroughs connected to cyber-physical systems and motivating blockchain technology for such systems. To give readers a straightforward collection of chapters, we will introduce various architectures, frameworks, models, implementation scenarios, and a wide range of use cases in various CPS fields. In addition, we will explore and showcase the latest innovations and emerging trends in blockchain technology as it relates to cybersecurity in CPS. For those in the cybersecurity field that are interested in Blockchain for CPS security in a more general sense, this book is a valuable resource. It provides a thorough whitepaper as a breakthrough in blockchain-enabled cybersecurity for cyber-physical systems. The well-rounded approach and careful arrangement of this book should help its readers grasp blockchain technology for cybersecurity and associated CPS technologies and methods.

We hope you enjoy reading this book!

Sultan Moulay, Morocco — Yassine Maleh
Canberra, ACT, Australia — Mamoun Alazab
Edinburgh, UK — Imed Romdhani

Contents

Editors and Contributors

About the Editors

Yassine Maleh is an Associate Professor at the National School of Applied Sciences at Sultan Moulay Slimane University, Morocco. He received his Ph.D. degree in Computer Science from Hassan 1st University, Morocco, in 2017. He is a cybersecurity and Information Technology researcher and practitioner with industry and academic experience. He worked for the National Ports Agency in Morocco as an IT manager from 2012 to 2019. He is a Senior Member of IEEE and a Member of the International Association of Engineers IAENG and Machine Intelligence Research Labs. Dr Maleh has made contributions to information security and privacy, Internet of Things Security, and Wireless and Constrained Networks Security. His research interests include Information Security and Privacy, Internet of Things, Networks Security, Information systems, and IT Governance. He has published over 140 papers (book chapters, international journals, and conferences/workshops), 20 edited books, and 4 authored books. He is the Editor-in-Chief of the *International Journal of Information Security and Privacy* and the *International Journal of Smart Security Technologies* (IJSST). He serves as an Associate Editor for IEEE Access (2019 Impact Factor 4.098), the *International Journal of Digital Crime and Forensics* (IJDCF), and the *International Journal of Information Security and Privacy* (IJISP). He is a Series Editor of *Advances in Cybersecurity Management*, by CRC Taylor & Francis. He was also a guest editor for many special issues in reputed journals (IEEE, Springer, Elsevier, etc.). He has served and continues to serve on executive and technical program committees and as a reviewer of numerous international conferences and journals such as Elsevier Ad Hoc Networks, IEEE Network Magazine, *IEEE Sensor Journal*, *ICT Express*, and *Springer Cluster Computing*. He was the Publicity Chair of BCCA 2019 and the General Chair of the MLBDACP 19 symposium and ICI2C'21 Conference. He received Publons Top 1% reviewer award for the years 2018 and 2019.

Mamoun Alazab is an Associate Professor in the College of Engineering, IT and Environment at Charles Darwin University, Australia. He received his Ph.D. degree in Computer Science from the Federation University of Australia, School of Science, Information Technology and Engineering. He is a cybersecurity researcher and practitioner with industry and academic experience. Dr Alazab's research is multidisciplinary and focuses on cybersecurity and digital forensics of computer systems including current and emerging issues in the cyber environment like cyber-physical systems and the Internet of Things, by considering the unique challenges present in these environments, with a focus on cybercrime detection and prevention. He looks into the intersection of machine learning as an essential tool for cybersecurity, for example, for detecting attacks, analyzing malicious code or uncovering vulnerabilities in software. He has more than 100 research papers. He is the recipient of a short fellowship from Japan Society for the Promotion of Science (JSPS) based on his nomination from the Australian Academy of Science. He delivered many invited and keynote speeches, 27 events in 2019 alone. He convened and chaired more than 50 conferences and workshops. He is the founding chair of the IEEE Northern Territory Subsection: (Feb 2019–current). He is a Senior Member of the IEEE, Cybersecurity Academic Ambassador for Oman's Information Technology Authority (ITA), and Member of the IEEE Computer Society's Technical Committee on Security and Privacy (TCSP). He has worked closely with government and industry on many projects, including IBM, Trend Micro, the Australian Federal Police (AFP), the Australian Communications and Media Authority (ACMA), Westpac, UNODC, and the Attorney General's Department.

Imed Romdhani is a full-time Associate Professor in Networking at Edinburgh Napier University since June 2005. He was awarded his Ph.D. from the University of Technology of Compiegne (UTC), France, in May 2005. He also holds Engineering and a Master's degree in Networking obtained respectively in 1998 and 2001 from the National School of Computing (ENSI, Tunisia) and Louis Pasteur University (ULP, France). He worked extensively with Motorola Research Labs in Paris and authored 4 patents.

Contributors

Nawal Ait Aali National Institute of Posts and Telecommunication, Rabat, Morocco

Toheeb A. Adeleke Computer Engineering Department, Ladoke Akintola University of Technology, Ogbomoso, Nigeria

Mohammed M. Alani Seneca College of Applied Arts and Technology, Toronto, Canada

Sunday Adeola Ajagbe Department of Computer & Industrial Production Engineering, First Technical University, Ibadan, Nigeria

Kamorudeen A. Amuda Computer Science Department, University of Ibadan, Ibadan, Nigeria

Teresa Arauz University of Seville, Seville, Spain

Amine Baina National Institute of Posts and Telecommunication, Rabat, Morocco

Joseph Bamidele Awotunde University of Ilorin, Ilorin, Nigeria

Abdul Barek Kennesaw State University, Kennesaw, GA, USA

Kamel Barkaoui Centre d'études et de recherche en informatique et communications, Conservatoire National des Arts et Métiers, Paris, France

Aswani Kumar Cherukuri Vellore Institute of Technology, Vellore, India

Abdelghani Chibani Laboratoire Images, Signaux et Systmes Intelligents (LISSI), University Paris-Est Créteil, Créteil, France

Mauro Conti Department of Mathematics, University of Padua, Padua, Italy

Juan-Manuel Corchado University of Salamanca, Salamanca, Spain

Vinamra Das Vellore Institute of Technology, Vellore, India

Md Jobair Hossain Faruk Kennesaw State University, Kennesaw, GA, USA

Ana-Belén Gil-González University of Salamanca, Salamanca, Spain

Kamini Girdhar Pandit Deendayal Energy University, Gandhinagar, Gujarat, India

Qin Hu Indiana University Purdue University, Indianapolis, IN, USA

Keshanth Jude Jegathees York St John University, York, UK

Harshit Jha Department of Information Technology, Manipal University Jaipur, Jaipur, Rajasthan, India

Annapurna Jonnalagadda Vellore Institute of Technology, Vellore, India

Firuz Kamalov Canadian University Dubai, Dubai, United Arab Emirates

Hossain Kordestani Department of Research and Innovation, Maidis SAS, Chatou, France

Yogesh Kumar Pandit Deendayal Energy University, Gandhinagar, Gujarat, India

José M. Maestre University of Seville, Seville, Spain

Yassine Maleh University Sultan Moulay Slimane, Beni Mellal, Morocco

Emilio Marín University of Seville, Seville, Spain

Yassine Mekdad Department of Electrical and Computer Engineering, Florida International University, Miami, FL, USA

Yeray Mezquita University of Salamanca, Salamanca, Spain

Jai Mishra Department of Information Technology, Manipal University Jaipur, Jaipur, Rajasthan, India

Roghayeh Mojarad Laboratoire Images, Signaux et Systmes Intelligents (LISSI), University Paris-Est Créteil, Créteil, France

Soufyane Mounir University Sultan Moulay Slimane, Beni Mellal, Morocco

Ankit Mundra Manipal University Jaipur, Jaipur, Rajasthan, India

Natasha Nigar Department of Computer Science (RCET), University of Engineering and Technology, Lahore, Pakistan

Ehsan Nowroozi Faculty of Engineering and Natural Sciences, Sabanci University, Istanbul, Turkey

Michael ODea York St John University, York, UK

Yetunde J. Oguns The Polytechnic Ibadan, Ibadan, Nigeria

Kazeem M. Olagunju Computer Engineering Department, Ladoke Akintola University of Technology, Ogbomoso, Nigeria

Karim Ouazzane Metropolitan University in London, London, UK

Javier Prieto University of Salamanca, Salamanca, Spain

Bilash Saha Kennesaw State University, Kennesaw, GA, USA

Erkay Savaş Faculty of Engineering and Natural Sciences, Sabanci University, Istanbul, Turkey

Seyedsadra Seyedshoari Department of Mathematics, University of Padua, Padua, Italy

Hossain Shahriar Kennesaw State University, Kennesaw, GA, USA

Chityanj Sharma Department of Information Technology, Manipal University Jaipur, Jaipur, Rajasthan, India

Chamkaur Singh Pandit Deendayal Energy University, Gandhinagar, Gujarat, India

Manuel Sivianes University of Seville, Seville, Spain

Aminu Bello Usman York St John University, York, UK

Wagdy Zahran Department of Research and Innovation, Maidis SAS, Chatou, France

Cryptocurrency Wallets: Assessment and Security

Ehsan Nowroozi, Seyedsadra Seyedshoari, Yassine Mekdad, Erkay Savaş, and Mauro Conti

1 Introduction

Since bitcoin's introduction, the number of cryptocurrencies has increased to thousands, with a market valuation of over $1.72 trillion as of March 2022 [1]. Meanwhile, the percentage of people owning or using cryptocurrencies is growing significantly in many countries as shown in Fig. 1. A digital wallet as a software application for cryptocurrencies, keeps private/public keys and properly operates on various blockchains, allowing users to transfer currencies to each other and monitor their currency balance eliminating need for a physical wallet [27]. Blockchain-based cryptocurrencies are built on the blockchain concept, which is a decentralized open database with entries that may be verified but not modified [17]. Various currencies could be stored, sent, and received using a digital wallet. Within the wallet, cryptocurrencies are not kept like real money. The blockchain captures and archives every transaction [18]. A wallet transaction involves sending currency

E. Nowroozi (✉)
Computer Engineering Department, Faculty of Engineering and Natural Sciences, Bahçeşehir University (Bahcesehir University), Istanbul, Turkey
e-mail: ehsan.nowroozi@eng.bau.edu.tr

S. Seyedshoari · M. Conti
Department of Mathematics, University of Padua, Padua, Italy
e-mail: seyedsadra.seyedshoari@studenti.unipd.it; mauro.conti@unipd.it

Y. Mekdad
Cyber-Physical Systems Security Lab, Department of Electrical and Computer Engineering, Florida International University, Miami, FL, USA
e-mail: ymekdad@fiu.edu

E. Savaş
Faculty of Engineering and Natural Sciences, Sabanci University, Istanbul, Turkey
e-mail: erkays@sabanciuniv.edu

Y. Maleh et al. (eds.), *Blockchain for Cybersecurity in Cyber-Physical Systems*, Advances in Information Security 102, https://doi.org/10.1007/978-3-031-25506-9_1

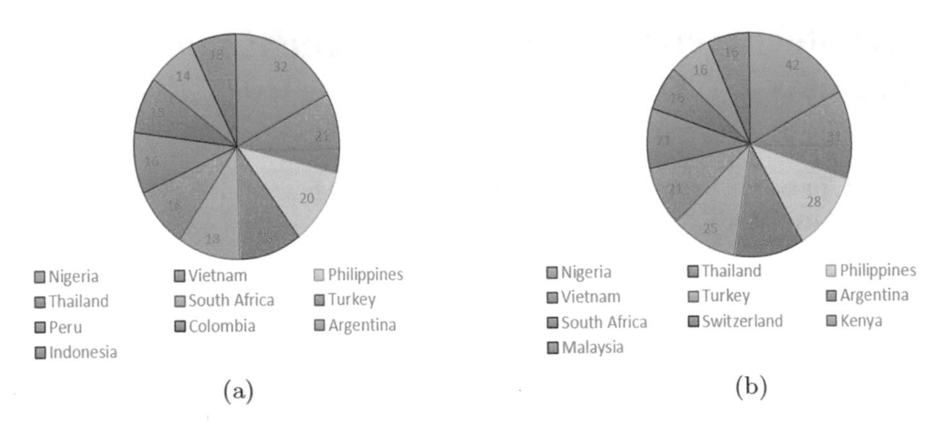

Fig. 1 The percentage of people owning or using cryptocurrencies in different countries. (**a**) Shows the percentage in 2020, and (**b**) Shows the percentage in 2021 [3]

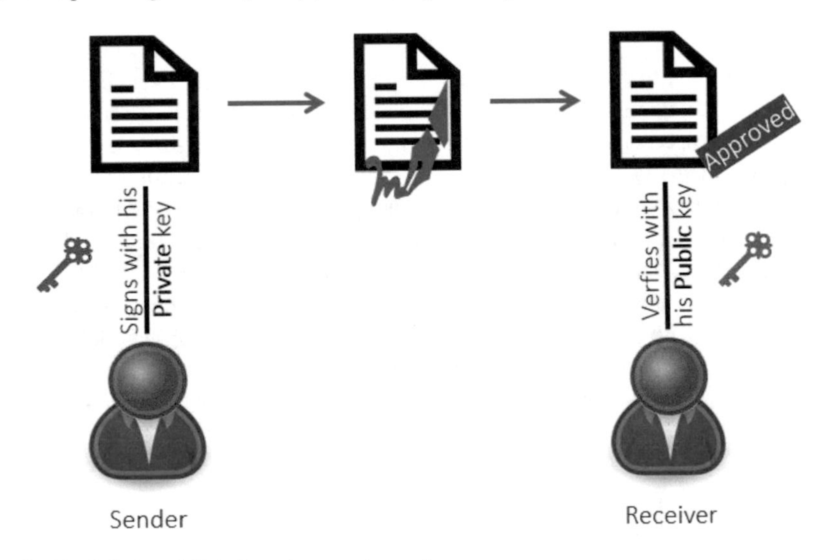

Fig. 2 Blockchain public/private key cryptography

between two addresses. Private key of the sender and public key of the receiver are required for a transaction to take place [2]. Any quantity of coins owned by the sender can be transferred to public key (or address) of the receiver. To verify that the transaction was started and performed by the sender, it digitally signs the transactions using its private key as shown in the Fig. 2. The mainnet includes both the sender and the recipient, is where transactions take place. There is a separate network called testnet that is utilized for testing, however testnet coins have no actual worth. Users cannot transmit cryptocurrencies between the mainnet and the testnet since they are two independent networks. In principle, bitcoin wallet applications establish new addresses, securely keep private keys, and assist in the automating of transactions. Several wallets only accept one type of cryptocurrency (for instance, Bitcoin), whereas others such as Exodus and Jaxx support a wide variety of cryptocurrencies.

Cryptography is a field of study that deals with above mentioned keys. Cryptocurrency consists of two security approaches, symmetric and asymmetric. First one has a secret key whereas the second model contains public and private key. Encryption and decryption in symmetric mode are simply done by utilizing traditional symmetric encryption techniques like Data Encryption Standard (DES), where the same key is used for encryption and decryption [10].

Asymmetric encryption often used in cryptocurrency exchanges, is an encryption technique that employs two keys (public and private key) paired with distinct encryption and decryption methods. Although the sender must have a duplicate of the recipient's public key in order to transmit a coin, it should be assumed that the adversary has the exact copy. In this case, the sender encrypts the message with the proper encryption mechanism, which the recipient of the message may decode using its private key. The purpose of the asymmetric approach is to prevent an attacker from utilizing the public key to decrypt an encrypted communication [18, 33].

1.1 Crypto Wallets' Categories

Digital wallets based on their features like online/offline working mode, can be divided in two categories: hot and cold wallets. Their main distinction is that a connection to the internet is required for hot wallets, whilst cold wallets do not. Users of a hot wallet typically use it to do online purchase and for that reason, users should not allocate a large amount of money, but a cold wallet functions similarly to a bank vault for storing various digital assets. It's advisable to have both wallets, mostly for security purposes [18].

There are some different types of hot and cold wallets. Desktop wallets, hardware wallets, mobile wallets, online wallets, and paper wallets are all available. Hot wallets include multisignature, desktop, mobile, and online wallets, whereas cold wallets include paper and hardware wallets. Each cryptocurrency wallet has its own level of safety and privacy to ensure that the private key is kept securely.

Each kind of digital wallets and their advantages and disadvantages have been described as follows [18]:

Mobile Wallets Mobile wallets are more efficient and simple than using other kinds of crypto wallets since they can be accessed from anywhere via an internet connection. Despite the fact that new mobile wallets take advantage of the security mechanisms of smartphones like ARM TrustZone to protect users [15], it is susceptible to viruses and hacking. This method allows users to use the TOR network for increased security. TOR is a common anonymous communication network with low rate of delay allows users to connect online resources without disclosing their network id [29]. Another fantastic feature is the ability to scan QR codes. Mobile phones, on the other hand, could be considered as unsafe equipment.

As a result, if the phone gets compromised, the user's crypto tokens may be lost. They are also vulnerable to viruses, malwares, and key loggers.

Online Wallets This form of wallet may be accessed through any web browser without need to download or install any software. Since these wallets are susceptible, it is not suggested that users keep a high number of cryptocurrencies in them. Cryptocurrency transfers are conducted in a timely manner. It is advised to hold a little quantity of cryptocurrencies in these wallets.

Several number of this digital wallet are capable of holding multiple cryptocurrencies as well as transferring funds amongst them. It allows customers to use the TOR network for more confidentiality and privacy. However, a third entity or centralized administration has complete control over the digital wallet. It is suggested to utilize a personal computer (PC) with a necessary security application pre-installed in order to use an online wallet. Users are vulnerable to different online scams due to a lack of awareness in information technology (IT).

Desktop Wallets Desktop wallets are assumed to be more secure than mobile and online wallets, however this might vary depending on the level of protection for online crypto wallets' security. Although a desktop wallet can create addresses for receiving cryptocurrency offline, it requires the use of the internet to send them out. Transaction logs will not be refreshed if there is no internet connection [30]. Although these wallets are simple to use and keep the private keys on the user's device, a machine connected to the internet becomes insecure and demands additional protection and security. Furthermore, frequent backups are required because the system may fail at any time, resulting in the loss of all data. Otherwise, the user needs to export the related private key or seed phrase. As a result, users will be able to access digital content on several devices [28].

Multisignature Wallets Based on level of protection, two or three private keys are required to access money after conducting a transfer using a multisignature wallet. This method is beneficial to businesses because it allows them to delegate responsibility to many staff, who must all provide their own private key in order to access assets. BitGo is an instance of a multisignature wallet, where the users store first key, a trusted third party stores the second key, and the firm itself stores the third key. Transactions might be slow due to the number of signatures required. Multisignature relies on the signing of the transaction by additional devices or a third party.

Paper Wallets This is one of the most secure wallets available. They fall in the category of cold crypto wallets. A paper wallet, as the name implies, is a printed sheet of paper containing both private and public keys.

A QR code is printed on the paper, which indicates the keys of the user and may be used for almost any kind of transaction. The user's principal attention should be retaining that paper safe, as the result this wallet is the safest. They are kept in physical wallet or pocket of the users without requiring a connection to the computer; however, the transaction takes longer to be completed.

Fig. 3 Ledger Nano X, portable hardware wallet [4]

Hardware Wallets These wallets are specialized devices of cryptography for generating, storing the private keys and authenticating transactions [24]. In most instances, they are safer wallets because transaction signing occurs on the hardware wallet, and the private key does not leave the safe hardware wallet system, it prohibits malware from stealing digital wallets [31]. Hardware wallet is commonly a USB flash memory (Fig. 3) with software installed and ready to use. Some of these devices contain a screen, allowing the user to conduct a transaction without the need of a computer. This kind of wallet provides the user with additional control over their cryptocurrency and is an appropriate option for long-term storage of crypto assets. The majority of secured USB wallets have a screen. They are safer than all other sorts of digital wallets. They are, however, quite tough to get and are not suggested for novices.

A comparison of different wallets is provided in Table 1.

Users who intend to trade in several currencies may consider multi-currency wallets. Although Bitcoin is the most well-known currency, there are large number of other cryptocurrencies on the market, each with its own infrastructure network [5].

1.2 Available Digital Wallets

It is important to keep in mind that the cryprtocurrency is outlawed or restricted in certain states and countries prior to deciding on a digital currency, while its usage and exchange is permitted in others. It is advisable to select a multi-currency wallet which supports several cryptocurrencies [28]. It is possible to lose money by selecting the incorrect wallet for a certain digital currency. Users should spend some time learning about the various types of crypto currency wallets and their functionality. In this section some of most common wallets are listed as follows: Exodus (online wallet), Coinpayments (online wallet), Ledger Nano S (hardware wallet), Jaxx (mobile wallet), and Ledger Blue (hardware wallet) [18].

Table 1 Comparison of different categories of cryptowallets

Wallet	Advantages	Disadvantages
Mobile wallets	– Efficiency and simplicity of use - Supporting TOR network – Using QR code	– Loss of crypto tokens due to compromising the phone – Prone to key logger, viruses, and malware
Online wallets	– Fast transactions – Supporting TOR network – Supporting multiple cryptocurrencies and transactions between them	– Fully controlled by central authorities or third party – Demanding a personal computer and installing specific application
Desktop wallets	– Simplicity of use – Storing private key on user's system	– Susceptible and requiring more security – Regular backup required
Multisignature wallets	– Dedicating responsibility to employees of a company	– Slow transactions
Paper wallets	– Kept in user's packet or physical wallet	– Slow transactions
Hardware wallets	– LCD screen on USB wallets – Safer than others	– Hard to purchase – Not suggested for beginners

Exodus is a web-based electronic wallet with a user-friendly interface shown on Fig. 4, a stylish design, and a reporting mechanism. When compared to other online wallets, Exodus provides comparable functionality, with some being better than others. In this type of wallets registration is free of charge so that anyone may submit the form and become the owner of a cryptowallet of this type. The cryptocurrency swap, where users can trade several crypto assets without incurring extra charges, is one of its best features It is a fantastic place for inexperienced traders. Although it is an online wallet, it is also an offline wallet since the data is kept on the computer of user when the wallet is generated [18].

Coinpayments is a digital wallet that can be accessed online. They become popular after proving that their wallet could hold at least 300 various digital currencies as illustrated on Fig. 5. They only get paid when a user finalizes a transaction using their wallet. Because this wallet accepts multiple currencies and is accepted by so many online retailers, it is feasible to shop online using this wallet. The BitGo services have been integrated within this wallet to provide a higher level of security and transaction speed. Moreover, safety function is added to keep money of the users safe from criminals. This wallet allows users to store several currencies in the same place. In addition, a lot of online retailers utilize it for online shopping.

Ledger Nano S is a digitized USB wallet for cryptocurrencies introduced in 2016. Due to the fact that hardware wallets are significantly more expensive than other digital wallets, but they are a cost-effective investiture with a variety of capabilities such as enabling the users to securely trade and monitor digital assets as well as supporting more than 1100 cryptocurrencies and tokens [6].

Fig. 4 Exodus platform

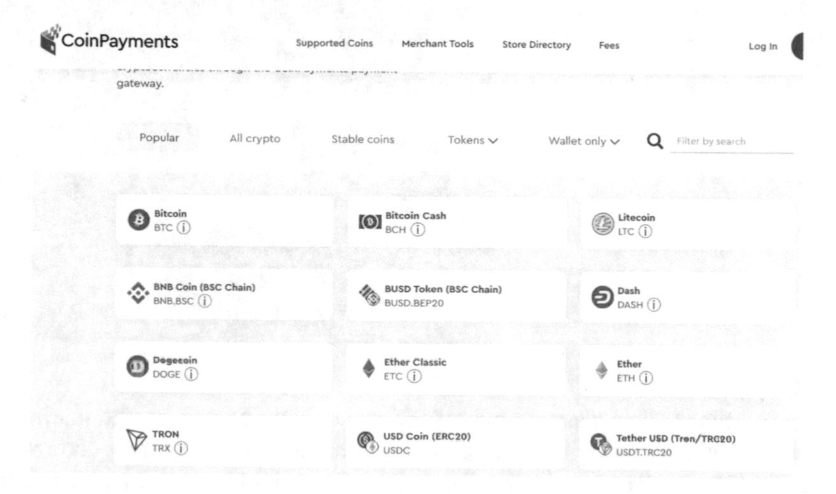

Fig. 5 Coinpayments platform

The private key's backup and the security are given special consideration. This gadget can be started without the need of a computer. It includes a little LCD screen on the front of the USB so that the users can use it easily as shown on Fig. 6. It makes possible moving money from one account to another as well as exchanging cryptocurrencies. There are two sizes of Ledger Nano S, the larger device is 98 mm, while the smaller device is 60 mm. This wallet can hold a variety of famous cryptocurrencies. The user may keep an eye on current transactions and utilize the button to double-chseck them. Several security features are available, as well as the possibility to lock the wallet using a password. Regardless of how little the gadget is, it can be conveniently utilized by the users.

Ledger Blue is also a hardware wallet designed by the same company. It outperforms the Ledger Nano S and adds plenty of additional features which could

Fig. 6 Ledger Nano S device

Fig. 7 Ledger Blue device [7]

be seen on Fig. 7. This wallet is one of the most costly wallets on the market due to these qualities. To prohibit external access, the users can specify a code with 4 to 6 digits. The Ledger blue wallet uses dual chip technology and includes built-in software for digital currency safety. It is completely immune to harmful malware which means attackers cannot hack it.

Jaxx Liberty is one of the mobile and web digital wallets (illustrated on Fig. 8) that may also be referred to as a desktop wallet since it operates on both Windows and mobile platforms that allows the user to trade digital assets using third-party services like Changelly as shown on Fig. 9. For all digital assets, Jaxx was intended to keep them safe from cybercriminals. New mobile wallets offer a variety of security measures if a user's phone is lost. If this is the case, they'll let users swap accounts. Jaxx is compatible with all main operating systems such as Android, IOS, Windows, Linux, and Mac OS. Jaxx enterprise is not able to view the user's digital currency since a private key is produced and saved on the computer of the user. In most cases, making a transaction with an online wallet requires a number of procedures. The Jaxx concept is focused on the Nada privacy model. Nada is responsible of protecting confidentiality and privacy.

Fig. 8 Jaxx Liberty platform

Fig. 9 Changelly exchange platform

Main features of discussed wallets are summarized in Table 2.

2 Overview of Digital Wallets' Security

The crypto wallets' security goals, include availability, integrity, and confidentiality, are compatible with most security standards. Adversary makes use of vulnerabilities in wallet libraries to create a distinctive impression of wallet finger that is linked to the user's identity for further monitoring. Although, Android and iOS provides a variety of tools for programmers and customers, some of these features can be

Table 2 Main features of some of most common wallets

Wallet	Main features
Exodus	**Safe**: Information are saved on user's system while being created
	Multi-currency: Supports diverse currencies in the same wallet
	Free registration: Possessing this wallet by simply filling out a form
Coinpayments	**Safe**: Money of user is protected against stealing
	Multi-currency: Supports diverse currencies in the same wallet
	Integrated with BitGo services: Increases speed and security of transactions
	Common: Used by thousands of online shops
Ledger Nano S	**User-friendly**: Could be used comfortably
	Multi-currency: Supports diverse currencies in the same wallet
	Small screen: Monitoring current transactions and confirming them by a button
	Backup and recovery: Fast restoring process if digital money is lost
	Safe: Provides many security features like a password to lock the wallet
Ledger Blue	**Pin code**: Restricts external access using a password with 4-6 characters
	Resistant to malicious software: Cannot be violated by malwares
	Safe: Benefits from dual chip design and includes a firmware for security
Jaxx	**Acceptable**: Could be implemented on any OS
	Easily operated: Does not need a lot of steps to execute a transaction
	Full control: Private key is stored and accessed only on user's computer

abused by hackers to violate the security of the framework of cryptocurrencies that runs on the platform [32].

There are many data transmission functions in the bitcoin cryptocurrency, the most prominent crypto money. These features may pose a security risk, but Bitcoin has a very strong security system, implying that they should be used correctly. The security of the platform should be a priority while investing in online platforms. When purchasing a wallet for this digital money, two-factor verifications is suggested which is a process of authentication that requires two assets of the users including what they know such as login credentials and what they have like a mobile phone to receive one-time password (OTP) [9]. In comparison to a physical wallet, a smart approach of storing money in the wallet is possible. This implies that a little quantity of currency should be kept in the digital cryptowallet for daily utilization.

Backup wallets can help eliminate issues such as information theft or errors of the computer; however, this condition can only be met if the data has been encoded. The security of data saved on the network is not completely guaranteed. Malicious softwares are able to infect any machine connected to the internet. Encryption of the data is essential to eliminate any risk of being compromised, which is a vital safety measure. Data should be kept in a multitude places like a backup wallet. It's not just about cloud storage when it comes to various places, but also regarding physical devices like CD, external hard disk, USB, and so on. Daily or frequent backups guarantee that the data is constantly updated. When it comes to digital wallets, encryption plays a critical role. Thus, encrypting user's cryptycurrency wallet is a quite effective method to protect the money saved within that wallet. When someone attempts to enter the digital wallet, a password is created. The password should not be forgotten or lost, as this would result in the loss of funds. The distinction between actual money and cryptocurrency is that if a user's password is lost, may obtain a new one. The user has complete responsibility in cryptocurrency and blockchain. It's critical to combine characters, numbers, and letters to set up a secure password.

2.1 Cryptocurrency Wallet's Backup

A backup wallet is simply another term for transferring money to another location or producing a replica [18]. There should be two wallets during the backup process: primary wallet and backup wallet which is working offline [25]. To keep safe cryptocurrencies, encrypted local backups of the funds can be saved in a hardware wallet that is not connected to the internet (backup wallet). Existing hardware wallets' backup and recovery procedure is a significant problem, since most of them employ a word list (mnemonics) to create a duplication of the private keys and restore them while required. These words must be written on a paper and kept secure by the user [24]. This approach pushes the problem outside the wallet by converting the private key's seed from digitized to physical type. Rezaeighaleh et al. [24] proposed a novel framework for backup and recovery process implementing Elliptic-Curve Diffie-Hellman (ECDH) algorithm, which could be easily used by users since they no long need to write the word list and save it.

2.2 Cryptocurrency Wallet's Encryption

Encrypting the confidential information and digital money has always been a robust and reliable method of protecting. In cryptocurrency, hashing is a way of transforming huge amount of data into little digits. It is used in the Bitcoin network for encoding the wallet's address, encoding transactions between two wallets, confirming the balance in a specific wallet. The Bitcoin network employs a safe hash algorithm such as SHA-256. One of the most important features of this technique is that modification of one bit of incoming data will totally alter the output. This is related to the Avalanche effect, which reflects the behavior of traditional mathematical algorithms such as Data Encryption Standard (DES) and Advance Encryption Standard (AES), in which a little variation in the input causes the entire hash value to alter considerably [22]. A slight modification in the plaintext may result in a large shift in the ciphertext when using symmetric ciphers. Otherwise, an error will appear while decrypting the encrypted text [23]. Public key cryptography is a technique of proving identity of the person by using a pair of cryptographic keys (a private key and a public key). A digital signature is created by combining both keys. The blockchain wallet, which connects with the blockchain network, stores these private keys, public keys, and blockchain addresses as well as keeps track of the coins that could be transmitted via digital signature [32].

Since possession of the private key entails complete control over the related cryptocurrency account, managing private key is critical for security. Before saving in the wallet, the private key must be encoded, and when used, must be decoded into plaintext. The plaintext of the private key in Ethereum, for instance, is a 256-bit binary integer that is usually displayed encoded as a hexadecimal number. Before being saved in the wallet, the private key must be encrypted and then decrypted if needed.

2.3 Cold Wallets as Another Solution

Cold wallets are another solution for storing and protecting data. These are hardware wallets which do not demand an online connection and use a USB stick to transfer transactions and keys [24]. Two computers share some parts of the same digital wallet while signing transactions offline. Only the first computer must be disconnected from all networks, as it is the only one with an entire digital wallet and permission for signing transactions. The other computer is connected to the internet and holds the digital wallet, which can only be used to observe and execute unsigned transactions. Only a few steps are required to complete the transaction:

- **Step 1**: A new transaction should be created on the computer with connection to the internet and save it to a USB device.
- **Step 2**: Transaction must be signed with the computer which does not have internet connection.
- **Step 3**: Signed transaction should be sent with the computer which is connected to the network.

2.4 Cryptocurrency Wallets and QR Code

Ghaffar Khan et al. [21] employed QR codes for cross verification across hot and cold wallets to keep digital currencies. Cold wallets are more safe against cyber-attacks due to their offline nature; This approach is like an additional protection layer of bitcoin transactions [14]. All the cryptocurrency investors should understand the differences between hot and cold wallets in order to ensure safe and secure digital money transactions. Online wallets can send the funds and distribute in network only after confirming private key of the cold wallet scanning the QR code.

The version of the digital wallet application must be upgraded on a regular basis since every time the program is updated, the users receive vital security upgrades. Updates may provide new capabilities for cryptowallets, as well as the prevention of a number of issues with different degrees of intensity. Numerous signatures can be used in crypto wallets, requiring several confirmations before a transaction can be funded. This form of security may be employed in larger businesses like banks with staff who have access to the government coffers. Multi-signature feature is also available in some web wallets such as BitGo, and Coinbase [5].

3 Cryptocurrency Wallets' Security Objectives

Cryptocurrency wallets have security goals that are similar to those of other security structures, including availability, integrity, and confidentiality [17].

Availability The purpose of availability is to guarantee that the legal use of data is not inhibited, which means that the information must be usable and available while demanded by a valid authority. It's critical for wallet applications to make sure that keys can be produced, saved, and retrieved appropriately. In addition, transactions should be properly signed, transmitted, and accessed in response to user queries [17].

The wallets can become unavailable if any failure, overload, or attack occurs. Important features of availability are fail-safe, reliability, scalability, fault-tolerance, up-time, and recoverability. The system could be called fail-safe if the attack or failure has least impact such as data loss. Reliability is known as probability of operating as expected if no outside source attempts to interrupt the system. A scalable system allow increasing the number of available resources without modifying the system architecture. A system can be assumed as a fault-tolerance system if it is able to continue operating properly even with a decreased level of functionality. Up-time is referred to the period of time that the system is actively working and accessible to users. Finally ,the term "recoverability" refers to the ability of a system to recover its data in an acceptable time frame in the event of a breakdown [12].

Integrity Integrity refers to the ability to prohibit illegitimate entities from altering data in order to ensure its completeness and correctness. When it comes to blockchain wallets, ensuring the integrity of the private key is critical. The user will lose his/her account's control if the private key kept in the wallet gets modified or deleted in an illegal way, resulting in the loss of the account's assets. Blockchain has employed cryptographic methods like hashes and signatures to verify that transaction data has not been changed before being transmitted to the blockchian. The integrity feature, on the other hand, is critical for a recently launched transaction. Even if the transaction's data has been altered before being signed by user with the private key, the transaction will be validated by the blockchain system since it carries the signature of the legal owner. It's also possible to tamper with historical transactions once they have been retrieved from the blockchain system.

Confidentiality The goal of confidentiality is to keep sensitive information safe from unwanted access. A digital currency account's private key grants complete control over the account and any digital assets held within it. As a result, the wallet's primary security feature is to guarantee that the private key is not accessible for an illegal manner. Because all the information is publicly available on the blockchain, transaction information is not assumed to be confidential.

4 Cryptocurrency Wallets' Adversary Model

Various sorts of digital money wallets have different adversary models like application oriented adversary model and physical access adversary model [25]. In this section, the adversary model for cryptocurrency wallets based on software has been discussed. Purpose of the adversary is to compromise the availability, integrity, or confidentiality of the wallet's data. This involves tampering with earlier transactions, preventing the initiation of new transactions, accessing the private key, manipulating newly launched transactions, refusing transaction information queries, etc [17].

The attacker lacks private information specific to a target wallet's owner, like the list of wallet transaction passwords or the user's account's private key. The attacker, however, has the potential to install and execute any program that is installed on the same system as the wallet operates. All the permissions requested by the installed program have been granted. Any option on the device where the wallet operates can be changed by the attacker. The attacker can also execute any program on the user's other devices who utilizes the wallet. The wallet's communication can be listened and modified by the adversary, even if they don't have access to the encrypted traffic's key. The servers connected to the wallets can be attacked by the adversary, but blockchain network cannot be controlled by attackers.

The adversary approach described above is realistic since the users might be persuaded to install a new program and then provide it all the necessary permissions. The program can imitate the appearance of a standard program. Furthermore, tactics like accessibility services, USB debugging, as well as other smartphone functions might provide attackers with extra possibilities to exploit [17].

5 Vulnerabilities in Cryptocurrency Wallets

Transaction management and private key management are two of the most important functionalities of cryptocurrency wallets. Transaction management comprises sending and gathering tokens, as well as querying balances and transactions, while key management covers a private key's creating, saving, importing, and exporting; however, if these capabilities are used incorrectly, attack points may be introduced into the attack surfaces. Furthermore, because an operating system (OS) hosts the digital wallet, attacker might be able to exploit the OS's properties, arising a danger to the digital wallet's security [17].

The attack surface from the perspective of the cryptocurrency wallet and its underlying operating system have been discussed in the following.

5.1 Cryptocurrency Wallets' Attack Surface

Transaction Management While a user intends to withdraw money from an account, the wallet creates a transaction and signs it using the user's private key. Then, it sends the signed transaction to the blockchain system for confirmation in order to accomplish operation of the transaction. When a user has to perform a collection process, must present the payer address of its account, that might contain the currency and amount.

Users can access the related account balances and account transaction logs using transaction records of the wallet application and balance inquiry services. This approach may need a connection to server of the wallet devoted to the service, instead of a blockchain network, because certain blockchain systems do not support direct queries of this information.

When sending or receiving money, information of the transaction provided by the user or shown by the wallet might be altered, causing a security risk and potentially resulting in user's money being moved to account of the adversary. If user's password input screen and the keyboard are observed during the money transfer, the encoded password might be thieved, which violates confidentiality. Diao et al. [13], derived unlock pattern of the user and the status of the foreground program without any authorization, revealing the intensity of security weaknesses in the transmission procedure. If an intruder can disrupt the money transfer or query of the balances or transactions by blocking the link between the wallet and its server or the blockchain network, postures a vulnerability to availability and may result in serious operations like the user extracting the private key to gain back administration of the account, resulting in more impairment. While looking up payments and account balances, an attacker also might deceive the users by falsifying the transmitted information between the wallet server and client, displayed data on the wallet, or data kept on the wallet's server. In this case, the wallet's integrity will be compromised, which results in display of incorrect transaction registers or incorrect balances on the wallet, consequently deceiving the users [17].

Key Management If the user has not created a cryptocurrency account, the wallet will randomly produce a couple of private and public keys for a new account on the local device. If the user owns an account, can import the account's private key into the wallet, which enables control the account from the wallet. Then, the created or imported private key gets encrypted by the digital wallet using user's encryption password. The users might lose full control of their account forever if they lose the private key which leads in loss of their funds. As a result, the private key should frequently be extracted for backup purposes.

If the random seed employed for producing a private key can be anticipated or retrieved during the creation process, the created private key is potential to be compromised. If the saved private key gets decrypted or retrieved in plaintext during the storage process, it can be stolen and exploited, putting the confidentiality at risk. Another way of violating the confidentiality occurs when the attacker observes input of the user and gains the key when the users is manually typing or copying and pasting the key. Moreover, the wallet may show information related to the key on its screen while importing and exporting keys so that an attacker could watch the data in order to achieve the key, endangering the confidentiality. Furthermore, when the password for key encryption is configured, an attacker can obtain it by eavesdropping the user's input, posing a danger to confidentiality. On the other hand, the account's integrity and availability may be at risk if a third party can manipulate or remove the saved key [17].

In the following section, some of the security threats against mobile wallets have been discussed.

5.2 Digital Wallet's Common Threats

Inappropriate Usage of Platform Android and Apple IOS, for example, supply a group of functions of host operating system. Abusing these services may cause security risks. All of the host system's services have presented implementation rules, and breaking these instructions is the most typical manner of imposing a recognized threat. For instance, using App local Storage instead of utilizing IOS Keychain to store confidential information in IOS apps. The data stored in app local storage may be exposed to other parts of the program, but the data kept in the Keychain is protected from illegal access by the operating system [26].

Unsafe Data Storage Unintended information disclosures and risky data storage fall under this category. If an attacker obtains access to the system, data saved locally in SQL databases and log files may be at risk. External storage of crucial data is recognized as unsafe and can be misused. Detection of unintentional data leaking is not as easy as detection of intentional leaks.

Data leaking might be caused by flaws in rooted devices, hardware, or frameworks. Data leakage vulnerabilities can be exploited in applications that lack sufficient monitoring measures of data leaking.

Inadequate Cryptography Cryptographic functions are frequently used in programs that require encryption. Inadequate cryptography can be exploited by two sorts of threats such as weakness in encryption process and damaged cryptography functionalities. The first is gaining access to confidential information by exploiting a flaw in the construction of encryption/decryption procedure. The second risk derives from the use of compromised functions of cryptography.

Reverse Engineering Like data, reverse engineering targets encryption keys and hardcoded passwords. This approach entails extracting source code from a digital wallet as well as numerous resources from an APK file. This attacks can be accomplished only by hackers who have a deep knowledge of digital wallets [16].

Public Wi-Fi Using public Wi-Fi such as in to conduct digital wallet money transfers can allow third parties to disrupt communication and possibly disrupt payment via MITMF, Wi-Fi sniffing, and DNS spoofing [11, 20]. For instance, an attacker could steal sensitive information of users who are connected to a public Wi-Fi such as in cafes.

Social Engineering Instead of breaching or employing practical hacking strategies, social engineering is a technique for gaining control over a computer or information of the users by exploiting human psychology. Attackers might sell the information in black markets or use them to make illegal payments. In addition, they can utilize the obtained information as their identity.

Phishing Attacks This kind of attack is one of the most frequent attacks where phishing link is a type of fraudulent access point that attackers exploit to get critical information and private data from users, such as credit card number, a financial lottery, or SMS. In phishing attacks, attackers try to acquire login information of the user and personal information, putting digital wallet accounts at risk of theft. For example, the Singapore Police Force (SPF) warned people about growth of the phishing attack in recent months and it has observed about 1200 cases from December 2021 till January 2022. In most cases, victims were called via messaging application like WhatsApp. During the conversation they were asked to provide some private information based on belief that the caller is from one of Government agencies [8].

6 Conclusion

Cryptocurrency wallet is a software application or a hardware device provides users the possibility to execute several transactions. Users aiming to buy a digital wallet should recognize their needs and objectives before choosing which type to obtain. Data organization as well as speed, security, and the possibility to execute transaction between two clients are pushing digital wallets in more demand. As these wallets become more popular, security and safety of the wallets become crucial [19]. In this study, we have seen that creating a backup of the private key and also encrypting the digital money using hash functions help diminish privacy and security threats as well as system errors. Employing QR code as cross-verifying cold wallets is another technique of keeping digital currencies safe. Security of digital wallets has the same objectives of other security systems including availability, integrity, and confidentiality. Moreover, the adversary model for cryptocurrency wallets has been discussed in this study where the adversary or attacker aims to violate security objectives of the digital wallets. Transaction management and key management as two principal features of crypto wallets provide several functionalities such as sending and collecting the tokens, creating and saving the private key. Exploiting these capabilities by attackers may vulnerabilities to blockchain-based wallets. It's critical to reinforce cryptocurrency wallets with the system's updated security standards, avoid infection of application supply chain, and mitigate repackaging threats in order to ensure wallet security.

References

1. https://coinmarketcap.com/, accessed: 2022-03-13
2. https://www.bitcoin.com/get-started/how-bitcointransactions-work, accessed: 2022-03-13
3. https://www.statista.com/statistics/1202468/global-cryptocurrency-ownership/, accessed: 2022-03-12
4. https://www.criptovaluta.it/hardware-wallet, accessed: 2022-03-11
5. https://hobowithalaptop.com/crypto-wallets, accessed: 2022-03-11
6. https://www.investopedia.com/ledger-nano-s-review-5190302, accessed: 2022-03-16
7. https://www.ledger.com/ledger-blue-an-enterprise-grade-security-device, accessed: 2022-03-12
8. https://www.channelnewsasia.com/singapore/police-warn-phishing-scams-2433296, accessed: 2022-03-15
9. Ali, G., Ally Dida, M., Elikana Sam, A.: Two-factor authentication scheme for mobile money: A review of threat models and countermeasures. Future Internet **12**(10), 160 (2020)
10. Aydar, M., Cetin, S.C., Ayvaz, S., Aygun, B.: Private key encryption and recovery in blockchain. arXiv preprint arXiv:1907.04156 (2019)
11. Bosamia, M.P.: Mobile wallet payments recent potential threats and vulnerabilities with its possible security measures. In: Proceedings of the 2017 International Conference on Soft Computing and its Engineering Applications (icSoftComp-2017), Changa, India. pp. 1–2 (2017)
12. Chaeikar, S.S., Jolfaei, A., Mohammad, N., Ostovari, P.: Security principles and challenges in electronic voting. In: 2021 IEEE 25th International Enterprise Distributed Object Computing Workshop (EDOCW). pp. 38–45 (2021)
13. Diao, W., Liu, X., Li, Z., Zhang, K.: No pardon for the interruption: New inference attacks on android through interrupt timing analysis. In: 2016 IEEE Symposium on Security and Privacy (SP). pp. 414–432 (2016)
14. Dikshit, P., Singh, K.: Efficient weighted threshold ecdsa for securing bitcoin wallet. In: 2017 ISEA Asia Security and Privacy (ISEASP). pp. 1–9 (2017)
15. Gentilal, M., Martins, P., Sousa, L.: Trustzone-backed bitcoin wallet. In: Proceedings of the Fourth Workshop on Cryptography and Security in Computing Systems. pp. 25–28 (2017)
16. Hassan, M.A., Shukur, Z.: Review of digital wallet requirements. In: 2019 International Conference on Cybersecurity (ICoCSec). pp. 43–48 (2019)
17. He, D., Li, S., Li, C., Zhu, S., Chan, S., Min, W., Guizani, N.: Security analysis of cryptocurrency wallets in android-based applications. IEEE Network **34**(6), 114–119 (2020)
18. Jokić, S.: Analysis and security of crypto currency wallets. ZBORNIK RADOVA UNIVERZITETA SINERGIJA **19**(4) (2019)
19. Jørgensen, K.P., Beck, R.: Universal wallets. Business & Information Systems Engineering pp. 1–11 (2022)

20. Kanimozhi, G., Kamatchi, K.: Security aspects of mobile based e wallet. International Journal on Recent and Innovation Trends in Computing and Communication **5**(6), 1223–1228 (2017)
21. Khan, A.G., Zahid, A.H., Hussain, M., Riaz, U.: Security of cryptocurrency using hardware wallet and qr code. In: 2019 International Conference on Innovative Computing (ICIC). pp. 1–10 (2019). https://doi.org/10.1109/ICIC48496.2019.8966739
22. Muthavhine, K.D., Sumbwanyambe, M.: An analysis and a comparative study of cryptographic algorithms used on the internet of things (iot) based on avalanche effect. In: 2018 International Conference on Information and Communications Technology (ICOIACT). pp. 114–119. IEEE (2018)
23. Palattella, M.R., Accettura, N., Vilajosana, X., Watteyne, T., Grieco, L.A., Boggia, G., Dohler, M.: Standardized protocol stack for the internet of (important) things. IEEE communications surveys & tutorials **15**(3), 1389–1406 (2012)
24. Rezaeighaleh, H., Zou, C.C.: New secure approach to backup cryptocurrency wallets. In: 2019 IEEE Global Communications Conference (GLOBECOM). pp. 1–6 (2019)
25. Rezaeighaleh, H., Zou, C.C.: Multilayered defense-in-depth architecture for cryptocurrency wallet. In: 2020 IEEE 6th International Conference on Computer and Communications (ICCC). pp. 2212–2217 (2020)
26. Sai, A.R., Buckley, J., Le Gear, A.: Privacy and security analysis of cryptocurrency mobile applications. In: 2019 Fifth Conference on Mobile and Secure Services (MobiSecServ). pp. 1–6 (2019)
27. Singh, G.: A review of factors affecting digital payments and adoption behaviour for mobile e-wallets. International Journal of Research in Management & Business Studies **6**(4), 89–96 (2019)
28. Suratkar, S., Shirole, M., Bhirud, S.: Cryptocurrency wallet: A review. In: 2020 4th International Conference on Computer, Communication and Signal Processing (ICCCSP). pp. 1–7 (2020)
29. Tan, Q., Gao, Y., Shi, J., Wang, X., Fang, B., Tian, Z.: Toward a comprehensive insight into the eclipse attacks of tor hidden services. IEEE Internet of Things Journal **6**(2), 1584–1593 (2019)
30. Taylor, S.K., Ariffin, A., Zainol Ariffin, K.A., Sheikh Abdullah, S.N.H.: Cryptocurrencies investigation: A methodology for the preservation of cryptowallets. In: 2021 3rd International Cyber Resilience Conference (CRC). pp. 1–5 (2021)
31. Uddin, M.S., Mannan, M., Youssef, A.: Horus: A security assessment framework for android crypto wallets. In: International Conference on Security and Privacy in Communication Systems. pp. 120–139. Springer (2021)
32. Varghese, H.M., Nagoree, D.A., Anshu, Jayapandian, N.: Cryptocurrency security and privacy issues: A research perspective. In: 2021 6th International Conference on Communication and Electronics Systems (ICCES). pp. 902–907 (2021)
33. Veinović, M., Adamović, S.: Kriptologija 1. Beograd: Univerzitet Singidunum (2013)

Cyber-Physical Systems Security: Analysis, Opportunities, Challenges, and Future Prospects

Joseph Bamidele Awotunde, Yetunde J. Oguns, Kamorudeen A. Amuda, Natasha Nigar, Toheeb A. Adeleke, Kazeem M. Olagunju, and Sunday Adeola Ajagbe

1 Introduction

The advancement of efficient technological approaches and techniques, such as period and regularity territory means, prediction, state-space exploration, system detection, sieving, vigorous and stochastic regulator, and optimization, has been pioneered by systems and control researchers over the years [1, 2]. Simultaneously, computer scientists have made significant strides in new programming languages,

J. B. Awotunde
Computer Science Department, University of Ilorin, Ilorin, Nigeria
e-mail: awotunde.jb@unilorin.edu.ng

Y. J. Oguns
Department of Computer Studies, The Polytechnic Ibadan, Ibadan, Nigeria
e-mail: oguns.yetunde@polyibadan.edu.ng

K. A. Amuda
Computer Science Department, University of Ibadan, Ibadan, Nigeria
e-mail: kamuda2883@stu.ui.edu.ng

N. Nigar
Department of Computer Science (RCET), University of Engineering and Technology, Lahore, Pakistan
e-mail: natasha@uet.edu.pk

T. A. Adeleke · K. M. Olagunju
Computer Engineering Department, Ladoke Akintola University of Technology, Ogbomoso, Nigeria
e-mail: taadeleke50@student.lautech.edu.ng; kmolagunju@student.lautech.edu.ng

S. A. Ajagbe (✉)
Department of Computer & Industrial Production Engineering, First Technical University, Ibadan, Nigeria
e-mail: saajagbe@pgschool.lautech.edu.ng

Y. Maleh et al. (eds.), *Blockchain for Cybersecurity in Cyber-Physical Systems*,
Advances in Information Security 102, https://doi.org/10.1007/978-3-031-25506-9_2

real-time computational methods, visualization, compiler prototypes, embedded system and virtual system architectures, and ground-breaking methods to maintain the reliability of computer systems, cybersecurity, and fault tolerance. A range of effective exhibiting formalisms and confirmation methods have also been developed by computer science researchers in the Cyber-Physical System (CPS) area [1, 3]. Research on CPS seeks to incorporate the concepts of information and engineering through computational and engineering disciplines to establish new science and technology [4]. So, CPS can combine cyber and physical parts well by using computers, the internet, and new network technologies [5].

Moreover, considering the new technological advancement globally, CPS, as the core component, is expected to be equipped with multiple sensors. The actuators can store and analyze data and be interconnected via the internet of things (IoT) sensor networks to provide interoperability, adaptability, operational efficiency, and knowledge integration of the smart of things. Nevertheless, adopting this novel technological advancement in various areas like medicine, security, and commerce is still challenging because their application will enforce a complete reconfiguration of the entire system so that CPS can effectively interface with the current system. Among notable CPS technology are cloud computing, IoT, Big Data, smart devices, and edge computing. The CPS has widely been used in the following areas: smart healthcare systems, smart cities, smart transportation, smart defense system, smart mobile system, meteorology system, smart agricultural systems, and many other areas [6]. Fig 1 depicts the general application of CPS.

But the development of CPS in recent years has raised new challenges for individuals regarding security, piracy, and confidentiality [3]. Among the most baffling obstacles in a wide range of cyber-attack defenses is maintaining the information security of cyber-physical networks. There is an increasing need to build information protection systems because of the pervasive use of wireless technologies for data processing, storage, and control devices, as in the case of a wireless sensor network (WSN). The remote location and autonomy of CPS systems result in the possibility of intrusions and attacks. The use of huge groups of devices in CPS can create various compromises within the system. There are several modern challenges associated with CPS [7]:

- The increase in the number of IoT-based devices can lead to an increase in CPS vulnerability to cyber-attacks like in the case of DDoS;
- Robust cybersecurity threat modeling is needed;
- The CPS needs an official manner for the development of weaknesses assessments;
- The development of an efficient error architecture for dealing with rapidly changing cyber and physical challenges;
- Designing dependable failure frameworks for grappling with rapidly changing cyber and physical challenges.

Hence, there is an urgent need for modern technologies and methodologies like people-centric quantum sensors, networks, sensing, multi-sensor systems, and biosensors to meet the CPS requirements in terms of reliability, privacy, confi-

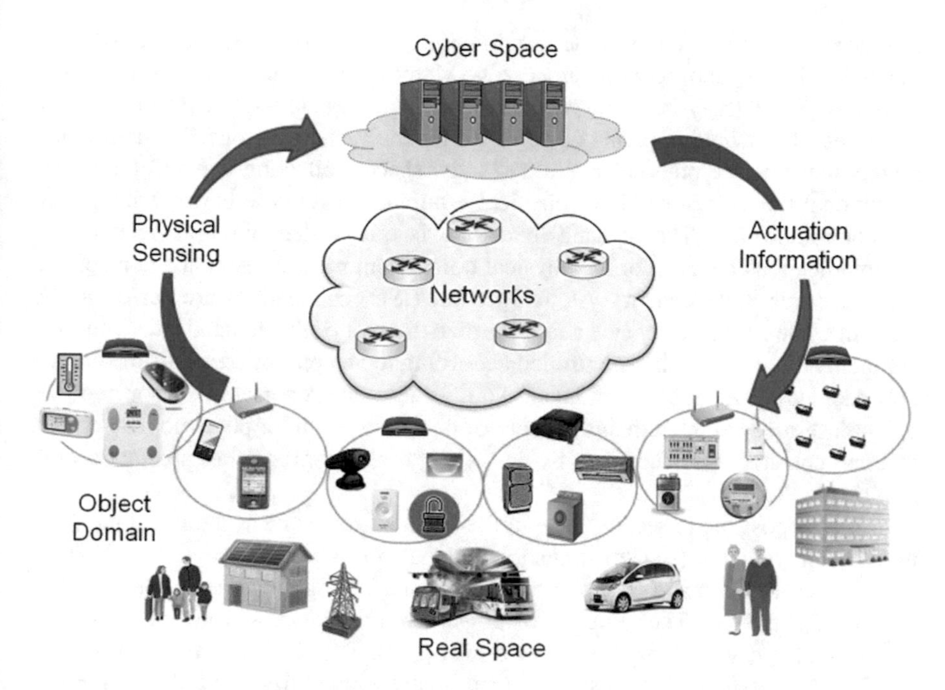

Fig. 1 Cyber-Physical System applications

dentiality, and security of individuals' data [1]. Therefore, this paper discusses the security threats in CPS to better understand the challenges in the entire system. It provides an overview that will help CPS researchers and practitioners know the security, privacy, shortcomings, and limitations in modern CPS architecture, future potential, intrusion detection, and the system's applicability in real-life projects.

The remaining sections of this chapter are organized as follows: Sect. 2 explains the overview of the cyber-physical system. Section 3 discusses CPS Security, Threats, and Attacks, while Sect. 4 presents the Risk Assessment. Section 5 concludes the chapter by discussing future work for achieving efficient CPS.

2 Overview of the Cyber-Physical System (CPS)

The research communities have recently seen booming scientific research on cyber-physical structures (CPS). By combining processing, communications, and physical systems, CPS is characterized as systems where computer systems usually monitor and track physical processes, whereas the latter influence computations and networks [2]. It should be remembered that incorporation does not always entail that the physical world and cyberspace merely converge but that the physical and cyber components connect profoundly. The research and design of the CPS are therefore focused on an understanding of the common dynamics of physical, machine, device, and network processes. Many devices, like smart grid request response [8–10], may be categorized as CPS, with demand-side consumers, such as various household appliances, constituting the physical properties, and smart meters that link the physical world and cyberspace, collecting demand load data. Traditional power systems have physical equipment as a core element integrated for communication. The demand load data is transmitted through the two-way communication channels for the physical component measurement and control. On the cyber side, Independent System Operator (ISO) calculations are performed to maximize the user's utility side and minimize the cost of the demand side (physical components), which is then controlled according to an appropriate real-time energy cost decided by the application layer. Another instance is a grid of sensor nodes, a network of medical instruments capable of detecting health or physiological details that can enhance body functions by delivering drugs or promoting prosthetic limb motion [11, 12].

It is also possible to see a CPS, a multi-agent system such as many autonomous underwater vehicles (AUVs). Sensors and networking systems allow AUVs to control their location and operation while communicating with one another to track a moving object [9]. Therefore, CPS ranges from miniature sensors such as body sensors to large-scale sensors such as power grids. The CPS report may provide a detailed and interdisciplinary structure for analyzing and designing these practical systems. With current prominent information and communications technology (ICT) systems, CPS shares some features with other related technology systems like embedded systems, Networked Control Systems (NCSs), IoT, and the industrial internet.

CPS can be chronological compared with other emerging technologies in the following ways:

1. Generic embedded systems and NCSs are not CPSs. It can be regarded as a networked embedded system [12].
2. CPSs are not IoTs, though they are used interchangeably sometimes. IoT typically refers to a hierarchical communication infrastructure with application-driven functionalities for sensing, processing, and transmitting information. While CPS emphasizes the interaction between physical processes and cyber dynamics, IoT is more like a platform for implementing some applications [4, 13]. It can, in other words, be considered an extension of the internet. In contrast to the IoT, CPS is a way to understand and build in the real world.
3. The upcoming industrial internet refers to global industrial environment convergence, widespread sensing, advanced computing, and ubiquitous network access, which will allow the global economy to reap growing benefits [14, 15]. It can also be shown that CPS is the technical foundation for this.

CPS dynamically reorganizes and reconfigures control structures with a high degree of automation at various spatial and temporal dimensions due to the close relation and synchronization between the cyber and physical worlds [16, 17]. The implementation of CPS relies on the closed-loop consideration and design of the entire system to allow seamless integration [18].

2.1 Principle of CPS Operations

There are also two primary layers in the CPS architecture [19, 20]: the cyber and physical layers. Variables representing data collected by the current situation of the CPS are represented by sensors and communication parameters that reflect control signals [21]. The controls in CPS calculate the distance between the values of the method parameters and the associated control levels. The controllers produce a local movement using complicated formulas after measuring this offset and calculating fresh actuation and NCSs. To hold the process closer to a given setpoint, the obtained controlling function is transmitted to the associated actuator [22].

Controllers often send the calculations to the central control servers and then perform the commands sent to them. The actual state of the items being handled should be known to system supervisors in the CPS. Thus, the graphical user interface (GUI), also known as the human-machine interaction (HMI), informs the human operator of the actual status of the controlled object. The CPS process is separated into the following stages [19]: (1) tracking; (2) communication; (3) computation processing; and (4) actuation. The cyber layer also employs an operational protocol, such as DNP3, to communicate with physical-layer applications [23, 24].

2.2 Cyber-Physical System and Internet of Things

Because of the vast gray field of overlap between the two, the words "CPS" and "IoT" have been used interchangeably in recent years. Educational establishments favor CPS while administrative entities and the private sector desire IoT [25, 26]. However, some big data analytics researchers use the name IoT as well. Generally, IoT is described as a communication network that connects objects with naming, sensing, and processing capabilities [27]. Furthermore, IoT allows for flexible, decentralized networks of interacting physical sensors that serve as expert systems, exchanging data with users via a communication medium such as WSN [28]. Common communication procedures such as Bluetooth, 6LoWPAN, RFID, and ZigBee are usually used to link smart objects such as smartphones, actuators, sensors, embedded computers, and RFID tags [4, 29]. However, CPS refers to a real-time system, such as distributed real-time management structures that merge computational and networking capabilities with physical entity tracking and control [30]. Actual control mechanisms and distributed networks with limited human intervention have also been described as CPS [31]. CPS is often the new generation of integrated, intelligent telecommunications technology networks. It computes, coordinates, and monitors/controls physical component functions in various applications, including transportation, electricity, medical, and production [32, 33].

While both IoT and CPS aim to increase the link between cyberspace and the real world and have distinctive features when they use knowledge sensing and digital technologies to communicate with the real world, the Internet of Things stresses networking. It aims to link everything in the real world, making it an open network interface and infrastructure. CPS focuses on knowledge and seeks to link everything in the real world, culminating in a closed network platform and technology [34]. CPS systems also aim to manage organizational and physical processes, addressing a tight human-machine relationship ignored mainly in the IoT [35]. CPS includes IoT, which typically relies on access control systems, while traditional control systems are sealed. CPS, for example, includes complex pricing for indirect/human-in-the-loop load management and closed-loop microgrid control [36]. The distinction between "IoT" and "CPS" is not made clear in most academic and project activities, and it is difficult to find a source that draws a clear difference between the two concepts. Most researchers consider the two concepts separate explanations for the same concept and interchangeably use the terms [37]. Since IoT reports noticing things in the real world, leveraging communication capacities, and gathering data required to handle things that aren't effectively handled today, it overlaps with CPS. Even though IoT was initially focused on identification and monitoring technologies, it is now used to regulate physical systems by incorporating RFID structures and sensor networks, specifically RFID sensor networks [37]. The physical and virtual worlds are connected via CPSs to form an IoT of data and services [38]. In what is known colloquially as the "IoT," CPS focuses more on control technologies. CPSs provide close computer-and information-centric physical and constructed frameworks and a relative balance and coordination of the system's functional and physical elements. The IoT is a common concept among grant-making organizations in Europe and Asia [39]. The IoT focuses on physical world sensing and (internet) networking. Things transmit data through the internet to direct (generally organizational) operations. Even though detecting physical data and sending it out, maybe not over the internet, is important for CPS, these devices often try to track both operational and physical processes, focusing on close user relationships in a way that the Internet of Things doesn't [40].

IoT is primarily concerned with unique identification, Internet connectivity, and the usability of "things." Nonetheless, defined objects in an IoT system may be networked together to monitor a specific scenario in a coordinated manner, in which case the IoT system can be considered to have grown to the level of a CPS. In general, CPS is concerned with the collective operation of sensors or actuators to achieve a specific objective and to do so efficiently. CPS employs an IoT framework to facilitate distributed system collaboration. Ashton [41], CPSs are computer-aided design systems that rely on seamlessly integrating computational algorithms and physical components. The IoT refers to the networked physical objects, computers, and systems that number in the billions (and growing). The new generations of CPS that have resulted, as well as their evolving networks, such as the IoT and the

Industrial Internet (II), have significant consequences for the future of smart and connected environments [42]. Industry Control Systems, Critical Infrastructures, the IoT, and Embedded Systems are all merged and integrated into Cyber-Physical Systems [35]. Also, CPS doesn't need to be connected to anything or have an address, and IoT is a subset of CPS [39].

2.3 Cyber-Physical System Layers

While there is global agreement on the definition of the CPS, there is no agreement on the CPS's basic components or communication models [43]. The physical and cyber layers of the CPS architecture are generally two main layers. The physical layer collects and executes data; the physical layer evaluates and analyses the data; and the cyber layer sends related commands [44]. Wu et al. [45] have discussed and identified that a three-layer structure cannot differentiate all of the CPS's characteristics. New CPS architectures, some of which are non-hierarchical and have ambiguous definitions, have been suggested [44]. According to Khan et al. [46], and Zhang et al. [47], more than three levels of architecture have been proposed to address serviceability coordination, service composition, service scheduling, and object abstraction, which are all aspects to consider. Given the prerequisite for handling and monitoring the CPS, the framework should include more stages and information. Besides, a new architecture has been modified to take into account the changes introduced, as described in Wu et al. [45], which has five layers: market, application, processing, transmission, and perception. CPS operates at three levels: awareness, transmission, and implementation, even though contradictory assumptions regarding the number of layers exist [48, 49]. The devices within these layers and the relevant functions that should be introduced describe each of these layers [49, 50]. This chapter discusses a three-level CPS architecture, based on the functions of each layer. The first three layers are the visualization (physical) layer, communication (system) layer, and interface (device) layer. The awareness layer, the uppermost surface [51], is also labeled as the identification or detector surface [29]. Transmitters, actuators, sensors, GPSs, laser scanners, intelligent machines, RFID tags with 2-D bar code stickers, and readers are part of the system [46]. This layer's devices can collect real-time data for various purposes, including monitoring and tracking, then interpret the data for what would be received from the real environment and perform application layer instructions. Data can be collected from sound, light, mechanics, heat, electricity, and position [32, 52]. In a wide and local network domain, sensors will produce real-time data with node cooperation [51], which will be compiled and evaluated at the application level. Sensors can collect temperature, acceleration, humidity, vibration, and location data, depending on their nature [46]. The transmission (transport) layer, or network layer [46], is the second layer, and it is in charge of exchanging and data processing knowledge and implementation. Data connectivity and storage in this layer are achieved using smaller networks, communications systems, the internet, or other existing networks using a combination of techniques such as Bluetooth, 4G, and 5G, Wi-Fi, Infrared, and ZigBee, based on the sensor hardware [46]. However, due to several reasons, including availability and cost-effectiveness, most interconnections are made via the internet. This implies that the networks in use should be capable of supporting real-time operations. Since handling and processing large amounts of data is critical, the transmitting level can initially access and handle a vast volume of data to achieve actual communication [44]. It is also responsible for providing efficient communication support [47].

Many protocols and functions, such as Internet Protocol version 6 (IPv6), can be found at this layer to handle a larger number of artifacts [45]. Data forwarding and delivery through multiple devices and nodes through the channels in use are also functions of this layer [51]. At this layer, Wi-Fi, LTE, Bluetooth, 4G/5G, and ZigBee technologies are used by cloud computing networks, routing systems, switching, and Internet Gateways. The network gateway connects various nodes by collecting, filtering, transmitting, and receiving data between them [51] and other layers of the CPS. Another problem in CPS is traffic and storage, which are exacerbated by the growing number of connected devices [46]. This will have an impact on CPS security. While protocols such as firewalls can handle this traffic, the protection of products with limited capabilities cannot be assured because their computing and storage capacities are finite [51]. The third and most accessible layer is the user layer. Its mission is to process data gathered at the transmission stage and send commands to physical systems, sensors, and actuators [32]. This layer makes the right decisions using sophisticated choice techniques on aggregated results [52], and instructions used in disciplinary acts will be regulated. Furthermore, before deciding on the required automatic response, this layer collects and processes data from the perception layer behavior [46]. This layer handles data aggregation from various sources, intelligent analysis of large amounts of data, and object control and management. To manage mobile computers, cloud services, middleware, and data mining algorithms can all be used [47]. This layer is also in charge of system control, observing the behavior of physical systems and issuing orders to change the behavior of hardware devices to ensure that the work environment runs correctly and appropriately. The framework layer also keeps track of prior activities to evaluate previous behavior for potential operational advancements. This layer combines CPS and industry-specialist applications to build a smart environment [51]. This has resulted in a wide range of intelligent applications in Smart Power systems, Home Automation and Towns, and Intelligent Automobiles, to name only a few examples [32], Smart Auto, environmental monitoring, industry regulation [44], Smart Health, and Smart Farming, all of which may involve private and safe data. Users' confidential knowledge, like medical information and financial records preferences, can be collected by such applications. As a result, it is important to implement data protection mechanisms. On the contrary, application systems are unique and require specific security policies. As a result, addressing a security policy for each application device separately is challenging. As the use of CPS grows, so does the number of security concerns that must be addressed.

2.4 Cyber-Physical System Model Types and Components

There are three main types of CPS models, each of them has its component to work with, and they are discussed as follows:

Timed Actor CPS This ideal emphasizes the non-functional aspects of success and pacing and the functional aspects of behavior and correctness. Geilen et al. [53] implemented a theory with practical and standard modification constraints, a collection of behaviors, increasing performance while reducing complexity. The enhancement centered on the "earlier-the-better" principle is the primary target since it facilitates the recognition of deterministic abstract concepts of non-deterministic structures [54]. These time-deterministic structures are less susceptible to state eruption, making analytical bounds easier to derive [55].

In Event-Based CPS An occurrence must be identified and observed by the appropriate CPS mechanisms before actuation resolutions are taken. On the other hand, individual part-timing constraints vary based on the non-deterministic device delay caused by various CPS behaviors such as sensing, actuating, communication, and computing [56]. According to Hu et al. [57], time constraints can be addressed using an event-driven methodology that uses CPS activities to ensure device connectivity, control processes, and computation. As a result, the CPS is better suited and more efficient for spatiotemporal data. The CPS activities are characterized by the event structure and the public and private activity parameters in the lattice-based event structure [58]. When such activities are mixed, they could be used to define any event's spatial-temporal structure and identify all of its elements.

Hybrid-Based CPS Blended CPS structures are heterogeneous systems that integrate two kinds of interacting systems: continuous state (physical dynamic) and discrete state (discrete computing) [59, 60]. The solution of discrete transient sequences triggered by finite state machines and the complex action generated by differential equations (s) is important for development and evolution [61]. Hybrid CPS, unlike other CPS types, is linked through a connection, rendering it prone to delays. Moreover, hybrid CPS models do not support hierarchy models and are not incompatible with concurrent systems modeling. As a result, Benveniste et al. [62] addressed the CPS issues associated with hybrid device simulation. Kumar et al. [63] used an actual hybrid verification approach to resolve and overcome CPS device network latency problems, while Tidwell et al. [64] proposed a configurable actual hybrid physical analysis for CPS. Finally, a hybrid automata-based event-driven monitoring of CPS was presented in Jianhui [65].

2.5 Challenges of Cyber-Physical System

Cyber-Physical Systems revolutionize the relationship with the physical world, but this revolution is not open. As even existing embedded systems must follow higher standards than general-purpose computing, we must pay particular attention to the physically-aware engineering application device specifications of the next generation if we want to put our faith in them. As a result, this chapter clarifies the meanings of common CPS system-level requirements. This challenge was linked to their characteristics.

Dependability The ability of a device to execute the essential task when in service without substantial deterioration in efficiency; the result is called dependability. The degree of confidence placed in the entire system is reflected in reliability. A highly reliable system should operate without interruption, deliver requested services as specified, and not fail. The terms "trustworthiness" and "reliability" are frequently used interchangeably [66]. Before actual system operation, ensuring reliability is a cumbersome task. Timing uncertainties in sensor readings and prompt actuation, for example, can reduce dependability, resulting in unforeseen effects. The cyber and physical elements of the project are inherently interrelated, and those underlying components, which are deliberately intertwined during system activity, make dependability analysis difficult. During the design phase, a common language should be announced for expressing information about dependability across the different parts of a system [67, 68].

Maintainability The ability of a scheme to be revamped in the event of a failure is referred to as "maintainability." A highly maintainable system should be able to be restored quickly and easily with minimal support services, and without causing additional faults during the maintenance process. Because of the close association of device components (e.g., detectors, actuators), automated forecasting diagnostic methods may be suggested for the physical mechanisms that make up the CPS infrastructure. Those mechanisms may be used to conduct continuous infrastructure monitoring and checking. The results of monitoring and testing facilities aid in determining the units that need repair. Some components with a high failure rate may be redesigned or scrapped and replaced with higher-quality alternatives [69, 70].

Availability The capacity of a system to be accessible even though mistakes occur is referred to as availability. A highly accessible system should be able to isolate a malfunctioning component from the rest of the system and keep running in the absence of it. Cyber-attacks that are harmful (such as denial of service assaults) significantly reduce the accessibility of device facilities. Medical data, for instance, sheds light on appropriate steps to be done on time to save a patient's life in Cyber-Physical Medical Systems. Malware activities or device failure may cause services that provide this data to become inaccessible, putting the patient's life at risk [71].

Safety The possession of a device to not cause any damage, danger, or danger, both within and without, during its service is called "safety." A precise, secure scheme would adhere to both common as well as implementation protection legislation to the most significant degree possible, as well as employee safety assurance mechanisms in the event of a failure. For example, tracking efficient output at a single point in time and real-time supervision of operations in the manufacturing process plant are two priorities for smart manufacturing (SM). Manufacturing plant protection can be significantly improved by using intelligent process management using integrated information systems and data collection processes in the production process. Smart sensors that are connected to a network can help find malfunctions and prevent disasters that could happen because of them [72].

Reliability It indicates the level of correctness with which a device executes its work. The certification of a system's capability to do things properly does not imply that they are performed appropriately. As a result, an extremely stable scheme ensures everything is done correctly. CPSs are required to function with ambiguity in details, features (e.g., timing), or the result of a system in the CPS infrastructure, effectively in the open, evolving, and volatile environments, which necessitates quantifying uncertainties throughout the CPS design stage. Successful CPS reliability characterization will result from this uncertainty study. Furthermore, the CPS reliability is limited by the reliability of physical and cyber components. Potential errors in the architecture flow and ad hoc cross-domain network links are all factors to consider [73, 74].

Robustness refers to a system's ability to maintain its robust specification and survive deficiencies. A highly robust solution should be designed to operate during disruptions while maintaining its initial structure and not be hampered or stopped by those failures. In addition to failures, disturbances such as sensor noises, actuator inaccuracies, defective communication channels, hardware errors, or software bugs can compromise the CPS's overall robustness. Other non-negligible variables that may occur in run-time include a lack of advanced system dynamics simulation (e.g., real-world environmental environments in which CPSs operate), a changing operating climate, or unusual incidents [33, 75].

Predictability This applies to the extent to which a system's state or performance can be predicted qualitatively or quantitatively. A highly predictable system can largely guarantee the defined result of the system's performance at any given time while meeting all system requirements. Smart medical devices fitted with sophisticated control systems are supposed to be well matched to the patient's symptoms, predict the patient's movements, and change their characteristics based on context knowledge from the external environment in Cyber-Physical Medical Systems (CPMS). Many medical devices operate in real-time, meeting various timing constraints and exhibiting varying degrees of awareness of pacing ambiguities (e.g., delays, jitters, etc.). Not all aspects of the CPMS can be predicted in advance. As a result, new resource allocation and scheduling policies, new programming and communication abstractions, and new programming and networking abstract concepts must be built to ensure transparent edge timing requirements [1].

Reconfigurability This refers to a system's ability to modify its settings in the event of a malfunction or in response to internal or external requests. A highly reconfigurable system should be self-configurable, which means it should be capable of fine-tuning and coordinating the action of its elements at a finer granularity. Self-configuring designed networks are what CPSs are. Remotely controlled systems may be required in some CPS deployment situations, such as international border surveillance, wildfire emergency response, and natural gas surveillance. In these situations, organizational requirements change. For example, safety threat level changes, regular code reviews, energy quality maintenance, and so on. This means that a lot of transceiver nodes or the whole system need to be changed to make better use of facilities and resources [76].

Security This system resource allows it to monitor access to its properties and safeguard confidential data from unofficial disclosure. An extremely protected scheme should have safeguards to prevent unauthorized information modification and resource withholding and be free of confidential data disclosure to a large degree. Due to their scalability, complexity, and dynamic existence, CPSs are vulnerable to physical and cyber failures and attacks. Malicious attacks (eavesdropping, denial-of-service, inserting false sensor dimensions or actuation demands, and so on) may be aimed at the cyberinfrastructure (for example, data collection, connectivity networks, verdict processes, and so on) or physical components to disrupt the system in operation or steal sensitive information. Other factors that make CPSs vulnerable to security threats include using a large-scale network (like the internet) and insecure networking protocols; dependency on legacy networks; or accelerated introduction of industrial off-the-shelf (COTS) innovations [77, 78].

Confidentiality This just extends to the resources of having approved parties' access to sensitive data provided by the device. The most reliable strategies for safeguarding against unwanted access, revelation, or alteration can be used in a highly confidential framework. In most CPS applications, data protection is a critical issue that must be addressed. Attacks on the confidentiality of data transmitted in an emergency management sensor network, for example, can reduce the system's effectiveness. The secrecy of data exchanged by infiltrated sensor nodes may be impeded, allowing data flow across the network to be redirected through contaminated detectors, sensitive data to be spied on, and bogus node personalities to be established. Furthermore, those fake nodes can be used to insert false or malicious data into the network. As a result, data circulation confidentiality must be maintained to a fair degree [79].

Heterogeneity refers to a system's ability to integrate various interacting and interconnected components into a dynamic whole. Because of the physical mechanics, computing structures, communication modules, and networking technology deployment, CPSs are inherently mixed. As a result, CPSs demand that all device components be heterogeneous in composition. For future medical devices, for instance, with the ability for various computations and connectivity, there is a possibility of increasing the interconnection complexity of open networks in a plug-and-play manner, necessitating the use of a diverse monitoring system and an integrated system locked control. The design of specific instruments can be very complicated based on the patient's healthcare issues. Future healthcare networks would be far more competent and challenging than today, with autonomous components and cooperation, actual assurance, and heterogeneous customized setting systems, thanks to advances in science and emerging technologies [80].

Scalability refers to a system's ability to continue to perform well after a shift in scale or elevated burden and to capitalize on it fully. The growth in machine capital should be proportional to the growth in system workload throughput. To improve efficiency, a highly scalable system should include scatter and gather mechanisms for workload balancing and efficient communication protocols. CPSs may have tens of thousands of integrated processors, sensors, and actuators that would all communicate with one another efficiently, depending on their size. In Gunes and Givargis [81], scalable embedded many-core architectures with programmable interconnect networks can be used to meet rising computing demand. Also, a high-performance and highly scalable infrastructure is required to dynamically enable CPS entities to enter and exit the existing network. Flexible software upgrades (i.e., updating the software program in real-time) can aid in continuously upgrading CPS programs and making better use of CPS tools through the productive space of regular data diffusion among those entities [82].

3 CPS Security, Threat, and Attacks

Security incidents show that attacks on the CPS, especially the cyber layer, may result in significant losses in people's necessities. As a result, CPS protection is becoming more critical at any time, and it should be considered first in the design procedure. Furthermore, sophisticated CPS protection techniques are better for securing these increasingly complex interconnected networks [83]. Most security efforts are focused on existing solutions explicitly developed for traditional information technology (IT) systems to build or construct advanced security technologies. These solutions, on the other hand, are not intended for CPS [84, 85]. Furthermore, most physical device performance, reliability, and productivity are the subjects of study and not protection, which is often overlooked due to restricted factors such as limited processing, communication, and storage capacity. CPS, on the other hand, will not work securely if protection is ignored [86]. The linkages between physical and cyber governing elements provide a near-coupling needed in reaction to the real need for protection methods. Security concerns are not recent; however, technological advancements have necessitated the development of new techniques to protect data from threats. Also, CPS protection is a very important issue that should be taken into account in every possible security solution [87].

3.1 Cyber-Physical System Security Issues

CPS are networked devices that incorporate cyber (computation and communication) and physical (sensors and actuators) elements in a feedback loop with the assistance of user intercession, interaction, and use. As the framework for new and future smart networks, these technologies can empower our essential infrastructure and have a colossal effect on daily human routines. On the other hand, the increased use of CPS introduces new risks that could have serious implications for users. Since security issues in computing have become a global problem, developing a stable, reliable, and productive CPS is a hot research topic. Security concerns are not recent, but technological advancements necessitate the development of new methods to protect data from unintended consequences. The vulnerability of the latest threats and cyber-attacks is inevitable, necessitating new approaches to defend CPS. Uncertainty in the network, security attacks, and physical device failures make maintaining overall system security difficult for CPS [77, 88]. Besides, cyber-physical pairing enables refined opponents to launch attacks that jeopardize another critical device characteristic, most notably safety [89, 90].

3.2 Cybersecurity

CPSs are systems that link computers, networking units, detectors, and actuators from the physical substratum in heterogeneous, transparent systems-of-systems or hybrid configurations. When systems become more interconnected, their sophistication rises [91]. Computer systems have now entered liquid, fuel, transportation, and electricity as vital resources for a country's economy. The use of CPS can be used in a variety of industries. The growth of CPS is critical to developing oil and gas, electricity supply processing, defense, and national services. Due to the vast number of mobile devices connected by network connectivity, CPS safety has become a public, technological, and economic issue for every nation in the world. "Wang and Lu" [92]. According to recent studies, cyber-attacks are aimed at destroying nation-state structures critical to the country's growth. CPS begins by attempting to damage infrastructures rather than just disrupting a single company or damaging a single machine [93].These attacks can potentially destroy vital infrastructure systems in security, banking, health, and the public sector [94]. Convicts, protesters, and terrorists are still looking for new and creative strategies and targets to achieve their objectives. Cyber-physical networks continue to be one of the most dangerous targets for hackers [95]. Additional security vulnerability awareness and sufficient security-related knowledge processing all contribute to maintaining a dependable system [96, 97]. The cyber system collects and analyzes data, impacting the physical system's activity through monetary and corrective behavior. While the convergence of cyber and physical schemes is crucial, the close coupling of the physical and cyber systems introduces new types of risks. However, as cyber-attacks occur, the cyber system can harm the physical system. Untimely and/or forged orders, for example, can cause damage to the facilities or even set off a chain of events. On the other hand, many of the CPS's essential functions depend on precise data and measurements from the physical system. Sensor, computer, and communication line failures result in incomplete data, computing delays, and failures to deliver critical commands. As a result, the physical system's stability is jeopardized. For several reasons, according to Alguliyev et al. [3], the focus of industrial IoT defense has mostly been on cyberattacks instead of direct attacks. This entails converting the power grid into an Advanced Metering Infrastructure (AMI), culminating in the proliferation of previously undisclosed cyber-attacks and SCADA vulnerabilities [98]. Electronic threats, as opposed to physical assaults, including physical appearance and devices, can now be performed from any device. Furthermore, the intelligent meter's interaction and connectivity with other devices in the Near-me Area Network (NAN) and Home Area Network (HAN) makes it vulnerable to a wide range of remote attacks. Lastly, cyber-attacks are difficult to hinder and overcome without the proper prevention and protective countermeasures. For more information on cyber threat intelligence [99], a brief look at how CPS security is handled was given.

Since cyber security is not limited to a particular feature, it can be viewed from a range of viewpoints, including:

Data Flow Protection Data flow safety is important during the collection, distribution, and processing stages.

Oriented Function The need for cyber-physical components to be integrated into the general CPS.

Oriented Threat The data confidentiality, honesty, availability, and transparency are all impacted [100].

This cyber intrusion incident demonstrates how attackers can cause significant damage to a large-scale ICT network in a short period. Compared with physical intrusion cases, cyber intruders are challenging to track down. Cybercriminals can operate from any location with a network connection. By analyzing packet information, many Internet Protocol (IP) traceback technologies can be used to pinpoint the origins of the assault [101, 102]. On the other hand, many websites include instructions on manipulating network packets and hijacking a victim's device. As a result, rather than detecting the attack source, cyber defense systems concentrate on blocking unknown links from WANs, such as cellular, the internet, and worldwide mobile interoperability for microwave access (WiMAX). Cybersecurity breaches, on the other hand, are often linked to the control unit interface settings in an electricity network.

3.3 Cyber-Physical System Vulnerabilities

A weakness in defense can lead to economic espionage (reconnaissance or active attacks). Consequently, a weakness evaluation entails recognizing and analyzing existing CPS flaws and determining effective remedial and preventive measures to minimize, diminish, or even remove any exposure [103]. As shown in Table 1, there are three major types of CPS vulnerabilities.

Table 1 CPS Major vulnerabilities, descriptions, and references

S/N	Major vulnerabilities	Descriptions	References
1	Network Vulnerabilities	This includes flaws in preventive security mechanisms, complimentary wired/wireless connectivity, and connections, jeopardized Eavesdropper, spoofing, sniffers, and interaction (network/transport/application layer) attackers, back-doors DoS/DDoS, and packet-skillful threats.	[104, 105]
2	Management Vulnerabilities	This includes a lack of security protocols, regulations, and guidelines.	[106]
3	Platform Vulnerabilities	Vulnerabilities in hardware, software, formation, and databases are all included.	[107]

Vulnerabilities may arise for a variety of reasons. Three significant factors cause vulnerabilities:

Assumption and Isolation In most CPS designs, it is founded on the concept of the "protection by anonymity" pattern. As a result, the emphasis here is on designing a dependable and stable system while also taking into account the process of required safety facilities without implying that systems are entirely disconnected from the external domain.

Increasing Connectivity The attack surface grows as the network becomes more linked. Manufacturers have established CPS through implementing and using public networks and open wireless technology, as CPS systems have become more connected in recent years. Up until 2001, the majority of external threats became the target of ICS attacks. This was before the invention of the internet, which moved threats to the outside world [77].

Heterogeneity CPS systems are made up of various third-party modules combined to create CPS applications. As a result, CPS has evolved into a multi-vendor scheme, with each product having its security issues [108].

Homogeneity Similar cyber-physical machine types share vulnerabilities that, once activated, can affect all computers in their immediate area; the Stuxnet worm attack on Iran's nuclear power stations is a prime example [109].

Suspicious Employees By thwarting and modifying the scripting code, allowing attackers remote access by accessing closed ports, or trying to plug in an unsafe USB or similar device, users may deliberately or unintentionally damage or hurt CPS devices. Consequently, three main types of CPS vulnerabilities are combined: electronic, physical, and cyber-physical hazards.

3.4 Cyber Attacks

There has been a recent increase in the number of cyber-attacks on CPS and IoCPT, with severe penalties. CPS is extremely vulnerable to malicious code injection attacks, according to recent studies [109–111] and code-reuse threats [112], in addition to subsequent false data threats [113], zero-control data [114], and, finally, Control-Flow Attestation (C-FLAT) threats [109]. An assault on CPS industrial instruments and equipment of this kind will result in a complete blackout. Table 2 presents major cyber-attacks and their descriptions in CPS.

4 Risk Assessment

With the engorged use of CPS in many delicate domains (such as medicine and smart homes), protection has become a pressing concern, necessitating the development of a suitable risk assessment process [115]. With so much reliance on the internet, the security emphasis is the focus of risk management that has changed from system risk to network risk [32]. The aim of evaluating CPS security is to create a quantifiable risk that can be used to secure potential systems. However, most of the efforts and studies are focused on business processes unrelated to CPS [116]. CPS security differs from conventional IT security in several ways, so the security features also vary. For example, standardized interfaces and techniques, untrustworthy linkages, and information sharing are all major ICS risk factors [117]. The three stages of the CPS risk analysis method are: (1) evaluating what might happen to the system; (2) reviewing the possibility of the incident; and (3) predicting the repercussions. Furthermore, three factors should be considered when assessing CPS risk: asset (value), hazard, and vulnerability identification [118].

4.1 Asset Identification

A tangible presence (e.g., medical devices, educational facilities, business facilities, operations, or information) or an intangible presence (e.g., medical devices, educational facilities, business facilities, operations, or information) may be used to secure an asset (for example, details about a business or the prestige of an organization). In reality, most assets are intangible, but as a result, assets have a clear value in several everyday transactions and services and should be safeguarded. Furthermore, direct and indirect monetary losses, as well as the resulting destruction, can be used to estimate asset quantification [33, 73].The value evaluation process involves determining the asset value rating and deriving the security layers, essential properties, and system key activities [33]. Physical properties, cyber assets, and communications with other networks are the three types of CPS assets. CPS assets

Table 2 Major cyber-attacks and descriptions in CPS

S/N	Cyber attacks	Descriptions
1	Eavesdropping	Snooping is the surveillance of non-secure CPS internet traffic to obtain sensitive material (passwords, usernames, or other CPS information). Snooping can be either passive or active. Passive eavesdropping is listening to CPS system data communication without inspecting, searching, or interfering with the packet.
2	Cross-Site Scripting	By inserting malignant coding, third-party web tools are used to execute malicious programs in the internet browser of the targeted user (primarily a targeted CPS developer, contractor, staff, etc.). XSS will hijack a victim's session, sometimes record keystrokes, and gain remote access to their computer.
3	SQL Injection or SQLi	SQLi is designed to deliver and/or change delicate data on CPS database-driven websites and perform administrative tasks like databank closure, mainly when CPS schemes still rely on SQL for managing data.
4	Phishing	This includes message phishing, vishing, phishing attacks, and whale hunting, which imitates business colleagues or service providers to threaten any or all CPS consumers (such as developers, experts, businesses, chief executive Officers (CEO), chief financial officers (CFO), or chief operating officers (COO).
5	DoS/DDoS	DoS attacks are conducted from many infected computers and target cyber-physical machine resources. Botnets normally carry out DDoS attacks, which use many infected computers to initiate a DDoS attack simultaneously from various positions. DoS assaults come in a variety of forms and dimensions.
6	Malicious Third Party	These involve malware that infiltrates data through a CPS cryptographic platform from an internal process (i.e., RTU or PLC) to a botnet instruction network, primarily using botnets, Trojans, or worms to infiltrate data through a CPS encrypted connection from an internal database (i.e., PLC, ICS, or RTU) using a Trusted Third Party in camouflage. As a result, CPSs and AMIs are being targeted.
7	Watering-hole Attack	An attacker looks for some vulnerabilities in cyber-physical defense. Once a vulnerability has been found, a "watering hole" will be set up on the chosen CPS website, where malware will be distributed by manipulating the targeted CPS framework, primarily via backdoors, rootkits, or zero-day exploits.

are different from traditional IT assets because they communicate with each other in ways that are dynamic, intangible, and use different structures.

4.2 Threat Identification

This move is used to aid in identifying threats that are a top priority in the domain of CPS, which is a daunting challenge. Statistical data can be used to analyze the hazard, while sampling reports and logs from Intrusion Detection Systems (IDS) can be used to analyze the danger [119]. While IDS strategies are beyond the scope of this article, Mitchell and Chen [120] provide a comprehensive literature review that categorizes new CPS IDS strategies, describes research strategies, and reviews the most frequently discussed CPS IDS techniques in the field.

4.3 Vulnerability Identification

An attacker could use all current vulnerabilities for espionage purposes to hear or harm the value of an item, which is referred to as a vulnerability. It can also be described as a situation or environment that an adversary can use to attack or harm systems [115]. A vulnerability evaluation analyzes a scheme and its purposes to identify deficiencies and determine suitable remedial measures or reductions that could be planned and enforced to minimize or remove any vulnerabilities [121]. CPS weaknesses are commonly classified into three categories: network, portal, and management. Security issues in network configuration, hardware, and logging are also examples of security flaws [122]. Security flaws in the specification, electronics, applications, and a lack of security mechanisms contribute to platform vulnerability. The lack of a security policy is the most common source of management weakness. Vulnerability quantification can be attained through various mechanisms, including past professional assessment procedures, historical evidence analyses, or industry best practices [123, 124]. It's a tough, if not impossible, task to eliminate or avoid all risks. Because of this, the least expensive ways to reduce risks to a good level are often used.

5 Conclusion and Future Directions

The component in transforming how humans interact with their physical environment is the cyber-physical system by integrating it with the cyber world. Either within or outside the IoT-based system (IoCPT), implementing CPS systems aims to enhance systems' availability, reliability, and product quality. Cyber and physical networks are connected in CPSs to provide essential services. For example, the smart grid, as an example of a CPS, extensively uses data gathered from the physical system. However, these have not been met in CPS because of privacy and security issues, which reduce their efficiency, safety, and reliability, thus obstructing their wide implementation. Privacy and security concerns have become a global problem in CPSs, so the design of stable, healthy, and effective CPSs is a popular field of research. Therefore, this chapter presents the CPS security vulnerabilities, attacks, and threats in terms of securing information on the internet. The limitations of existing security measures are also presented, and their effect on people's lives is discussed. The CPS's main types of security attacks and threats are presented and analyzed. Finally, the challenges of CPS, possible solutions, and areas for future research are discussed. The best practices, services, and security aspects must be put in place to ensure irrepressible and secure CPS systems by maintaining the quality of service and the required performance. Analyzing the currently available security methods and solutions will give future researchers direction on which security measures to take when designing CPS systems.

References

1. Baheti, R., & Gill, H. (2011). Cyber-physical systems. The impact of control technology, 12(1), 161–166.
2. Jazdi, N. (2014, May). Cyber-physical systems in the context of Industry 4.0. In 2014 IEEE international conference on automation, quality, and testing, robotics (pp. 1–4). IEEE.
3. Alguliyev, R., Imamverdiyev, Y., & Sukhostat, L. (2018). Cyber-physical systems and their security issues. Computers in Industry, 100, 212–223.
4. Tao, F., Qi, Q., Wang, L., & Nee, A. Y. C. (2019). Digital twins and cyber–physical systems toward smart manufacturing and industry 4.0: Correlation and comparison. Engineering, 5(4), 653–661.
5. Zeadally, S., & Jabeur, N. (2016). Cyber-physical system design with sensor networking technologies. The Institution of Engineering and Technology.
6. Bamimore I. & Ajagbe S. A., (2020) Design and implementation of smart home for security using Radio Frequency modules, *International Journal of Digital Signals and Smart Systems (Inderscience Journal) Vol.4, Issue 4, Pp 286–303*
7. Liu, C. H., & Zhang, Y. (Eds.). (2015). Cyber-physical systems: architectures, protocols, and applications (Vol. 22). CRC Press.
8. Affum, E. A., Ajagbe, S. A., Boateng, K. A., Adigun, M. O., (2022), Response Analysis of Varied Q-Power Values of Cosine Distribution in Spatial Correlation, *2022 IEEE International Symposium on Antennas and Propagation and USNC-URSI Radio Science Meeting (AP-S/URSI), 2022, pp. 2070–2071,*https://doi.org/10.1109/AP-S/USNC-URSI47032.2022.9887202.
9. Li, N., Chen, L., & Low, S. H. (2011, July). Optimal demand response based on utility maximization in power networks. In 2011 IEEE power and energy society general meeting (pp. 1–8). IEEE.
10. Deng, R., Chen, J., Cao, X., Zhang, Y., Maharjan, S., & Gjessing, S. (2013). Sensing-performance tradeoff in cognitive radio enabled smart grid. IEEE Transactions on Smart Grid, 4(1), 302–310.
11. Calhoun, B. H., Lach, J., Stankovic, J., Wentzloff, D. D., Whitehouse, K., Barth, A. T., ... & Zhang, Y. (2011). Body sensor networks: A holistic approach from silicon to users. Proceedings of the IEEE, 100(1), 91–106.

12. Guan, X., Yang, B., Chen, C., Dai, W., & Wang, Y. (2016). A comprehensive overview of cyber-physical systems: From perspective of feedback system. IEEE/CAA Journal of Automatica Sinica, 3(1), 1–14.
13. Hehenberger, P., Vogel-Heuser, B., Bradley, D., Eynard, B., Tomiyama, T., & Achiche, S. (2016). Design, modelling, simulation and integration of cyber physical systems: Methods and applications. Computers in Industry, 82, 273–289.
14. Hatzivasilis, G., Fysarakis, K., Soultatos, O., Askoxylakis, I., Papaefstathiou, I., & Demetriou, G. (2018). The industrial internet of things as an enabler for a circular economy Hy-LP: a Novel IIoT protocol, evaluated on a wind park's SDN/NFV-enabled 5G industrial network. Computer communications, 119, 127–137.
15. Basir, R., Qaisar, S., Ali, M., Aldwairi, M., Ashraf, M. I., Mahmood, A., & Gidlund, M. (2019). Fog computing enabling industrial internet of things: State-of-the-art and research challenges. Sensors, 19(21), 4807.
16. Dey, N., Ashour, A. S., Shi, F., Fong, S. J., & Tavares, J. M. R. (2018). Medical cyber-physical systems: A survey. Journal of medical systems, 42(4), 1–13.
17. Penas, O., Plateaux, R., Patalano, S., & Hammadi, M. (2017). Multi-scale approach from mechatronic to Cyber-Physical Systems for the design of manufacturing systems. Computers in Industry, 86, 52–69.
18. Nikolakis, N., Maratos, V., & Makris, S. (2019). A cyber physical system (CPS) approach for safe human-robot collaboration in a shared workplace. Robotics and Computer-Integrated Manufacturing, 56, 233–243.
19. Hahn, A., Thomas, R. K., Lozano, I., & Cardenas, A. (2015). A multi-layered and kill-chain based security analysis framework for cyber-physical systems. International Journal of Critical Infrastructure Protection, 11, 39–50.
20. Krotofil, M., & Larsen, J. (2014, August). Are you threatening my hazards?. In International Workshop on Security (pp. 17–32). Springer, Cham.
21. Jiang, Y., Yin, S., & Kaynak, O. (2018). Data-driven monitoring and safety control of industrial cyber-physical systems: Basics and beyond. IEEE Access, 6, 47374–47384.
22. Orojloo, H., & Azgomi, M. A. (2017). A method for evaluating the consequence propagation of security attacks in cyber–physical systems. Future Generation Computer Systems, 67, 57–71.
23. Ajagbe, S. A., Ayegboyin, M. O., Idowu, I. R., Adeleke, T. A., & Thanh, D. N. H. (2022) Investigating Energy Efficiency of Mobile Ad-hoc Network (MANET) Routing Protocols, *An International Journal of Computing and informatics, Vol 46, no. 2, pp.* 269–275, https://doi.org/10.31449/inf.v46i2.3576
24. Ozansoy, C. R., Zayegh, A., & Kalam, A. (2008, December). Time synchronisation in a IEC 61850 based substation automation system. In 2008 Australasian Universities Power Engineering Conference (pp. 1–7). IEEE.
25. Modbus, I. D. A. (2004). Modbus application protocol specification v1. 1a. North Grafton, Massachusetts (www. modbus. org/specs. php).
26. Gładysz, B. (2015). An assessment of RFID applications in manufacturing companies. Management and Production Engineering Review, 6(4), 33–42.
27. Jeschke, S., Brecher, C., Meisen, T., Özdemir, D., & Eschert, T. (2017). Industrial internet of things and cyber manufacturing systems. In Industrial internet of things (pp. 3–19). Springer, Cham.
28. Younan, M., Houssein, E. H., Elhoseny, M., & Ali, A. A. (2020). Challenges and recommended technologies for the industrial internet of things: A comprehensive review. Measurement, 151, 107198.
29. Kumar, J. S., & Patel, D. R. (2014). A survey on internet of things: Security and privacy issues. International Journal of Computer Applications, 90(11).
30. Al-Sarawi, S., Anbar, M., Alieyan, K., & Alzubaidi, M. (2017, May). Internet of Things (IoT) communication protocols. In 2017 8th International conference on information technology (ICIT) (pp. 685–690). IEEE.

31. Suo, H., Wan, J., Zou, C., & Liu, J. (2012, March). Security in the internet of things: a review. In 2012 international conference on computer science and electronics engineering (Vol. 3, pp. 648–651). IEEE.
32. Yaacoub, J. A., Salman, O., Noura, H. N., Kaaniche, N., Chehab, A., & Malli, M. (2020); Cyber-physical systems security: Limitations, issues and future trends. Micro process Microsyst. 2020 Sep;77:103201. doi: 10.1016/j.micpro.2020.103201. Epub 2020 Jul 8. PMID: 32834204; PMCID: PMC7340599.
33. Ashibani, Y., & Mahmoud, Q. H. (2017). Cyber physical systems security: Analysis, challenges and solutions. Computers & Security, 68, 81–97.
34. Ma, H. D. (2011). Internet of things: Objectives and scientific challenges. Journal of Computer science and Technology, 26(6), 919–924.
35. Schätz, B., Törngren, M., Passerone, R., Pfeifer, H., Bensalem, S., McDermid, J., ... & Cengarle, M. V. (2015). CyPhERS-cyber-physical European roadmap and strategy. Fortiss GmbH, Munich, Germany, Tech. Rep, 611430.
36. Yeboah-Ofori, A., Abdulai, J., & Katsriku, F. (2019). Cybercrime and Risks for Cyber Physical Systems. International Journal of Cyber-Security and Digital Forensics (IJCSDF), 8(1), 43–57.
37. Minerva, R., Biru, A., & Rotondi, D. (2015). Towards a definition of the Internet of Things (IoT). IEEE Internet Initiative, 1(1), 1–86.
38. Broo, D. G., Boman, U., & Törngren, M. (2020). Cyber-physical systems research and education in 2030: Scenarios and strategies. Journal of Industrial Information Integration, 21, 100192.
39. Zheng, X., & Julien, C. (2015, May). Verification and validation in cyber physical systems: Research challenges and a way forward. In 2015 IEEE/ACM 1st International Workshop on Software Engineering for Smart Cyber-Physical Systems (pp. 15–18). IEEE.
40. Greer, C., Burns, M., Wollman, D., & Griffor, E. (2019). Cyber-physical systems and internet of things.
41. Ashton, K. (2009). That 'internet of things' thing. RFID journal, 22(7), 97–114.
42. Gunes, V., Peter, S., Givargis, T., & Vahid, F. (2014). A survey on concepts, applications, and challenges in cyber-physical systems. KSII Transactions on Internet & Information Systems, 8(12).
43. La, H. J., & Kim, S. D. (2010, August). A service-based approach to designing cyber physical systems. In 2010 IEEE/ACIS 9th International Conference on Computer and Information Science (pp. 895–900). IEEE.
44. Lu, T., Lin, J., Zhao, L., Li, Y., & Peng, Y. (2015). A security architecture in cyber-physical systems: security theories, analysis, simulation and application fields. International Journal of Security and Its Applications, 9(7), 1–16.
45. Wu, M., Lu, T. J., Ling, F. Y., Sun, J., & Du, H. Y. (2010, August). Research on the architecture of Internet of Things. In 2010 3rd international conference on advanced computer theory and engineering (ICACTE) (Vol. 5, pp. V5–484). IEEE.
46. Khan, R., Khan, S. U., Zaheer, R., & Khan, S. (2012, December). Future internet: the internet of things architecture, possible applications and key challenges. In 2012 10th international conference on frontiers of information technology (pp. 257–260). IEEE.
47. Zhang, B., Ma, X. X., & Qin, Z. G. (2011). Security architecture on the trusting internet of things. Journal of Electronic Science and Technology, 9(4), 364–367.
48. Bajeh, A. O., Mojeed, H. A., Ameen, A. O., Abikoye, O. C., Salihu, S. A., Abdulraheem, M., ... & Awotunde, J. B. (2021). Internet of robotic things: its domain, methodologies, and applications. In Emergence of Cyber Physical System and IoT in Smart Automation and Robotics (pp. 135–146). Springer, Cham.
49. Chang, W., Burton, S., Lin, C. W., Zhu, Q., Gauerhof, L., & McDermid, J. (2020). Intelligent and connected cyber-physical systems: A perspective from connected autonomous vehicles. In Intelligent Internet of Things (pp. 357–392). Springer, Cham.
50. Cao, L., Jiang, X., Zhao, Y., Wang, S., You, D., & Xu, X. (2020). A survey of network attacks on cyber-physical systems. IEEE Access, 8, 44219–44227.

51. Awotunde, J. B., Jimoh, R. G., Folorunso, S. O., Adeniyi, E. A., Abiodun, K. M., & Banjo, O. O. (2021). Privacy and security concerns in IoT-based healthcare systems. In The Fusion of Internet of Things, Artificial Intelligence, and Cloud Computing in Health Care (pp. 105–134). Springer, Cham.
52. Ali, S., Al Balushi, T., Nadir, Z., & Hussain, O. K. (2018). Cyber Security for Cyber Physical Systems (Vol. 768, pp. 11–33). Springer.
53. Geilen, M., Tripakis, S., & Wiggers, M. (2011, April). The earlier the better: A theory of timed actor interfaces. In Proceedings of the 14th international conference on Hybrid systems: computation and control (pp. 23–32).
54. Vicaire, P. A., Hoque, E., Xie, Z., & Stankovic, J. A. (2011). Bundle: A group-based programming abstraction for cyber-physical systems. IEEE Transactions on Industrial Informatics, 8(2), 379–392.
55. Canedo, A., Schwarzenbach, E., & Faruque, M. A. A. (2013, April). Context-sensitive synthesis of executable functional models of cyber-physical systems. In 2013 ACM/IEEE International Conference on Cyber-Physical Systems (ICCPS) (pp. 99–108). IEEE.
56. Zhang, Z., Eyisi, E., Koutsoukos, X., Porter, J., Karsai, G., & Sztipanovits, J. (2014). A co-simulation framework for design of time-triggered automotive cyber physical systems. Simulation modelling practice and theory, 43, 16–33.
57. Hu, F., Lu, Y., Vasilakos, A. V., Hao, Q., Ma, R., Patil, Y., ... & Xiong, N. N. (2016). Robust cyber–physical systems: Concept, models, and implementation. Future generation computer systems, 56, 449–475.
58. Tan, Y., Vuran, M. C., Goddard, S., Yu, Y., Song, M., & Ren, S. (2010, April). A concept lattice-based event model for cyber-physical systems. In Proceedings of the 1st ACM/IEEE International Conference on Cyber-physical Systems (pp. 50–60).
59. Alur, R., Courcoubetis, C., Halbwachs, N., Henzinger, T. A., Ho, P. H., Nicollin, X., ... & Yovine, S. (1995). The algorithmic analysis of hybrid systems. Theoretical computer science, 138(1), 3–34.
60. Antsaklis, P. J., Stiver, J. A., & Lemmon, M. (1992). Hybrid system modeling and autonomous control systems. In Hybrid systems (pp. 366-392). Springer, Berlin, Heidelberg.
61. Yalei, Y., & Xingshe, Z. (2013, June). Cyber-physical systems modeling based on extended hybrid automata. In 2013 International Conference on Computational and Information Sciences (pp. 1871–1874). IEEE.
62. Benveniste, A., Bourke, T., Caillaud, B., & Pouzet, M. (2013). Hybrid systems modeling challenges caused by cyber-physical systems. Cyber-Physical Systems (CPS) Foundations and Challenges. Available on-line: http://people. rennes. inria. fr/Albert. Benveniste/pub/NIST2012. pdf.
63. Kumar, P., Goswami, D., Chakraborty, S., Annaswamy, A., Lampka, K., & Thiele, L. (2012, June). A hybrid approach to cyber-physical systems verification. In DAC Design Automation Conference 2012 (pp. 688–696). IEEE.
64. Tidwell, T., Gao, X., Huang, H. M., Lu, C., Dyke, S., & Gill, C. (2009, March). Towards configurable real-time hybrid structural testing: a cyber-physical system approach. In 2009 IEEE International Symposium on Object/Component/Service-Oriented Real-Time Distributed Computing (pp. 37–44). IEEE.
65. Jianhui, M. (2011). Event driven monitoring of cyber-physical systems based on hybrid automata. National University of Defense Technology Changsha.
66. Wan, K., & Alagar, V. (2011, November). Dependable context-sensitive services in cyber physical systems. In 2011IEEE 10th International Conference on Trust, Security and Privacy in Computing and Communications (pp. 687–694). IEEE.
67. Denker, G., Dutt, N., Mehrotra, S., Stehr, M. O., Talcott, C., & Venkatasubramanian, N. (2012). Resilient dependable cyber-physical systems: a middleware perspective. Journal of Internet Services and Applications, 3(1), 41–49.
68. Höfig, K., Armbruster, M., & Schmidt, R. (2014). A vehicle control platform as safety element out of context.

69. Shcherbakov, M. V., Glotov, A. V., & Cheremisinov, S. V. (2020). Proactive and predictive maintenance of cyber-physical systems. In Cyber-Physical Systems: Advances in Design & Modelling (pp. 263–278). Springer, Cham.
70. Napoleone, A., Macchi, M., & Pozzetti, A. (2020). A review on the characteristics of cyber-physical systems for the future smart factories. Journal of manufacturing systems, 54, 305–335.
71. Haque, S. A., Aziz, S. M., & Rahman, M. (2014). Review of cyber-physical system in healthcare. international journal of distributed sensor networks, 10(4), 217415.
72. Wei, M. (2016). Modeling, Evaluation and Enhancement of Threats-Induced Reliability in Cyber-Physical Systems.
73. Chen, K. C., Lin, S. C., Hsiao, J. H., Liu, C. H., Molisch, A. F., & Fettweis, G. P. (2020). Wireless networked multirobot systems in smart factories. Proceedings of the IEEE.
74. Hoffmann, M. (2019). Smart Agents for the Industry 4.0: Enabling Machine Learning in Industrial Production. Springer Nature.
75. Letichevsky, A. A., Letychevskyi, O. O., Skobelev, V. G., & Volkov, V. A. (2017). Cyber-physical systems. Cybernetics and Systems Analysis, 53(6), 821–834.
76. Misra, S., & Eronu, E. (2012). Implementing reconfigurable wireless sensor networks: The embedded operating system approach. Embedded Systems-High Performance Systems, Applications and Projects, Intechopen, 221–232.
77. Yaacoub, J. P. A., Salman, O., Noura, H. N., Kaaniche, N., Chehab, A., & Malli, M. (2020). Cyber-physical systems security: Limitations, issues and future trends. Microprocessors and Microsystems, 77, 103201.
78. Gunduz, M. Z., & Das, R. (2020). Cyber-security on smart grid: Threats and potential solutions. Computer networks, 169, 107094.
79. Maglaras, L. A., Kim, K. H., Janicke, H., Ferrag, M. A., Rallis, S., Fragkou, P., ... & Cruz, T. J. (2018). Cyber security of critical infrastructures. Ict Express, 4(1), 42–45.
80. Jimenez, J. I., Jahankhani, H., & Kendzierskyj, S. (2020). Health care in the cyberspace: Medical cyber-physical system and digital twin challenges. In Digital Twin Technologies and Smart Cities (pp. 79–92). Springer, Cham.
81. Gunes, V., & Givargis, T. (2015, August). XGRID: A scalable many-core embedded processor. In 2015 IEEE 17th International Conference on High Performance Computing and Communications, 2015 IEEE 7th International Symposium on Cyberspace Safety and Security, and 2015 IEEE 12th International Conference on Embedded Software and Systems (pp. 1143–1146). IEEE.
82. Park, M. J., Kim, D. K., Kim, W. T., & Park, S. M. (2010, November). Dynamic software updates in cyber-physical systems. In 2010 International Conference on Information and Communication Technology Convergence (ICTC) (pp. 425–426). IEEE.
83. Jalali, S. (2009). Trends and implications in embedded systems development. TCS white paper.
84. Konstantinou, C., Maniatakos, M., Saqib, F., Hu, S., Plusquellic, J., & Jin, Y. (2015, May). Cyber-physical systems: A security perspective. In 2015 20th IEEE European Test Symposium (ETS) (pp. 1–8). IEEE.
85. Wang, E. K., Ye, Y., Xu, X., Yiu, S. M., Hui, L. C. K., & Chow, K. P. (2010, December). Security issues and challenges for cyber physical system. In 2010 IEEE/ACM Int'l Conference on Green Computing and Communications & Int'l Conference on Cyber, Physical and Social Computing (pp. 733–738). IEEE.
86. Kim, N. Y., Rathore, S., Ryu, J. H., Park, J. H., & Park, J. H. (2018). A survey on cyber physical system security for IoT: issues, challenges, threats, solutions. Journal of Information Processing Systems, 14(6), 1361–1384.
87. Humayed, A., Lin, J., Li, F., & Luo, B. (2017). Cyber-physical systems security—A survey. IEEE Internet of Things Journal, 4(6), 1802–1831.
88. Kumar, C., Marston, S., & Sen, R. (2020). Cyber-physical Systems (CPS) Security: State of the Art and Research Opportunities for Information Systems Academics. Communications of the Association for Information Systems, 47(1), 36.

89. Hassan, M. U., Rehmani, M. H., & Chen, J. (2019). Differential privacy techniques for cyber physical systems: a survey. IEEE Communications Surveys & Tutorials, 22(1), 746–789.
90. Griffioen, P., Weerakkody, S., Sinopoli, B., Ozel, O., & Mo, Y. (2019, June). A Tutorial on Detecting Security Attacks on Cyber-Physical Systems. In 2019 18th European Control Conference (ECC) (pp. 979–984). IEEE.
91. Friedberg, I., McLaughlin, K., Smith, P., Laverty, D., & Sezer, S. (2017). STPA-SafeSec: Safety and security analysis for cyber-physical systems. Journal of information security and applications, 34, 183–196.
92. Wang, W., & Lu, Z. (2013). Cyber security in the smart grid: Survey and challenges. Computer networks, 57(5), 1344–1371.
93. Ali, N. S. (2016). A four-phase methodology for protecting web applications using an effective real-time technique. International Journal of Internet Technology and Secured Transactions, 6(4), 303–323.
94. Al-Mhiqani, M. N., Ahmad, R., Abdulkareem, K. H., & Ali, N. S. (2017). Investigation study of Cyber-Physical Systems: Characteristics, application domains, and security challenges. ARPN Journal of Engineering and Applied Sciences, 12(22), 6557–6567.
95. Ten, C. W., Manimaran, G., & Liu, C. C. (2010). Cybersecurity for critical infrastructures: Attack and defense modeling. IEEE Transactions on Systems, Man, and Cybernetics-Part A: Systems and Humans, 40(4), 853–865.
96. Ali, N. S., & Shibghatullah, A. S. (2016). Protection web applications using real-time technique to detect structured query language injection attacks. International Journal of Computer Applications, 149(6), 26–32.
97. Sridhar, S., Hahn, A., & Govindarasu, M. (2011). Cyber–physical system security for the electric power grid. Proceedings of the IEEE, 100(1), 210–224.
98. Coffey, K., Smith, R., Maglaras, L., & Janicke, H. (2018). Vulnerability analysis of network scanning on SCADA systems. Security and Communication Networks, 2018.
99. Bou-Harb, E. (2016, November). A brief survey of security approaches for cyber-physical systems. In 2016 8th IFIP International Conference on New Technologies, Mobility and Security (NTMS) (pp. 1–5). IEEE.
100. Cleveland, F. M. (2008, July). Cyber security issues for advanced metering infrasttructure (AMI). In 2008 IEEE Power and Energy Society General Meeting-Conversion and Delivery of Electrical Energy in the 21st Century (pp. 1–5). IEEE.
101. Cronin, John. "Automated IP tracking system and method." U.S. Patent Application 11/249,575, filed February 9, 2006.
102. Safa, H., Chouman, M., Artail, H., & Karam, M. (2008). A collaborative defense mechanism against SYN flooding attacks in IP networks. Journal of Network and Computer Applications, 31(4), 509–534.
103. Moteff, J. (2005, February). Risk management and critical infrastructure protection: Assessing, integrating, and managing threats, vulnerabilities and consequences. Library of Congress Washington DC Congressional Research Service.
104. Zhu, B., Joseph, A., & Sastry, S. (2011, October). A taxonomy of cyber attacks on SCADA systems. In 2011 International conference on internet of things and 4th international conference on cyber, physical and social computing (pp. 380–388). IEEE.
105. Nash, T. (2005). Backdoors and holes in network perimeters. Online]: http://ics-cert. us-cert. gov/controlsystems.
106. Amin, S., Litrico, X., Sastry, S., & Bayen, A. M. (2012). Cyber security of water SCADA systems—Part I: Analysis and experimentation of stealthy deception attacks. IEEE Transactions on Control Systems Technology, 21(5), 1963–1970.
107. Cerdeira, D., Santos, N., Fonseca, P., & Pinto, S. (2020, May). Sok: Understanding the prevailing security vulnerabilities in trustzone-assisted tee systems. In 2020 IEEE Symposium on Security and Privacy (SP) (pp. 1416–1432). IEEE.
108. Amin, S., Schwartz, G. A., & Hussain, A. (2013). In quest of benchmarking security risks to cyber-physical systems. IEEE Network, 27(1), 19–24.

109. Abera, T., Asokan, N., Davi, L., Ekberg, J. E., Nyman, T., Paverd, A., ... & Tsudik, G. (2016, October). C-FLAT: control-flow attestation for embedded systems software. In Proceedings of the 2016 ACM SIGSAC Conference on Computer and Communications Security (pp. 743–754).
110. Francillon, A., & Castelluccia, C. (2008, October). Code injection attacks on harvard-architecture devices. In Proceedings of the 15th ACM conference on Computer and communications security (pp. 15–26).
111. Roemer, R., Buchanan, E., Shacham, H., & Savage, S. (2012). Return-oriented programming: Systems, languages, and applications. ACM Transactions on Information and System Security (TISSEC), 15(1), 1–34.
112. Alemzadeh, H., Chen, D., Li, X., Kesavadas, T., Kalbarczyk, Z. T., & Iyer, R. K. (2016, June). Targeted attacks on teleoperated surgical robots: Dynamic model-based detection and mitigation. In 2016 46th Annual IEEE/IFIP International Conference on Dependable Systems and Networks (DSN) (pp. 395–406). IEEE.
113. Hu, H., Shinde, S., Adrian, S., Chua, Z. L., Saxena, P., & Liang, Z. (2016, May). Data-oriented programming: On the expressiveness of non-control data attacks. In 2016 IEEE Symposium on Security and Privacy (SP) (pp. 969–986). IEEE.
114. Lu, T., Xu, B., Guo, X., Zhao, L., & Xie, F. (2013, March). A new multilevel framework for cyber-physical system security. In First international Workshop on the Swarm at the Edge of the Cloud.
115. Zalewski, J., Drager, S., McKeever, W., & Kornecki, A. J. (2013, January). Threat modeling for security assessment in cyberphysical systems. In Proceedings of the Eighth Annual Cyber Security and Information Intelligence Research Workshop (pp. 1–4).
116. Khattak, H. A., Shah, M. A., Khan, S., Ali, I., & Imran, M. (2019). Perception layer security in Internet of Things. Future Generation Computer Systems, 100, 144–164.
117. Corallo, A., Lazoi, M., & Lezzi, M. (2020). Cybersecurity in the context of industry 4.0: A structured classification of critical assets and business impacts. Computers in industry, 114, 103165.
118. Zhong, M., Zhou, Y., & Chen, G. (2021). Sequential Model Based Intrusion Detection System for IoT Servers Using Deep Learning Methods. Sensors, 21(4), 1113.
119. Mitchell, R., & Chen, I. R. (2014). A survey of intrusion detection techniques for cyber-physical systems. ACM Computing Surveys (CSUR), 46(4), 1-29.
120. Chen, D. D., Woo, M., Brumley, D., & Egele, M. (2016, February). Towards Automated Dynamic Analysis for Linux-based Embedded Firmware. In NDSS (Vol. 1, pp. 1–1).
121. Upadhyay, D., & Sampalli, S. (2020). SCADA (Supervisory Control and Data Acquisition) systems: Vulnerability assessment and security recommendations. Computers & Security, 89, 101666.
122. Abdelrahman, A. M., Rodrigues, J. J., Mahmoud, M. M., Saleem, K., Das, A. K., Korotaev, V., & Kozlov, S. A. (2021). Software-defined networking security for private data center networks and clouds: Vulnerabilities, attacks, countermeasures, and solutions. International Journal of Communication Systems, 34(4), e4706.
123. Abikoye, O. C., Bajeh, A. O., Awotunde, J. B., Ameen, A. O., Mojeed, H. A., Abdulraheem, M., ... & Salihu, S. A. (2021). Application of internet of thing and cyber physical system in Industry 4.0 smart manufacturing. In Emergence of Cyber Physical System and IoT in Smart Automation and Robotics (pp. 203–217). Springer, Cham.
124. Ajagbe, S. A., Oyediran, M. O., Nayyar, A., Awokola, J. A. and Al-Amri, J. F., (2022) "P-acohoneybee: a novel load balancer for cloud computing using mathematical approach," *Computers, Materials & Continua*, vol. 73, no.1, pp. 1943–1959, https://doi.org/10.32604/cmc.2022.028331

Cybersecurity-Based Blockchain for Cyber-Physical Systems: Challenges and Applications

Yassine Maleh, Soufyane Mounir, and Karim Ouazzane

1 Introduction

Today Industry 4.0 is the leading trend of the fourth industrial revolution. It is characterized by fully automated production. Developing and implementing appropriate high-performance systems – cyber-physical systems – is necessary to achieve full industrial automation. Industry 4.0 is based on four aspects: virtualization, interoperability, decentralization and real-time operation. These same aspects correspond to cyber-physical systems (CPS). In such systems, equipment, sensors, and information systems are interconnected and can function almost independently of humans. Cyber-physical systems require a high level of decentralization to make independent decisions independent of humans, which modern technologies such as the Internet of Things, big data, cloud computing and various methods of artificial intelligence enable and support. Thanks to this set of technologies, CPS can monitor and control actuators in real-time [19, 41].

CPS are described as systems with a decentralized control system, formed by merging the physical and virtual worlds with autonomous behavior, creating a common system with other similar systems and establishing deep cooperation with humans [43].

Cyber-physical systems use embedded software sensors, and actuators, establish connections with each other and with operators, exchange information through interfaces, and store and process data from sensors or networks. The first scientific

Y. Maleh (✉) · S. Mounir
University Sultan Moulay Slimane, Beni Mellal, Morocco
e-mail: y.maleh@usms.ma; s.soufyane@usms.ma

K. Ouazzane
Metropolitan University in London, London, UK
e-mail: ouazzank@staff.londonmet.ac.uk

Y. Maleh et al. (eds.), *Blockchain for Cybersecurity in Cyber-Physical Systems*,
Advances in Information Security 102, https://doi.org/10.1007/978-3-031-25506-9_3

sources about cyber-physical systems appeared in 2006. The notion was first introduced at the American National Science Foundation seminar. Over the past decade, several scientific articles have been written explaining the concept. According to one study, the number of scientific papers devoted to the term has increased by about 40% each year, demonstrating the rapidly growing interest in this concept in academia [13].

Furthermore, these statistics show that the field of cyber-physical systems is expanding rapidly. As cyber-physical systems' capabilities and applications grow, so make the scientific approaches to defining and interpreting them. Therefore, major differences exist between the original approaches in 2006 and those in 2021–2022. Cyber-physical systems, acting as coupled computing systems, establish intense connections to the surrounding physical world and the processes that occur there and make their data available and processed. Figure 1 provides a detailed illustration of how the CPS acts as a bridge between the cybernetic and physical systems.

Cyber-physical systems are more complex than traditional computer systems: they combine solutions at the digital and physical levels. On this basis, the approaches to identifying and assessing threats to cyber-physical systems are somewhat different. Most threats, in this case, originate in cyberspace; in the future, these threats will already reach the physical space level [39].

Blockchain BC technology is known to be the basis of cryptocurrencies. But it also offers potential for many other areas. Besides the financial sector, IoT is one of the two main application areas. A CB is essentially a distributed database, or Distributed Ledger, containing records of all digital events or transactions executed and exchanged between the parties involved. Once recorded in the blockchain, the data cannot be manipulated or deleted. Therefore, it contains verifiable records of all digital events that have been executed. In this way, the blockchain enables the creation of a network without intermediaries, in which the parties involved can trust each other without having to trust each other outside the network. Entities joined together in a BC network are called nodes. Full nodes store the entire blockchain, ensuring the accuracy of the data. Lightweight nodes do not store or process the whole blockchain, allowing them to use less storage and computing capacity. Because of its operating principle, BC technology addresses the above challenges of CPSs. In these systems, protection against data manipulation, cyber-attacks, and data integrity are fundamental areas in which BC technology has significant advantages over other solutions. The decentralized and redundant data storage also provides high transparency and security against technical failures.

The application of blockchain technologies in cybersecurity is generally less known, but no less promising. This chapter aims to study the different cybersecurity applications in cyber-physical systems based on blockchain technology.

This chapter will cover the following topics: Sect. 2 presents the research methodology adopted in this paper. Section 3 discusses the various case studies of blockchain technology for cyber-physical systems applications, such as IoT, healthcare, e-commerce, transportation, and cybersecurity. Section 4 analyses in depth the application of blockchain for cybersecurity in CPS. Section 5 discusses some

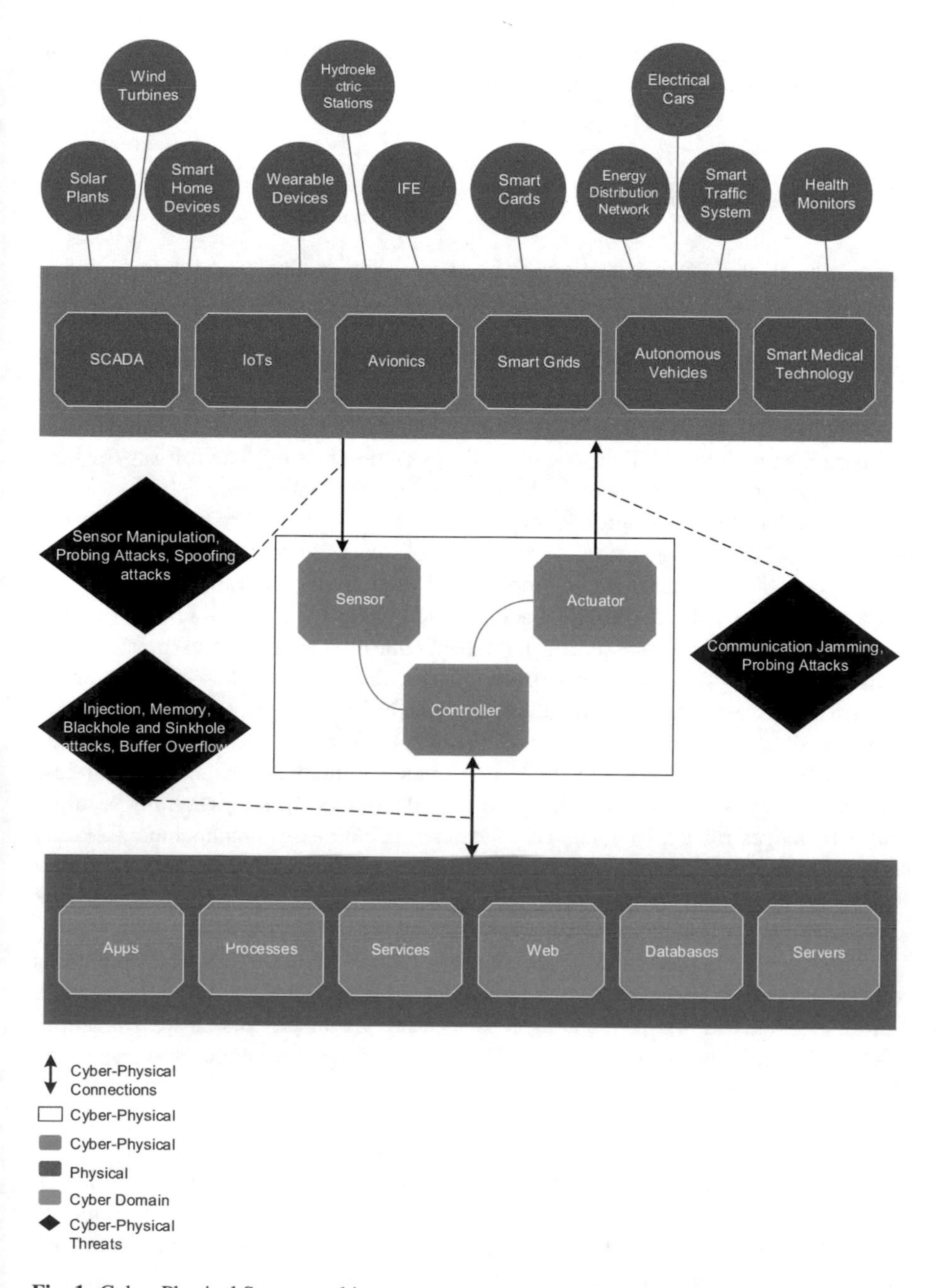

Fig. 1 Cyber-Physical System architecture

limitations and open research issues related to blockchain-enabled cybersecurity in CPS. The last section presents the conclusion.

2 Research Methodology

Dedeoglu et al. [17] discussed various challenges encountered while applying blockchain systems to CPS. For example, scalability and resources challenge. Blockchain needs miners and a lot of computational capacity to mine new blocks. Broadcasting transactions demand a lot of bandwidth; immutability property further increases the size of the database. Similarly, a high processing time is needed to validate the transactions. Similarly, other challenges include privacy concerns, as all participating nodes will have access to transaction history. The third is trust, an element of risk as data gathered by blockchain may be malicious.

A complete set of design challenges of blockchain in the health care domain is discussed by Rathore et al. [51]. Some of them relate to data exchange as health care data is sensitive. Others are related to incorporating big data for data mining, acquisition, storage, backup, and recovery. Solutions related to these challenges were also discussed. In industrial control systems, the associated challenges are storing the data from sensors, meters, smartphones, etc.—sharing device information, secure mutual authentication, scalability management of IoT devices, creation of secure virtual zones, etc. Some of the challenges in transportation zone include providing secure vehicle-to-vehicle communication, a secure wireless update system, a network security platform, distributed trust system for security, etc. Challenges related to smart grid systems include asset management, identity management, real-time data recording, data sharing, privacy, no manipulation by third parties, decentralized storage, etc. Rathore et al. [66] also discussed the application of blockchain on cyber-physical systems.

Khalil et al. [30] presented the application of blockchain for CPS from operations and security perspective. The studies related to operations and security are classified into five groups: Cyber-physical system security, Cyber-physical System Control, Cyber-physical system trust, Cyber-physical system performance or storage, and Cyber-physical system applications.

The following Fig. 2 presents a statistical review of research papers presented at conferences and journals on CPS security and operation that uses blockchain technology between 2018 and 2022. In Fig. 2, we see a visual representation of the statistical results in the literature; more precisely, the articles are displayed chronologically. From the stacked bar chart, it can be observed that although CPS security concentrated papers continued until 2022, the trend was transitioning from CPS security concentrated papers to CPS operation-focused papers as the volume of CPS blockchain research dramatically increased. There is an increasing volume of work outside of CPS security and operations that explore the possibilities presented by blockchain technology. Though more articles have been published in 2022 than

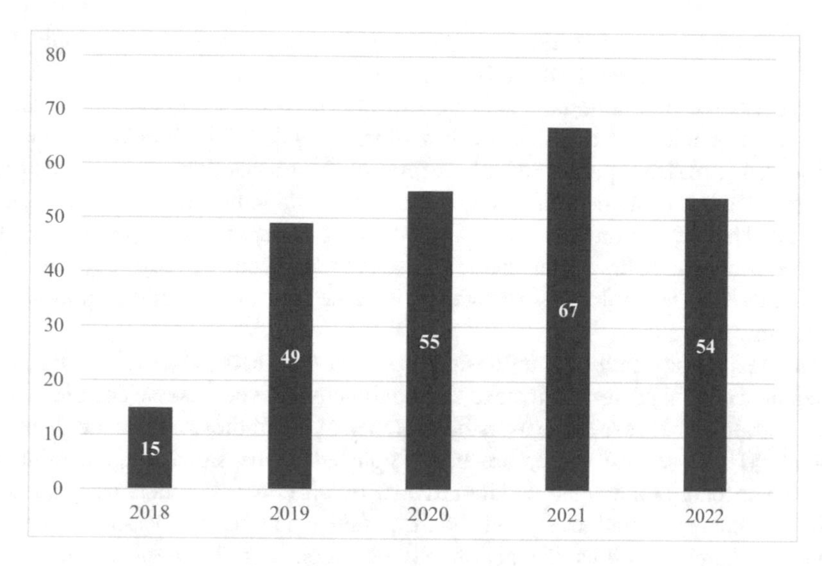

Fig. 2 Statistics of the papers published between 2018 and July 2022

in the previous several years, this study only accounts for those published up through the beginning of July.

3 Blockchain (BC) Applications for Cyber-Physical Systems (CPS)

The blockchain is a distributed ledger that prevents fraud and tampering by recording transactions in an unalterable public ledger that cannot be altered. The distributed ledger approach proposed by Abadi and Brunnermeier [1] would allow records to be held independently on any blockchain network. The distributed nature of the blockchain makes it challenging to scam or steal from the system. They also show that technological record-keeping is superior because of its accuracy, decentralization, and low cost. By contrasting it with the old centralized approach, they show how revolutionary blockchain is for maintaining records. The algorithms built into the blockchain allow for the redaction of previous transactions in the ledger in the event of an error. The system's use is not limited to the financial sector; other sectors where openness is essential can also benefit significantly from it. Successful companies are those that effectively communicate with their constituents.

The term "blockchain" describes a series of linked electronic blocks that may be used to record data of any type. "a method for storing and sending information that is transparent, secure, and functions without a central governing body" is how blockchain is typically described. Blockchain is a type of Distributed Ledger Technology (DLT), which are digital systems that keep track of and share information about financial asset transactions over a network of computers. When people think of DLT, they usually think of the blockchain. A blockchain is a distributed ledger that records the transactional activity of all of a network's users from the time it was launched until the moment the last block is added. The blockchain may be considered a massive public record storing all the transactions done by its users ever since it was created, which is why the image of a ledger is commonly used to depict it. That it is immutable is one of its defining features. It is impossible to alter the blocks once they have been created, thanks to many cutting-edge cryptographic safeguards. This technology is trustworthy because you can only add operations, not change or remove them.

This technology relies on a dispersed peer-to-peer network, where data is not stored on a central server but instead shared directly between users. The blockchain is decentralized because some users have copies of it and these copies are dispersed globally. These several copies are being updated at the same time, continually. Blockchain data is managed collaboratively, by all network nodes, as opposed to centrally managed databases. Each of these nodes adheres to the same computer protocol, which specifies the actions to be taken and the requirements to be completed before updating the database.

In its most basic form, blockchain technology is a database that stores records, often transactions, in groups called blocks. A block is more than a collection of data; the connections between them cannot be broken (by chains). Users, or "miners," validate each new block as it is added to the blockchain. The blocks are timestamped and added to the blockchain when verification is complete, making the data available to anybody with access to the network. This makes the transaction public to the recipient and the network at large. Tokens serve as the medium of exchange for miner compensation. Changing or modifying an existing block or the chains is not feasible; the only method to alter the blockchain is to add a new block.

There is no doubt that blockchain technology represents a major advancement. The number of businesses that are using blockchain technology has skyrocketed recently. Many businesses have been revolutionized, not only Airbnb and Dropbox. This shift in strategy will boost revenue and boost productivity for the company. Increased government employment equals better service delivery. It's been a game-changer for companies since it's a faster and cheaper alternative to cash transactions. The payment mechanism relies on blockchain technology to work. This technology has several potential uses, including digital identification, online shopping, insurance, real estate management, online payments, and crowdsourcing. The capacity to validate every transaction over a network makes it useful for facilitating the exchange of data and information. In less than 2 s, you can perform what might take days or weeks at a traditional bank or company—business and industrial innovation velocity. Consensus is the key to the success of Blockchain technology. Many sectors have been profoundly affected by blockchain's introduction. Accenture, the largest IT business in the United States, has indicated that it is considering a blockchain deployment to create blockchain-based applications. Like any other technology, blockchain can be used in many different contexts; for example, it could make certain types of transactions more secure and efficient. Or it could enable

Table 1 Cyber-physical systems applications

Domains	Applications	Advantages
IoT	Managing identities and attributes taking stock with the help of data. Safety in Internet of Things communications Communicating from the edge to the cloud safely	Advised a SOA design approach Facilitate improved company procedures smart contracts for controlling access Introduction to Internet of Things apps that make use of blockchain technology
Transportation	Vehicle electronics, transportation networks, aviation, and airspace control.	Management of complicated flows and equipment compliance is simplified. Reducing the complexity of the payment process Flows that can be traced Inverse logistics
e-commerce	Market transparency, monitoring and supply chain tracing. Redesigned payment gateway, safe online store, and verified user reviews of actual product.	Modalities of exchange besides conventional currency Faster order processing Strengthened protection for financial transactions Improved speed of financial dealings
Healthcare	E-health, electronic health records and health data collection, storage, and sharing.	Organization of Health Records Efficiency improvement in clinical trials Anti-counterfeiting measures and drug tracing
Finance	Trade finance, smart assets, and smart contracts to combat fraud and money laundering and expand access to banking services	The Bank as Uber: a new model for financial services Enabling of monetary transactions More reliable and productive financial dealings
Cybersecurity	Data authentication, anonymity of users, keyless signature infrastructure, and cyber-physical transaction validation	Complete, permanent protection of data Certifying agencies may be replaced by decentralized blockchain technology. High-tech password verification

automated, low-latency bidding through automated contract systems or systems of record for large, complex businesses.

Emerging blockchain applications for cyber-physical systems, including IoT, medical records, transportation, e-commerce, banking, and cyber security, are the primary emphasis of this chapter. The study's application domains and the systems they cover are listed in Table 1.

Blockchain technology's first use case was in the financial sector, but now it's being adopted in diverse industries like healthcare, education, banking, the Internet of Things, and more. This day and age, cyber-physical systems are all the rage. Blockchain solves the problems plaguing cyber-physical systems because of its inherent decentralized character. Some blockchain use cases in CPSs are outlined below.

3.1 Health Care

Health care is another domain in which blockchain can be extensively applied (Rathore et al. [51]). Preservation of medical records and avoiding modifications to them is vital. Blockchain can help modify medical records. It is a critical aspect of medical records. Any modification to the medical record will leave a digital trace, thus ensuring the document's integrity. IoT devices for medical purposes can be combined with Blockchain technology for real-time data processing in a secure manner. There is a use case where an Ethereum-based private blockchain communicates with smart devices using sensors.

Many sectors like banking and retail used Big Data (Bhuiyan et al. [11]). Health care also is one of the industries that can leverage Big Data. The data that gets generated every day is increasing. Healthcare data generated by healthcare devices, providers' computer systems, and wearable devices can be structured and unstructured. When transformed into Big Data, the data can help in clinical decision-making. Traditionally doctors used to depend on their judgment in providing treatment. With Big Data, decision-making takes into consideration of the best available information. However, there are concerns regarding the privacy and security of Big Data in health care.

Blockchain is considered a solution for privacy and security. There are different blockchains—Public, private, and Hybrid (Bhuiyan et al. [11]). For healthcare purposes, a patient can choose a private or hybrid blockchain where he/she can decide to whom to give access to their medical information. For example, to the health care provider, family, etc. This way, the patient will have complete control over his/her health information. Otherwise, if a patient goes to four different providers, he/she will have his medical information at four different places and cannot control any modification to his medical information.

In the case of Blockchain, to add information, it needs at least half of the blockchain participants (Bhuiyan et al. [11]). The Blockchain proposed for health-care data consists of three layers. The first layer is the interface to the Blockchain and is called the application layer. The second layer is the private blockchain consisting of nodes and an essential authoring entity responsible for generating public and private keys. The third layer is the encrypted database layer, where all the protected health information is saved.

When a health record gets created, it would be encrypted and stored in the database (Bhuiyan et al. 2018). Every time a new record is inserted, a pointer is made in the Blockchain along with the creator information. The patient would also get notified. The patient can access his health records from anywhere in the world. They can control who gives access to their health records. Figure 3 presents the blockchain layered architecture.

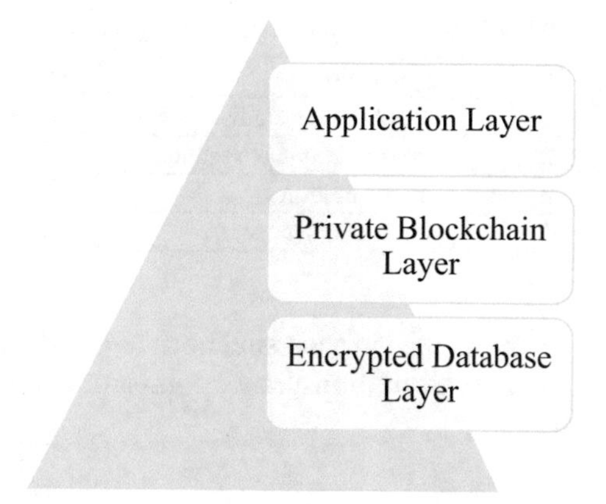

Fig. 3 Blockchain layered architecture

3.2 Industrial Control Systems

Industrial control systems (ICS) monitor and physical control entities that, in turn are used in various industries such as nuclear power plants and irrigation industries (Rathore et al. [66]). The data collected from various sensors are sent to controllers, which propagate it to actuators. A use case depicts how Ethereum-based blockchain can be used to communicate electricity usage. Depending on the blub and air conditioning usage, Ethereum communicates to the network to update the devices from a normal model to energy savings mode.

3.3 Transportation

Connected vehicles connected to the internet gather a lot of information inbuilt into the car (Rathore et al. [66]). They use that data to communicate with other cars. This technology created several programs like lane change warnings, collision avoidance, and cruise control. These programs assist the driver in providing a better safety and driving experience. Several protocols aid in this process. For example, Dedicated Short Range Communications (DSRC) is approved to handle communication between vehicle applications. Similarly, IEEE 1609.2-4 is the protocol used for V2V communication that includes the change of lane warnings, blind spots, forward collision warnings, etc.

Information sharing between vehicles is made possible by SAE J2735 Basic Safety Messages (BSMs) (Rathore et al. [66]). Basic Safety messages are trusted and not encrypted as they are broadcasted to all vehicles nearby. In contrast, messages regarding velocity, position, size, etc., information is trusted and encrypted using PKI certificates. Although BSM messages are encrypted, there is a possibility of tampering with the messages, leading to serious consequences. For example, sensors can be jammed to remove vehicles from navigation systems. Blockchain

Table 2 Smart grid applications using blockchain

No	Applications	Solutions
1	Health care systems	Ethereum based blockchain
2	Industrial control systems	Ethereum based framework
3	Transportation	Blockchain technology in the vehicular cloud
4	Smart grid systems	Blockchain with artificial agents

can be applied in such situations to avoid tampering with data. That data becomes a valuable asset for insurance companies and can help in other investigations.

3.4 Smart Grid Systems

A smart grid system is an intelligent energy network system that delivers electricity efficiently from the producer to the consumer (Rathore et al. [66]). Since existing systems do not defend against cyberattacks, blockchain technology can defend against cyber-attacks. Various frameworks and techniques have been discussed to use blockchain for smart grids. For example, a private blockchain with an artificial agent is proposed that can provide real-time pricing information. Similarly, blockchain-based power grid communications in smart communities are used to preserve privacy. Table 2 illustrates smart grid applications using blockchain.

Another promising use of blockchain in the industry is its symbiosis with Internet of Things (IoT) technologies [35]. The production and improvement of "smart" devices occur in almost all industries. This is the industrial Internet of Things, IoT in enterprise resource planning systems, self-diagnosing equipment performing self-service, creating networks of miniature devices that form a "web" of senses for powerful artificial intelligence, and much more. Given the low trust in such IoT devices, blockchain could be a solution to security problems, requiring, however, simultaneous high performance and scalability. If the third generation offers reasonable compromises to this issue, combining IoT and artificial intelligence with blockchain could spawn a new stream of revolutionary industrial innovation.

3.5 Cybersecurity

Using blockchain technology, confidential data may be maintained in an immutable ledger of transactions, accessible only to authorized parties and immune to tampering. The adoption of smart contracts is crucial for security improvement since they ensure that all the conditions necessary to carry out a command or action have been met in advance. Every node in the network that has been authenticated has access to an immutable copy of the blockchain, making decentralization a highly available solution [35]. Maintaining order in data storage is less difficult, and altering financial transactions is extremely difficult. The sheer number of nodes that make up a single blockchain makes it almost difficult for an outsider to compromise it without using enormous amounts of processing power trying to take over the network.

The state-of-the-art security methods provide a central repository from which to provide access. On the other hand, Blockchain is built on top of distributed ledger technology, making it more resistant to hacking (Taylor et al. [57]). Below are some areas where Blockchain can be applied from a cybersecurity perspective.

- ***IoT Devices***

  Temperature sensors and routers are just two examples of edge devices that hackers are increasingly targeting to access the system as a whole. Hackers may now more easily access broad systems like home automation via edge devices like "smart" switches because of the current fixation with artificial intelligence (AI). Most Internet of Things gadgets doesn't have full-fledged safety measures in place. By removing the need for a centralized authority figure, Blockchain may be utilized to secure these types of systems or devices. By taking this technique, devices can independently choose their level of security. Edge devices are more secure because they can identify and act on questionable orders from unfamiliar networks without depending on a centralized administration or authority.

  When hackers enter the device's central administration, they take over everything. Blockchain makes it harder for these assaults to succeed by decentralizing device authority structures (if they are possible at all).

  Data and harmful behavior on networked IoT devices may be monitored using blockchains. Also, IoT devices may safely install firmware updates through peer-to-peer transmission. It guarantees the identification and verification of devices and continuous, safe data transmission.
- ***Data Storage and Sharing***

  Blockchain employs hashing algorithms to guarantee the integrity of data saved in any medium, including the cloud. Because blockchains are distributed ledgers with client-side encryption, all participants must agree to any data modifications before being broadcast. By guaranteeing a decentralized data storage network, blockchain technology provides a safeguard for private information. Using this countermeasure would make it exceedingly difficult, if not impossible, for hackers to access computer systems containing sensitive information. Several cloud storage providers are investigating how blockchain may be implemented to keep customer information safe from hackers.
- ***Securing DNS and DDoS***

  When users of a network resource, server, or website are prevented from accessing or using the resource in question, this is known as a distributed denial of service (DDoS) assault. These assaults disable or hinder the underlying resource infrastructure.

  On the other side, hackers may easily break into the link between an IP address and a domain name by targeting a DNS that hasn't been patched or otherwise compromised. This attack may render a website inoperable, charged, and even lead users to malicious domains.

By distributing DNS records, blockchain technology offers a promising solution for reducing these attacks. Blockchain would have closed all security loopholes that hackers may have used by utilizing decentralized solutions.

- ***Protecting data transmission***

  In the future, blockchain technology might be used to secure the data while it is in transit. The technology's full encryption function allows for protected data transfer by shielding information from prying eyes.

  With this method, confidence in blockchain-based data would rise. Hackers with bad intentions frequently intercept data in transit to destroy or make unauthorized changes. As a result, there will be a significant void in poorly streamlined avenues of communication like email.

4 Blockchain-Enabled Cybersecurity in CPS

Using the latest tools in cybersecurity, a unified database may be used to control who has access. On the other hand, Blockchain is based on distributed ledger technology, making it more robust and less vulnerable to tampering. Blockchain can build trust among its users and not rely on a central authority because it uses cryptography and mathematical techniques. Blockchain technology's genuine, transparent, and immutable qualities make it useful in many fields. It has found use in fields as diverse as finance, medicine, the Internet of Things (IoT), schools, and cyber security. We then explore cyber security issues that blockchain can solve [31].

Blockchain technology can address privacy and security concerns (Tariq et al. [56]). The immutability, decentralization, and security make it ideal for fog-enabled IoT devices. Another essential feature of blockchain is smart contracts. They define rules based on which authentication is carried out. Smart contracts can also sense malicious actions and breached blockchains reject updates. Besides, a unique GUID and a symmetric key pair for each IoT device connected to the blockchain network are provided. However, there are also challenges in Blockchain as the data starts to grow bigger. Therefore, further research is needed to overcome the challenges of Blockchain. Using a database to handle enormous amounts of data and linking Blockchain might overcome the difficulties of storing big data in a blockchain. In the following section, we will organize recent research according to the various topics that have been explored. There is an undeniable connection between all the research, and several may be appropriate for use in more than one of the groups. As we organize the studies, however, we focus on the authors' most pressing aims, which helps us see the big picture of the most interesting current developments in the field.

4.1 Cyber-Physical Systems Security

There has been a recent shift toward using blockchain technology in creating security processes and systems to guard against cyberattacks. Without a trusted central repository, the Blockchain guarantees the authenticity of all transactions. The transactions involving system users' tangible and intangible assets are limited to a small set of preset actions. Hashes connect successive blocks that include transaction data to create a chain. A consensus algorithm ensures that all participants in the system have access to the same, immutable set of blocks, making it more difficult for an adversary to compromise the blockchain. In terms of data safety, blockchain's primary benefit is that it is exceedingly challenging to compromise previously recorded transactions. Due to the interconnected nature of the blocks, even a small alteration in one might cause the entire chain to fail. However, the computational difficulty of this endeavor significantly reduces the likelihood of a hack of a blockchain system.

Several applications of blockchain technology are already in use in cyber-physical systems. The key advantage of this technology is, as was previously said, the capacity to validate a wide variety of transactions that would otherwise be impossible in a non-trusted setting. Research has concluded that blockchain technology will play a significant role in the coming fourth industrial revolution (Industry 4.0) [6, 22]. In addition, as part of Industry 4.0, blockchain is being promoted together with other promising technologies of our time [5]. The Internet of Things [21], fog computing [10], big data [68] and augmented reality [21], are examples. Blockchain is widely considered a vital technology in the Industrial Internet of Things, helping to turn conventional factories into cutting-edge "smart" factories that employ the most recent advances in digital technology. Here, we'll discuss several ongoing studies that provide concrete instances of how blockchain technology is being applied to address security issues in real-world cyber-physical systems. One crucial aspect of the current effort is the safe administration of various assets, including those in cyber-physical systems. Bitcoin was the first digital currency to make use of blockchain technology. So, that's the way it turned out. The cryptocurrency market expanded alongside the development of blockchain technology, and it is now deeply embedded in daily life. There has been a significant increase in the use cases for blockchain technology recently. For example, recent research by Bhushan et al. [12] examined blockchain's potential for addressing security concerns in massive cyber-physical systems like smart cities. The authors analyze the potential of blockchain technology in connection to key components, such as transportation, healthcare, smart grids, financial systems, supply chain management, and data center networks. They offer future research areas based on their findings.

There are several distinct ways to categorize studies of blockchain technology. The initial body of research deals with logistics and distribution in blockchain technology. The researchers explain various parts of the problem and provide a generic solution for blockchain-based supply chain management. Still, they do not specialize in any sector or subset of cyber-physical systems. The offered methods may, in certain situations, be tailored for implementation in cyber-physical systems serving a variety of objectives. They aren't always given such a broad definition in the documentation. As a result, Saberi et al. [54] proposed a taxonomy of the obstacles to using blockchain technology in Supply Chain Management SCM. Some of the difficulties in addressing these obstacles were explored by Aceto et al. [5]. Neither option specifies the exact asset being dealt. This dataset also includes a large number of supply chain services and products. Kshetri et al. [33] detailed practical uses of blockchain technology for inventory and supply chain management.

Using blockchain and IoT technologies together is a significant focus of many cyber-physical systems. Research into the organization and administration of production and its related responsibilities would fall under the first category if it were not subdivided further. In the context of collaborative manufacturing, Yu et al. [65] provided an architectural method for ensuring data integrity in cyber-physical production systems. The second set of investigations, which involves supply chain management, is committed to doing away with everything that might endanger the smooth functioning of a particular asset or service. These kinds of programs have increased recently so you can easily choose among them. Electricity [36], gasoline [37], computer resources [55], and software may all be distributed and sold with the use of blockchain technology. All of the earlier research involves commodity exchange for monetary gain. Therefore, blockchain-based solutions significantly rely on the ideas behind cryptocurrencies. The following subset comprises studies addressing the challenge of coordinating reliable communication between several devices. Data integrity duties of such devices may vary depending on the data being secured. Many papers focus on the network dynamics of generic IoT devices without addressing the nuances of cyber-physical system interactions. The following are only a few recent instances of efforts in this area: [15, 32, 38].

The energy efficiency of architectural solutions for the Internet of Things is a common theme in this research, and many different approaches are proposed for achieving it. The duties at hand may be broken down into two categories: those that focus solely on ensuring transactional integrity, and everything else. Those that do so also protect customer information privacy during financial dealings. In one research [34], for instance, it is argued that information on the whereabouts of Internet of Things devices should be protected. Since the authors recognize the need to keep this information private, they suggest a system combining blockchain technology with encryption. When discussing specific use-cases for the use of blockchain technology for data security in Internet of Things-based cyber-physical systems, it is important to keep in mind the category of cyber-physical systems that includes linked cars, which provide for unmanned vehicles [14, 49, 50].

Xu et al. [63] presented an anonymous blockchain-based system for charging-connected electric cars, using blockchain technology and creating a multi-party security system between EVs and EVSPs. A novel blockchain-enabled batch authentication strategy for artificial intelligence-envisioned Internet of Vehicles-based smart city deployment was developed by Bagga et al. [9].

In recent works, to improve security and trustworthiness in UAV-enabled IoT applications, Abualsauod [4] presented a unique Hybrid blockchain-based approach. To provide robust authentication and secure communication, Gupta et al. [24] proposed a lightweight secured architecture for blockchain-based IoV.

There will be an explosive increase in the number of academic journal articles devoted to this topic between 2022 and 2024. Therefore, it is reasonable to conclude that adopting blockchain technology to secure this class of cyber-physical systems is a significant development toward resolving this issue.

4.2 Data Authentication in Cyber-Physical Systems

A forensic procedure may be applied to digital evidence, including the following steps: identification, gathering, inspection, analysis, documentation, and presentation. The notion of preserving digital evidence is crucial and should be considered at every stage of this procedure. To this end, blockchain is crucial in guaranteeing the authenticity and provenance of the evidence gathered. However, the security systems and methodologies employed must meet stringent non-functional criteria due to the complexity of Cyber-Physical Systems operating scenarios regarding scalability, computing performance, and communication network utilization. Several blockchain-based methods have been developed to safeguard data from cyber-physical threats and guarantee its authenticity.

When it comes to digital steganography, on the other hand, the methods used rarely have anything to do with protecting the authenticity of the material being hidden. It is important to differentiate between several initiatives aimed at enhancing the security of sensitive data in cyber-physical systems by inventing methods and algorithms for hiding information in digital photos and other digital objects [49, 58, 59].

While the authors of these solutions state that they are concerned with protecting user privacy in the IoT, they fail to offer illustrations of how their algorithms may be applied outside of that setting. The authors of much of this study are concerned about the lack of safety measures in telemedicine. Considering the prevalence of such research, it merits its category. However, this group of works does not push the boundaries of multimedia embedding and will not be the subject of future research. The works below also incorporate the incorporation of more conventional forms of data into multimedia forms. The authors highlight the applicability of their methods in cyber-physical systems while identifying and explaining the restrictions associated with data transmission conditions peculiar to such systems. Not as many works fall into this category, but they deserve to be treated independently of the first. The problem of insecure telemedicine image transmission is addressed in [48]. Confidential images are hidden in photographs that also include public information. The image container contains the secret picture's fingerprint (perceptual hash) for verification purposes. One unique feature of this method is that it keeps tabs on the order in which images are sent.

Hoang et al. [27] implement digital watermarks into wireless sensor network information. The embedding algorithm is predicated on a gamma-like transformation of the binary alphabet. Lightweight algorithms are believed to provide benefits. The algorithms reliably and independently incorporate the digital watermark components into the sensory data without regard to the values of the sensory data or any of their attributes. A digital watermark based on encrypted information can be generated using established techniques and newer ones that address the challenges of wireless sensor networks and the Internet of Things. Digital watermark parts are generated in the simplest case by simply using the values of sensor data items.

A good example is the embedding method described in [61]. With this method, the digital watermark bit is generated and embedded in the subsequent sensor value based on the values that came before them. While there are many benefits to using a digital watermark with several embedded components, the timing of this method is problematic. Removing a digital watermark will be complicated even if an attacker isn't actively intercepting your communications. With regards to this problem, Wang et al. [59] proposed that sensor data be partitioned into groups of varying lengths, with the key serving as a separator. Chains of digital watermarks are generated and embedded in neighboring sets. Digital watermarks are used to verify the sensor data. The second digital watermark chain provides separation and data synchronization, which encodes group separators. A digital watermark containing values for sensory quantities and some of their attributes is achievable in a more complex setting. For instance, look at Internet of Things device authentication [20]. It is possible to use the stochastic properties of data streams as the basis for digital watermarks. The use of spectrum expansion to embed digital watermarks into data streams is a common practice. Several features of the obtained data, including data length, frequency of occurrence, and time of capture, are discussed by Hameed et al. [26] to produce a digital watermark. To prevent sensor node clone attacks, Nguyen et al. [46] developed a digital watermark dependent on CSMA/CA collisions. Furthermore, the unique visualization of the sensor data sets this research apart. They resemble a computer image and are called a matrix when put together.

This setup generally permits using tried-and-true methods for processing digital images, including sensory input. All the research techniques above use digital watermarks representing binary sequences. It is also a topic of research to use watermarking on analog transmissions (most often modulated signals) to solve the problem of signal source authentication. The solutions to the puzzles presented there are, ideally, analogous to those offered by digital watermarking. The only difference is in how the signal is represented and, by extension, how it is processed. The study of sender authentication in NB-IoT (Narrow Band Internet of Things) compliant systems [66].

For this inquiry, we employ the concept of a radio-frequency watermark. The watermark is digitally generated rather than binary sequences for further embedding and then transformed to modulated signals. Since the usable and watermark signals do not interfere, the proposed method is more trustworthy. In some situations, watermarks can be used to prevent specific attacks. Rubio-Hernán et al. [53] proposed a new technique for identifying cyber replication efforts on networked control industrial systems based on adaptive control theory. An attacker attempt to gain command of the system by recreating stolen data sequences. The strategy for applying this algorithm to guard against an intruder is the significant contribution of this study, not the embedding technique derived from earlier papers.

A method for including reversible air indications in the signals transmitted by "hard" real-time industrial control systems is provided by Huang et al. [28]. The authors focused on ship control systems because of their potential impact. Before beginning the embedding process, a secret key must be transmitted through a safe communication channel. This method can identify assaults that attempt to delay or distort a signal. As was said before, some preliminary attempts have been made to combine blockchain with digital watermark technologies. Blockchain technology and digital watermarks are two solutions that handle various cyber-physical system security issues. When used together, these techniques could generate more safety than any of them could on their own.

To guarantee credibility amongst users of various blockchain-based CPS apps, Mohanta et al. suggested a signature storage system in 2020 [44]. Docker technologies and the Ethereum network were used to construct the solution, assuring security and significantly saving storage space and associated costs. A Cyber-Physical Trust System (CPTS) is an IoT-based CPS that incorporates trust as a system component, and blockchain was proposed by Beckmann et al. [43]. The CPTS motivated by blockchain was developed further by Milne et al. [43], who used the Tamarin Prover tool to provide formal proofs of attributes such as integrity, identity, authentication, and non-repudiation.

To enhance the security of SDN and blockchain-enabled networks, Neelam and Shinray [45] introduced a CPS model that incorporates IoT devices and makes use of a completely programmable recursive internetworking architecture (RINA) with secure authentication through RINA password authentication. Rathore and Park [52] addressed the challenges of centralized control, privacy, and security in deep learning (DL) for CPS. They suggested using blockchain for DL operations deployed at the edge layer to provide decentralized and secure operations in IoT CPS networks, which they called DeepBlockIoTNet.

A few studies from 2022 emphasize a combination of objectives, including security, control, performance, data storage, and privacy. Table 3 below summarizes the various applications of blockchain technology in cybersecurity.

5 Limitations and Open Research Issues

Blockchain technology has been experiencing meteoric growth in recent years. Because it paves the way for future applications that make use of interconnected devices, it has the potential to revolutionize how people work and communicate. But it does have some restrictions, such as:

Table 3 Blockchain Applications for Cybersecurity in CPS

Authors	Techniques	Applications	Contributions
Mei et al. [42]	Blockchain-enabled privacy-preserving authentication	Transportation CPS	Employs an elliptic curve to construct a pairing-free ring signature scheme, significantly reducing overhead resources in the transportation CPS with cloud-edge computing.
Abd El-Latif et al. [2]	Quantum-inspired Blockchain	Smart edge utilities in IoT-based smart cities	Protect yourself from attacks that might be made using digital or quantum computers.
Zou et al. [69]	Blockchain-based medical data sharing	Medical data sharing	Designing SPChain, a new data sharing, and privacy-preserving eHealth system.
Abdulkader et al. [3]	Lightweight Blockchain-based cybersecurity (LBC)	IoT environments	Deal with high-volume data transfers and demanding computations.
Abdulkader et al. [3]	Blockchain empowered cooperative authentication	Vehicular edge computing	Deal with large amounts of data and processing that demand a lot of bandwidth.
Badsha et al. [8]	BloCyNfo-share	Cybersecurity information exchange (CYBEX)	Specify the channels via which confidential data may be disseminated or accessed by third parties.
Xia et al. [60]	Trust-less medical data sharing	Electronic medical Records in Cloud Environments	Try to get health records out there where they won't be so closely watched.
Dagher et al. [16]	Privacy-preserving framework for access control	Electronic medical records	Protect the confidentiality of patients' personal data.
Patel [47]	Blockchain-based secure and decentralized sharing	Transferring images across sites Hyperledger fabric trees may be used to process individual health records in bulk	Keep patients' private data safe and secure.
Dorri et al. [18]	BC-based smart home framework	Interconnected smart vehicles	Take into consideration the privacy and security risks associated with autonomous cars.

- Due to its reliance on block size and the time required for hash computations, it cannot grow with the number of connected devices.
- In certain implementations, this necessitates the existence of transaction fees or some kind of miner compensation.
- Even while it's not as centralized as the idea of a single bank, it still relies on a small group of powerful participants (miners).
- Since they are responsible for storing the complete ledger and acting as endorsers or miners in the transaction verification process, those involved in the blockchain have significant computing and storage needs.

The use of blockchain technology has become crucial in today's battle against cybercrime. For instance, numerous actors in the cybersecurity industry provide Content Delivery Network CDN and Dynamic Deny of Service DDoS mitigation services using blockchain technology. Users can lease unused network capacity to aid content delivery and bot defense. Some players have understood this and are positioning themselves on this rapidly developing technology. Indeed, in response to a different type of Distributed Denial of Service (DDoS) attacks, only decentralized solutions can be used, not as an additional cybersecurity product, but as a parallel system to guarantee data protection. The blockchain offers the security of a history set in stone, eliminates grey areas and facilitates the traceability of operations. Each act is time-stamped and guarantees the identity of the person who generated it [40]. This makes it ideally suited for various future communication networks and service management. Research is needed on many fronts to evaluate the potential opportunities of blockchain in next-generation network and service management.

- **A multiplied attack surface:** The attack surface is the weak point in a network that hackers can use to infiltrate a system. The connection density of an IoT network with 5G is considerably high: millions of devices are capable of connecting in a single square mile. This means the attack surface will be five times larger than 4G. This means that all Internet of Things (IoT) devices and sensors inside buildings and cities will need to be secured... [7].
- **Telemedicine:** a sensitive sector: Telemedicine is another area where next-generation networks will play an important role [25]. It will now be possible to quickly transfer heavy patient files, perform operations remotely with surgical robots, and use connected objects to monitor, for example, patients' insulin levels. If the devices are not sufficiently protected against attacks, the risks in terms of computer security are worrying: identity theft, invasion of privacy, and attacks on individuals.
- **IoT Security:** It's becoming common for hackers to exploit vulnerabilities in "edge" devices like thermostats and routers to access more extensive networks. Hackers may now more easily access larger systems, such as home automation, via edge devices, such as smart switches, thanks to the current fixation with artificial intelligence (AI). The majority of these IoT gadgets typically have inadequate safety measures. This is where blockchain's decentralized nature may be utilized to safeguard large-scale networks or individual devices (Xu et al. [62]). This strategy will enable the gadget to choose how to protect itself

independently. The edge devices are more secure because they can identify and respond to questionable requests from unfamiliar networks without relying on a central admin or authority.

Normal hacking procedures involve intruding on the device's central administration and gaining complete control of all connected devices and systems. Blockchain's ability to decentralize authority over devices makes these kinds of assaults more challenging to carry out (if even possible).

- **A decentralized network:** 5G has a different architecture than 4G. It relies more on software applications than its predecessors. It uses complementary SDN and NFV technologies, which have been developed with the cloud and offer a decentralized approach to the network [64]. The advantage is that it can adapt to users' needs; the medical sector, for example, requires more security and reliability than a video download. Resources can be allocated to each department according to its challenges. The downside is that this decentralization makes the network more difficult to monitor, which is especially problematic for surveillance agencies engaged in the fight against terrorism.
- **Verification of Cyber-Physical Infrastructures:** Information created by cyber-physical systems has compromised its integrity due to data manipulation, system misconfiguration, and component failure. However, blockchain technology's strengths in information integrity and verification might be used to verify the condition of existing cyber-physical systems. A more secure chain of custody may be possible thanks to data recorded on the blockchain about the infrastructure's parts [67].
- **Protecting Data Transmission:** The information produced by cyber-physical systems has been compromised due to data manipulation, systems misconfiguration, and component failure. However, blockchain technology's information integrity and verification capabilities may be used to verify the standing of any cyber-physical infrastructure. There may be greater confidence in the whole chain of custody if the information on the infrastructure's components is generated using blockchain [23]. Data in transit can be hacked by malevolent actors who then use the data for their ends, such as erasing it or making unauthorized changes. This creates a significant bottleneck in ineffective means of communication like email.
- **Standardization and Regulatory Aspects:** The future of our technological solutions in CPS, IoT networks, and beyond will be influenced by regulatory and standardization efforts. Since nations have vastly varying viewpoints and rules in this field, it has been difficult for standards agencies to ensure the privacy and identification of their citizens. Many organizations are working on privacy frameworks, including international standards groups like JTC 1. [29].

Cyber-physical systems will therefore serve as the basis for developing new uses in various fields: health, entertainment, industry, education, etc. Its technical qualities will enable the development of systems that respond to high digital power, are hyperconnected, and have a concrete impact on the physical world. However, the development of the Internet of Things (or IoT) in these multiple sectors exposes its users to increased risks of cyber-attacks (intrusion, identity theft, espionage, security breach, etc.). Some national authorities are also concerned about the risk of backdoors allowing access to data, or even the risk of remote control.

6 Conclusion

Blockchain has qualities that allow it to be used for cyber defense. However, today, technology is not mature enough to move from the category of fad novelties to the mainstream. In this chapter, we have provided a comprehensive analysis of CPS using blockchain technology. We began by cataloging the many CPS activities that may be improved by incorporating blockchain technology. We next explore the use of blockchain for cybersecurity in CPS, after reviewing hundreds of operations described in the literature. This chapter contributes significantly to the theme by expanding our knowledge of an area that has received less attention in the literature. It became apparent throughout development that this needed more theoretical exploration, and it was given that opportunity. The results reveal that the topic has become increasingly important and has risen in importance each year as new blockchain-based applications for cyber-physical system security have emerged and developed. It is too soon to compare Blockchain technology to other technologies already in use and determine if it is more suited in the context of applications involving cyber-physical systems.

References

1. Abadi, J., & Brunnermeier, M. (2018). Blockchain economics. *National Bureau of Economic Research (No. W25407).*
2. Abd El-Latif, A. A., Abd-El-Atty, B., Mehmood, I., Muhammad, K., Venegas-Andraca, S. E., & Peng, J. (2021). Quantum-Inspired Blockchain-Based Cybersecurity: Securing Smart Edge Utilities in IoT-Based Smart Cities. *Information Processing & Management*, *58*(4), 102549. https://doi.org/10.1016/j.ipm.2021.102549
3. Abdulkader, O., Bamhdi, A. M., Thayananthan, V., Elbouraey, F., & Al-Ghamdi, B. (2019). A Lightweight Blockchain-Based Cybersecurity for IoT environments. *2019 6th IEEE International Conference on Cyber Security and Cloud Computing (CSCloud)/ 2019 5th IEEE International Conference on Edge Computing and Scalable Cloud (EdgeCom)*, 139–144. https://doi.org/10.1109/CSCloud/EdgeCom.2019.000-5
4. Abualsauod, E. H. (2022). A hybrid blockchain method in internet of things for privacy and security in unmanned aerial vehicles network. *Computers and Electrical Engineering*, *99*, 107847. https://doi.org/10.1016/j.compeleceng.2022.107847
5. Aceto, G., Persico, V., & Pescapé, A. (2019). A Survey on Information and Communication Technologies for Industry 4.0: State-of-the-Art, Taxonomies, Perspectives, and Challenges. *IEEE Communications Surveys & Tutorials*, *21*(4), 3467–3501. https://doi.org/10.1109/COMST.2019.2938259
6. Alladi, T., Chamola, V., Parizi, R. M., & Choo, K. R. (2019). Blockchain Applications for Industry 4.0 and Industrial IoT: A Review. *IEEE Access*, *7*, 176935–176951. https://doi.org/10.1109/ACCESS.2019.2956748

7. Althobaiti, O. S., & Dohler, M. (2020). Cybersecurity Challenges Associated With the Internet of Things in a Post-Quantum World. *IEEE Access*, *8*, 157356–157381. https://doi.org/10.1109/ACCESS.2020.3019345
8. Badsha, S., Vakilinia, I., & Sengupta, S. (2020). BloCyNfo-Share: Blockchain based Cybersecurity Information Sharing with Fine Grained Access Control. *2020 10th Annual Computing and Communication Workshop and Conference (CCWC)*, 317–323. https://doi.org/10.1109/CCWC47524.2020.9031164
9. Bagga, P., Sutrala, A. K., Das, A. K., & Vijayakumar, P. (2021). Blockchain-based batch authentication protocol for Internet of Vehicles. *Journal of Systems Architecture*, *113*, 101877. https://doi.org/10.1016/j.sysarc.2020.101877
10. Baniata, H., & Kertesz, A. (2020). A Survey on Blockchain-Fog Integration Approaches. *IEEE Access*, *8*, 102657–102668. https://doi.org/10.1109/ACCESS.2020.2999213
11. Bhuiyan, M. Z. A., Zaman, A., Wang, T., Wang, G., Tao, H., & Hassan, M. M. (2018, May). Blockchain and big data to transform the healthcare. In Proceedings of the international conference on data processing and applications (pp. 62–68). https://doi.org/10.1145/3224207.3224220
12. Bhushan, B., Khamparia, A., Sagayam, K. M., Sharma, S. K., Ahad, M. A., & Debnath, N. C. (2020). Blockchain for smart cities: A review of architectures, integration trends and future research directions. *Sustainable Cities and Society*, *61*, 102360. https://doi.org/10.1016/j.scs.2020.102360
13. Cardin, O. (2019). Classification of cyber-physical production systems applications: Proposition of an analysis framework. *Computers in Industry*, *104*, 11–21. https://doi.org/10.1016/j.compind.2018.10.002
14. Cebe, M., Erdin, E., Akkaya, K., Aksu, H., & Uluagac, S. (2018). Block4Forensic: An Integrated Lightweight Blockchain Framework for Forensics Applications of Connected Vehicles. *IEEE Communications Magazine*, *56*(10), 50–57. https://doi.org/10.1109/MCOM.2018.1800137
15. Chi, J., Li, Y., Huang, J., Liu, J., Jin, Y., Chen, C., & Qiu, T. (2020). A secure and efficient data sharing scheme based on blockchain in industrial Internet of Things. *Journal of Network and Computer Applications*, *167*, 102710. https://doi.org/10.1016/j.jnca.2020.102710
16. Dagher, G. G., Mohler, J., Milojkovic, M., & Marella, P. B. (2018). Ancile: Privacy-preserving framework for access control and interoperability of electronic health records using blockchain technology. *Sustainable Cities and Society*, *39*, 283–297. https://doi.org/10.1016/j.scs.2018.02.014
17. Dedeoglu, V., Dorri, A., Jurdak, R., Michelin, R. A., Lunardi, R. C., Kanhere, S. S., & Zorzo, A. F. (2020). A Journey in Applying Blockchain for Cyberphysical Systems. *2020 International Conference on COMmunication Systems & NETworkS (COMSNETS)*, 383–390. https://doi.org/10.1109/COMSNETS48256.2020.9027487
18. Dorri, A., Steger, M., Kanhere, S. S., & Jurdak, R. (2017). BlockChain: A Distributed Solution to Automotive Security and Privacy. *IEEE Communications Magazine*, *55*(12), 119–125. https://doi.org/10.1109/MCOM.2017.1700879
19. Evsutin, O., Meshcheryakov, R., Tolmachev, V., Iskhakov, A., & Iskhakova, A. (2019). Algorithm for Embedding Digital Watermarks in Wireless Sensor Networks Data with Control of Embedding Distortions. In V. M. Vishnevskiy, K. E. Samouylov, & D. V Kozyrev (Eds.), *Distributed Computer and Communication Networks* (pp. 574–585). Springer International Publishing.
20. Ferdowsi, A., & Saad, W. (2019). Deep Learning for Signal Authentication and Security in Massive Internet-of-Things Systems. *IEEE Transactions on Communications*, *67*(2), 1371–1387. https://doi.org/10.1109/TCOMM.2018.2878025
21. Fernández-Caramés, T. M., & Fraga-Lamas, P. (2018). A Review on the Use of Blockchain for the Internet of Things. *IEEE Access*, *6*, 32979–33001. https://doi.org/10.1109/ACCESS.2018.2842685

22. Fernández-Caramés, T. M., & Fraga-Lamas, P. (2019). A Review on the Application of Blockchain to the Next Generation of Cybersecure Industry 4.0 Smart Factories. *IEEE Access*, *7*, 45201–45218. https://doi.org/10.1109/ACCESS.2019.2908780
23. Gao, F., Zhu, L., Gai, K., Zhang, C., & Liu, S. (2020). Achieving a Covert Channel over an Open Blockchain Network. *IEEE Network*, *34*(2), 6–13. https://doi.org/10.1109/MNET.001.1900225
24. Gupta, M., Patel, R. B., Jain, S., Garg, H., & Sharma, B. (2022). Lightweight branched blockchain security framework for Internet of Vehicles. *Transactions on Emerging Telecommunications Technologies*, *n/a*(n/a), e4520. https://doi.org/10.1002/ett.4520
25. Hameed, K., Bajwa, I. S., Sarwar, N., Anwar, W., Mushtaq, Z., & Rashid, T. (2021). Integration of 5G and Block-Chain Technologies in Smart Telemedicine Using IoT. *Journal of Healthcare Engineering*, *2021*, 8814364. https://doi.org/10.1155/2021/8814364
26. Hameed, K., Khan, A., Ahmed, M., Goutham Reddy, A., & Rathore, M. M. (2018). Towards a formally verified zero watermarking scheme for data integrity in the Internet of Things based-wireless sensor networks. *Future Generation Computer Systems*, *82*, 274–289. https://doi.org/10.1016/j.future.2017.12.009
27. Hoang, T., Bui, V., Vu, N., & Hoang, D. (2020). A Lightweight Mixed Secure Scheme based on the Watermarking Technique for Hierarchy Wireless Sensor Networks. *2020 International Conference on Information Networking (ICOIN)*, 649–653. https://doi.org/10.1109/ICOIN48656.2020.9016541
28. Huang, H., & Zhang, L. (2019). Reliable and Secure Constellation Shifting Aided Differential Radio Frequency Watermark Design for NB-IoT Systems. *IEEE Communications Letters*, *23*(12), 2262–2265. https://doi.org/10.1109/LCOMM.2019.2944811
29. Jiang, W., Han, B., Habibi, M. A., & Schotten, H. D. (2021). The Road Towards 6G: A Comprehensive Survey. *IEEE Open Journal of the Communications Society*, *2*, 334–366. https://doi.org/10.1109/OJCOMS.2021.3057679
30. Khalil, A. A., Franco, J., Parvez, I., Uluagac, S., Shahriar, H., & Rahman, M. A. (2022). A Literature Review on Blockchain-enabled Security and Operation of Cyber-Physical Systems. *2022 IEEE 46th Annual Computers, Software, and Applications Conference (COMPSAC)*, 1774–1779. https://doi.org/10.1109/COMPSAC54236.2022.00282
31. Koh, L., Dolgui, A., & Sarkis, J. (2020). Blockchain in transport and logistics – paradigms and transitions. *International Journal of Production Research*, *58*(7), 2054–2062. https://doi.org/10.1080/00207543.2020.1736428
32. Koshy, P., Babu, S., & Manoj, B. S. (2020). Sliding Window Blockchain Architecture for Internet of Things. *IEEE Internet of Things Journal*, *7*(4), 3338–3348. https://doi.org/10.1109/JIOT.2020.2967119
33. Kshetri, N. (2018). Blockchain's roles in meeting key supply chain management objectives. *International Journal of Information Management*, *39*, 80–89. https://doi.org/10.1016/j.ijinfomgt.2017.12.005
34. Li, M., Hu, D., Lal, C., Conti, M., & Zhang, Z. (2020). Blockchain-Enabled Secure Energy Trading With Verifiable Fairness in Industrial Internet of Things. *IEEE Transactions on Industrial Informatics*, *16*(10), 6564–6574. https://doi.org/10.1109/TII.2020.2974537
35. Liang, G., Weller, S. R., Luo, F., Zhao, J., & Dong, Z. Y. (2019). Distributed Blockchain-Based Data Protection Framework for Modern Power Systems Against Cyber Attacks. *IEEE Transactions on Smart Grid*, *10*(3), 3162–3173. https://doi.org/10.1109/TSG.2018.2819663
36. Liu, Z., & Li, Z. (2020). A blockchain-based framework of cross-border e-commerce supply chain. *International Journal of Information Management*, *52*, 102059. https://doi.org/10.1016/j.ijinfomgt.2019.102059
37. Lu, H., Huang, K., Azimi, M., & Guo, L. (2019). Blockchain Technology in the Oil and Gas Industry: A Review of Applications, Opportunities, Challenges, and Risks. *IEEE Access*, *7*, 41426–41444. https://doi.org/10.1109/ACCESS.2019.2907695
38. Luo, J., Chen, Q., Yu, F. R., & Tang, L. (2020). Blockchain-Enabled Software-Defined Industrial Internet of Things With Deep Reinforcement Learning. *IEEE Internet of Things Journal*, *7*(6), 5466–5480. https://doi.org/10.1109/JIOT.2020.2978516

39. Maleh, Y., Lakkineni, S., Tawalbeh, L., & AbdEl-Latif, A. A. (2022). Blockchain for Cyber-Physical Systems: Challenges and Applications. In Y. Maleh, L. Tawalbeh, S. Motahhir, & A. S. Hafid (Eds.), *Advances in Blockchain Technology for Cyber Physical Systems* (pp. 11–59). Springer International Publishing. https://doi.org/10.1007/978-3-030-93646-4_2
40. Maleh, Y., Shojafar, M., Alazab, M., & Romdhani, I. (2020). *Blockchain for Cybersecurity and Privacy: Architectures, Challenges, and Applications*. CRC press. https://doi.org/10.1201/9780429324932
41. Maleh Yassine, Mohammad Shojaafar, Ashraf Darwish, A. H. (2019). Cybersecurity and Privacy in Cyber-Physical Systems. *CRC Press*. https://www.crcpress.com/Cybersecurity-and-Privacy-in-Cyber-Physical-Systems/Maleh/p/book/9781138346673
42. Mei, Q., Xiong, H., Chen, Y.-C., & Chen, C.-M. (2022). Blockchain-Enabled Privacy-Preserving Authentication Mechanism for Transportation CPS With Cloud-Edge Computing. *IEEE Transactions on Engineering Management*, 1–12. https://doi.org/10.1109/TEM.2022.3159311
43. Milne, A. J. M., Beckmann, A., & Kumar, P. (2020). Cyber-Physical Trust Systems Driven by Blockchain. *IEEE Access*, *8*, 66423–66437. https://doi.org/10.1109/ACCESS.2020.2984675
44. Mohanta, B. K., Satapathy, U., Dey, M. R., Panda, S. S., & Jena, D. (2020). Trust Management in Cyber Physical System using Blockchain. *2020 11th International Conference on Computing, Communication and Networking Technologies (ICCCNT)*, 1–5. https://doi.org/10.1109/ICCCNT49239.2020.9225272
45. Neelam, B. S., & Shimray, B. A. (2021). Applicability of RINA in IoT communication for acceptable latency and resiliency against device authentication attacks. *2021 6th International Conference for Convergence in Technology (I2CT)*, 1–7. https://doi.org/10.1109/I2CT51068.2021.9417992
46. Nguyen, V., Hoang, T., Duong, T., Nguyen, Q., & Bui, V. (2019). A Lightweight Watermark Scheme Utilizing MAC Layer Behaviors for Wireless Sensor Networks. *2019 3rd International Conference on Recent Advances in Signal Processing, Telecommunications & Computing (SigTelCom)*, 176–180. https://doi.org/10.1109/SIGTELCOM.2019.8696234
47. Patel, V. (2019). A framework for secure and decentralized sharing of medical imaging data via blockchain consensus. *Health Informatics Journal*, *25*(4), 1398–1411. https://doi.org/10.1177/1460458218769699
48. Peng, H., Yang, B., Li, L., & Yang, Y. (2020). Secure and Traceable Image Transmission Scheme Based on Semitensor Product Compressed Sensing in Telemedicine System. *IEEE Internet of Things Journal*, *7*(3), 2432–2451. https://doi.org/10.1109/JIOT.2019.2957747
49. Qian, Y., Jiang, Y., Hu, L., Hossain, M. S., Alrashoud, M., & Al-Hammadi, M. (2020). Blockchain-Based Privacy-Aware Content Caching in Cognitive Internet of Vehicles. *IEEE Network*, *34*(2), 46–51. https://doi.org/10.1109/MNET.001.1900161
50. Rathee, G., Sharma, A., Iqbal, R., Aloqaily, M., Jaglan, N., & Kumar, R. (2019). A Blockchain Framework for Securing Connected and Autonomous Vehicles. In *Sensors* (Vol. 19, Issue 14). https://doi.org/10.3390/s19143165
51. Rathore, H., Mohamed, A., & Guizani, M. (2020). A survey of blockchain enabled cyber-physical systems. Sensors, *20*(1), 282. https://doi.org/10.3390/s20010282
52. Rathore, S., & Park, J. H. (2021). A Blockchain-Based Deep Learning Approach for Cyber Security in Next Generation Industrial Cyber-Physical Systems. *IEEE Transactions on Industrial Informatics*, *17*(8), 5522–5532. https://doi.org/10.1109/TII.2020.3040968
53. Rubio-Hernan, J., De Cicco, L., & Garcia-Alfaro, J. (2018). Adaptive control-theoretic detection of integrity attacks against cyber-physical industrial systems. *Transactions on Emerging Telecommunications Technologies*, *29*(7), e3209. https://doi.org/10.1002/ett.3209
54. Saberi, S., Kouhizadeh, M., Sarkis, J., & Shen, L. (2019). Blockchain technology and its relationships to sustainable supply chain management. *International Journal of Production Research*, *57*(7), 2117–2135. https://doi.org/10.1080/00207543.2018.1533261
55. Seitz, A., Henze, D., Miehle, D., Bruegge, B., Nickles, J., & Sauer, M. (2018). Fog Computing as Enabler for Blockchain-Based IIoT App Marketplaces - A Case Study. *2018 Fifth International Conference on Internet of Things: Systems, Management and Security*, 182–188. https://doi.org/10.1109/IoTSMS.2018.8554484

56. Tariq, N., Qamar, A., Asim, M., & Khan, F. A. (2020). Blockchain and smart healthcare security: A survey. *Procedia Computer Science, 175*, 615–620. https://doi.org/10.1016/j.procs.2020.07.089
57. Taylor, P. J., Dargahi, T., Dehghantanha, A., Parizi, R. M., & Choo, K. K. R. (2020). A systematic literature review of blockchain cyber security. Digital Communications and Networks, *6*(2), 147–156. https://doi.org/10.1016/j.dcan.2019.01.005
58. Wang, B., Kong, W., Li, W., & Xiong, N. N. (2019a). A Dual-Chaining Watermark Scheme for Data Integrity Protection in Internet of Things. In *Computers, Materials \& Continua* (Vol. 58, Issue 3). https://doi.org/10.32604/cmc.2019.06106
59. Wang, S., Huang, C., Li, J., Yuan, Y., & Wang, F.-Y. (2019b). Decentralized Construction of Knowledge Graphs for Deep Recommender Systems Based on Blockchain-Powered Smart Contracts. *IEEE Access*, *7*, 136951–136961. https://doi.org/10.1109/ACCESS.2019.2942338
60. Xia, Q., Sifah, E. B., Smahi, A., Amofa, S., & Zhang, X. (2017). BBDS: Blockchain-Based Data Sharing for Electronic Medical Records in Cloud Environments. In *Information* (Vol. 8, Issue 2). https://doi.org/10.3390/info8020044
61. Xiao, Y., & Gao, G. (2019). Digital Watermark-Based Independent Individual Certification Scheme in WSNs. *IEEE Access*, *7*, 145516–145523. https://doi.org/10.1109/ACCESS.2019.2945177
62. Xu, L. Da, Lu, Y., & Li, L. (2021). Embedding Blockchain Technology into IoT for Security: A Survey. *IEEE Internet of Things Journal*, 1. https://doi.org/10.1109/JIOT.2021.3060508
63. Xu, S., Chen, X., & He, Y. (2021). EVchain: An Anonymous Blockchain-Based System for Charging-Connected Electric Vehicles. *Tsinghua Science and Technology*, *26*(6), 845–856. https://doi.org/10.26599/TST.2020.9010043
64. Yazdinejad, A., Parizi, R. M., Dehghantanha, A., & Choo, K. R. (2019). Blockchain-enabled Authentication Handover with Efficient Privacy Protection in SDN-based 5G Networks. *IEEE Transactions on Network Science and Engineering*, 1. https://doi.org/10.1109/TNSE.2019.2937481
65. Yu, C., Jiang, X., Yu, S., & Yang, C. (2020). Blockchain-based shared manufacturing in support of cyber physical systems: concept, framework, and operation. *Robotics and Computer-Integrated Manufacturing*, *64*, 101931. https://doi.org/10.1016/j.rcim.2019.101931
66. Zhao, B., Fang, L., Zhang, H., Ge, C., Meng, W., Liu, L., & Su, C. (2019). Y-DWMS: A Digital Watermark Management System Based on Smart Contracts. In *Sensors* (Vol. 19, Issue 14). https://doi.org/10.3390/s19143091
67. Zhao, W., Jiang, C., Gao, H., Yang, S., & Luo, X. (2021). Blockchain-Enabled Cyber–Physical Systems: A Review. *IEEE Internet of Things Journal*, *8*(6), 4023–4034. https://doi.org/10.1109/JIOT.2020.3014864
68. Zhaofeng, M., Lingyun, W., Xiaochang, W., Zhen, W., & Weizhe, Z. (2020). Blockchain-Enabled Decentralized Trust Management and Secure Usage Control of IoT Big Data. *IEEE Internet of Things Journal*, *7*(5), 4000–4015. https://doi.org/10.1109/JIOT.2019.2960526
69. Zou, R., Lv, X., & Zhao, J. (2021). SPChain: Blockchain-based medical data sharing and privacy-preserving eHealth system. *Information Processing & Management*, *58*(4), 102604. https://doi.org/10.1016/j.ipm.2021.102604

Trust Management in Cyber-Physical Systems: Issues and Challenges

Nawal Ait Aali and Amine Baina

1 Introduction

Cyber-physical systems (CPS) are cooperative systems, with decentralized control, resulting from the fusion between the real and virtual worlds, having autonomous behaviors and, depending on the context in which they are found, being able to form systems of systems with other CPS and leading to extensive collaboration with humans [1].

Since its appearance, the CPSs have been considered as an industrial revolution (industry 4.0) [2] given the significant role that these systems have brought to the industry in terms of the interaction between physical and cyber entities to achieve decentralized and autonomous controls. Also, these systems can make decisions without human intervention by integrating Artificial Intelligence [3] in their deployment. Indeed, the CPS makes it possible to control and command physical entities through the use of cyber ones. The thing that leads to personalized production according to the consumer's needs.

The principle of CPS is based on embedded software [4] that uses sensors and actuators, connects to human operators by communicating via interfaces and can store and process information from sensors or the network [5].

Although CPSs bring several advantages to the industry, they confront many attacks and threats [6] affecting physical and cyber entities and the interactions between collaborating CPSs. In this context, the security of CPSs is essential to protect all types of their entities. For this reason, we aim through this chapter to identify the security problems generated in a CPS, the different confronted attacks, and vulnerabilities. Also, we are interested in demonstrating the necessity of Trust Management [7] within the CPS.

N. Ait Aali (✉) · A. Baina
STRS Laboratory, National Institute of Posts and Telecommunication, Rabat, Morocco

Y. Maleh et al. (eds.), *Blockchain for Cybersecurity in Cyber-Physical Systems*,
Advances in Information Security 102, https://doi.org/10.1007/978-3-031-25506-9_4

For this purpose, we will organize this chapter as follows: in Sect. 2, we present the concept of Cyber-Physical System, its definitions, and its characteristics. In Sect. 3, we will describe the operation of the CPS and the different technologies involved in its deployment. Section 4 aims to identify the vulnerabilities and attacks affecting the CPS while reserving Sect. 5 to introduce Trust Management; we present the trust concept, its definitions, and its classifications. And also, we describe our motivations for integrating trust management into CPS. Section 6 is dedicated to discussing the Blockchain concept, especially, its effectiveness in the CPS security. Then, Sect. 7 concludes the chapter while introducing our future research works.

2 Backgrounds

2.1 Cyber Physical Systems

Industry 4.0 [8] is believed to include a significant portion of Cyber Physical Systems. This technology automates and streamlines production. Additionally, due to CPS, manufacturing lines automate while adjusting to environmental factors and customer demands.

Since its introduction, the idea of a CPS has caught the interest of numerous research initiatives, institutions, and scientific organizations. Additionally, it has been described by a number of definitions. The NSF (National Science Foundation) [6] offers the initial definition of the CPS [9]: "*the CPS is a system integrating physical and computational processes. Embedded and computers networks monitor and control physical processes, generally with feedback loops where the physical processes affect computations and vice versa*". Also, CPS received new definitions as a result of the development of this notion. In [10], the researchers define CPS as "*physical systems whose operations are monitored, controlled, coordinated and integrated by a computing and communications center*". However, the following definition [11] is thought to be clearer and more synthetic than the others offered in the literature: "*CPSs are systems made up of collaborative cyber entities, which are in intensive connection with the surrounding physical world and the phenomena taking place in it, while providing and using both the data provision and data processing services available on the network*".

Moreover, the CPSs are used in various industries, including meteorology, agriculture, transportation, health care, and energy. The many fields employing the CPS are shown in Fig. 1 [12].

As we have already presented, the CPS comprises a number of cyber entities interacting to control and supervise the physical components. This connection between the real and cyber environments is described in Fig. 2. Several entities, like sensors [13], and connected objects [14], use cyber entities to send the information they collect to the computation and processing units. The output from the computation units is then translated into instructions for the actuators, which are physical components. An information system is responsible for overseeing all of these interactions between physical and virtual entities [15];

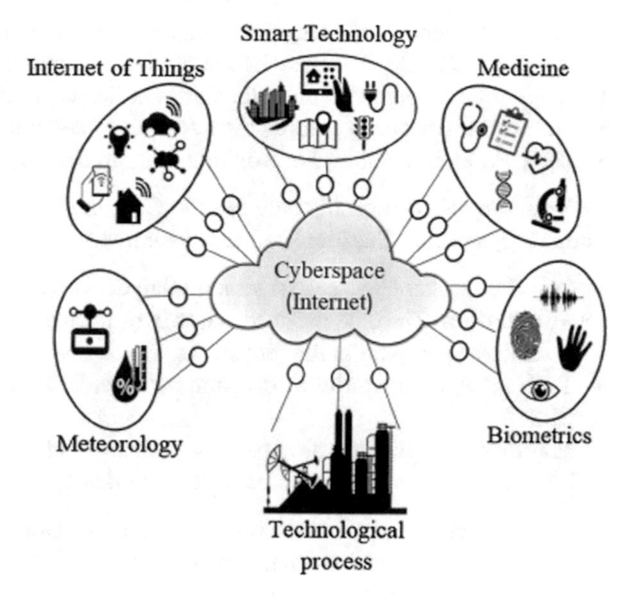

Fig. 1 Cyber-physical system [12]

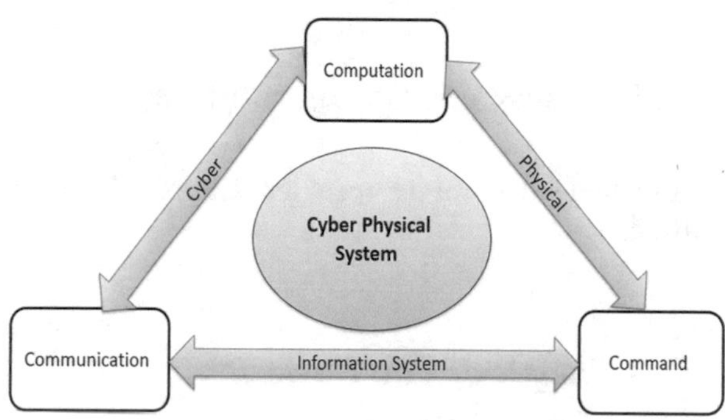

Fig. 2 Life cycle of CPS. (Derived from [15])

CPSs have particular features besides the interaction between physical and cyber entities.

2.2 *Different Features of CPSs*

Like all systems, CPSs have some features [15] that enable them to outperform existing systems (such as the embedded system). In this study, we list some of these features.

Regarding the connection between physical and cyber entities:

- The CPS stands out for its strong physical and cyber integration;
- The physical parts of the CPS have, all, a great computational capacity;

- The CPS offers a wide range of connection and communication options, including Bluetooth, GSM, GPS, wired and wireless networks, etc.;
- A dynamic reconfiguration of the CPS entities is possible;
- The CPS resilience is ensured by automated intellectual control;
- The CPS interoperability is provided through the internet.

Referring to the relationship between the CPS and humans, the CPSs are classified into four categories. These are defined by taking into account the decision-making:

- **Full**: In a Cyber-Physical System, human decision-making is not required;
- **Automation**: The CPS makes a majority of the decisions, and the role of humans is limited to adaptation functions;
- **Tool**: Now, the majority of decisions are made by humans, and they direct the CPS;
- **Manual**: All decisions are made by humans, and the CPS's only responsibility is to provide information (informant) to humans.

This classification shows that, within the CPS, decision-making in a manufacturing line is possible without human involvement [16].

3 CPS: How Does It Operate?

This section will cover the operation of the CPS as well as the different technologies utilized.

3.1 Principle of CPS Operation

The general operation of the CPS [6] can be divided into three sections which are characterized by a set of components as indicated in Fig. 3:

- Communication and networking;
- Computation and control;
- Monitoring and manipulation.

The CPS can communicate with other CPSs or control centers via wired or wireless connections. The intelligence is incorporated in the computation and control part. Control commands are transmitted, detected measurements are received, and decisions are produced. The monitoring and manipulation component's role is to connect the CPS and the physical environment using sensors to monitor the physical components and actuators to manipulate them.

The system operator in CPS should be aware of the situations around the controlled objects. As a result, the human operator can observe the present state of the controlled objects due to the Graphical User Interface (GUI), also known as the Human Machine Interface (HMI) [12].

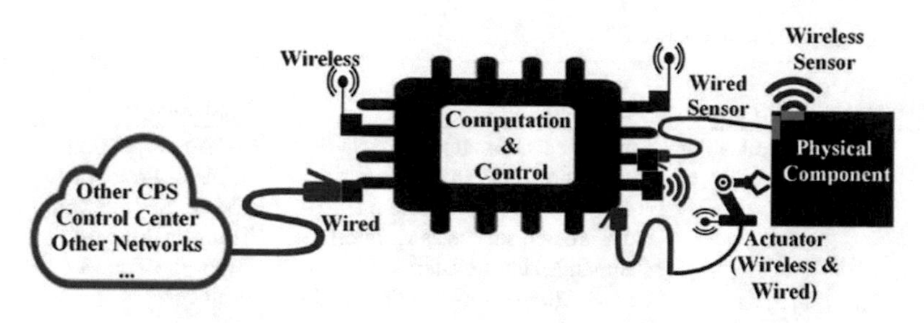

Fig. 3 Operation of CPS [6]

To better understand the operation of the CPS, we illustrate its architecture by detailing the components of each layer and its actions. Indeed, several types of architecture are proposed to the CPS [17–19]. In this chapter, we illustrate the architecture of CPS using the OSI model [20], which consists of seven layers, from the physical layer to the application layer. Table 1 regroups these layers while we describe the operation of each layer and its protocols.

As presented in Table 1, we observe that the operation of some layers (application and physical) depends on the context of the CPS; other layers (presentation, session, transport, network, data link) operate independently of this context. Additionally, we conclude that the CPS's establishment is based on several technologies, including sensor networks, the Internet of Things, artificial intelligence, collaborative systems, etc.

3.2 *Integrated Technologies in CPS*

As described in the previous sub-section, Cyber-Physical Systems are realized by integrating a set of technologies.

- **Internet of Things (IoT)**

Among the different definitions proposed for the concept of the Internet of Things [21–24], we base on the definition presented in [25] to describe this concept. The Internet of Things is a global network infrastructure with self-configuring capabilities based on standards and interoperable communication protocols. In the Internet of Things, physical and virtual objects, called: connected objects, have standardized digital identifications (IP address, protocols, etc.) These objects can be remotely controlled using a computer, smartphone, or tablet.

The principle of the IoT consists of assigning an identity to each object, most often the IP address. This identity is unique and recognizable in the network. This makes it possible to find the object and to communicate to it all the instructions to follow via a computer or telephone. Each connected object is characterized by a sensor, which enables data measurement.

Table 1 OSI model layers in CPS

CPS architecture	CPS layers	Description	Protocols
	Application	Makes decisions, updates information, and stores them It has systems management software and serves as an interface (Human Machine Interface) between the system operator and the CPS	HTTP: HyperText Transfer Protocol SMTP: Simple Mail Transfer Protocol FTP: File Transfer Protocol
	Presentation	Information is transformed to make it compatible with other communicating entities (code conversion, data compression, etc.) It consists of data transformation, data encryption, and decryption	ASCII: American Standard Code for Information Interchange. Unicode
	Session	Manages the communication session Creates a connection between two applications that must work together Monitors the order of message transmission in the network	SSH: Secure Shel
	Transport	Enables error-free, loss-free message transmission between two cooperating entities Each message/data (from the session layer) is divided into small fragments to be transmitted to the network layer	TCP: Transmission Control Protocol UDP: User Datagram Protocol
	Network	Routes the packets containing several information in CPS to the recipient entity	IP: Internet Protocol ARP: Address Resolution Protocol RARP: Reverse Address Resolution Protocol
	Data link	Transmits the received measurements to the network layer and sends the decisions to the physical layer It is based on a point-to-point or multipoint link	Ethernet Protocol Point to Point Protocol
	Physical	It consists of sensors, and actuators, which are connected via wireless or wired networks; Physical components have little memory and processing power	2G/3G/4G, ZigBee, Bluetooth WiMAX, Wifi

The IoT is actually used in several applications: healthcare, agriculture, smart homes, etc. For example, we can turn off a television remotely. We can start a washing machine at a specific hour, etc.

The various sensors used in IoT, as well as other components constitute a sensor network.

- **Sensor network**

A sensor network is a network that provides access to information independently of time and location by collecting, processing, analyzing and disseminating data [26].

Its operation is to route the data collected by the nodes using multi-hop routing to a node that is considered a "collection point", called a *sink node* [27]. The latter can be connected to the network user (via the Internet, a satellite, or another system). The user can address requests to the other nodes of the network, specifying the type of data required, and then collect the captured data.

Focusing on wireless sensor networks [28], these are characterized by the existence of many nodes distributed in the network. Their role is to collect and transmit data autonomously. The particularity of this type of network is that the nodes are dispersed randomly in their network without negatively impacting the primary operation of the wireless sensor network.

- **Embedded systems**

An embedded system is a system that performs a specific task, in real-time, within the device in which it is integrated. It is a system that combines software and hardware. Indeed, the software is executed on a microcontroller or a microprocessor [29]. In addition, the tasks within the embedded system must be carried out with great precision and within imposed deadlines.

As described in [30], an embedded system is divided into four interacting entities: the sensors, the operative part, the control and reconfiguration system, and the actuators. Sensors measure physical quantities. The control and reconfiguration system establishes the actions to be carried out based on these measurements. The actuators act on the operative part. Many industries use embedded systems, including transportation, medical technology, and IT. Additionally, it is utilized in everyday home devices like washing machines, refrigerators, microwaves, etc.

- **Collaborative systems**

From a computing point of view and in information systems, collaboration is identified by sharing a set of resources, services, and information between the different actors/entities to carry out certain activities [31]. In general, collaborative systems are defined by a set of entities that collaborate to achieve specific goals. These objectives generally include sharing resources, accessing various services, completing the actions necessary to operate the systems, etc. Collaborative systems are required in various fields, namely: environment [32], automatic systems based on connected objects and sensors [33], computer systems [34], etc.

The operation of the collaborative system is based on at least two entities that communicate and collaborate. The collaboration is based on a negotiation between entities that can be carried out from one entity to another or through a central authority, which manages communication and access between these entities. Therefore, the collaboration between the entities can be centralized (existence of central authority), or decentralized (peer-to-peer communication) [35].

In addition to the technologies described above, CPSs incorporate artificial intelligence as the decision maker. Thanks to artificial intelligence and its derivatives, Machine Learning, and Deep Learning. The decisions can be made within the CPS without necessarily having human intervention. These decisions are based on a set of classification [36], prediction [37], and clustering algorithms [38]. The choice of algorithms depends on the decision to be made and the application field of the CPS.

From what we have presented in this chapter concerning the architecture of the CPS and the technologies deployed there, we deduce that the CPS is rich in integrated components, actions to be carried out, and technologies involved. All of these are managed by computer system software. In this sense, the CPS is confronted by a set of attacks. For this reason, in the following section, we will describe the security vulnerability and all the attacks affecting the CPS in all its aspects.

4 Security Issues in CPS

Like any system (cyber or physical), Cyber-Physical Systems are confronted by several attacks that can have severe economic and social impacts. In this sense, the security of these systems becomes an obligation.

To secure a CPS, we must study and treat the security of its three parts: the cyber part, the physical part, and the cyber-physical part. Each is exposed to several attacks and confronts various vulnerabilities [6]. In addition, regarding the CPS architecture, each layer of the OSI model presented in Table 1 is confronted by several attacks. Therefore, the security of CPS presents a significant challenge; such a proposed security model must protect the three parts of the CPS against all possible attacks.

In addition, the interactions within the CPS vary from one CPS to another depending on its application domain (vehicle, health care, transport, smart grid, agriculture, etc. . . .). Therefore, other characteristics and factors must be considered in CPS security.

4.1 Security Vulnerabilities in CPS

We classify the CPS vulnerabilities into three categories according to their class: cyber, physical, and cyber-physical vulnerabilities.

Regarding cyber vulnerabilities, they affect the means of communication between a CPS and another CPS or a computing center, or another network. This means integrating network protocols (such as TCP/IP, ICCP (Inter-Control

Table 2 Security vulnerabilities in CPS

Cyber-physical system	Vulnerabilities
Cyber	Open communication protocols: TCP/IP, ICCP, RTC Wired and wireless communication Insecure protocols Insecure Database Insecure Operating System Software
Physical	Physical equipment stealing, sabotage, and aging
Cyber-physical	Insecure protocols: DNP3, Modbus Insecure communications between CPS components Direct Access to remote devices Insecure Real-Time Operating System Running software for controlling and monitoring the physical equipment

Center Communications Protocol), and RTC (Real-Time Transport Control)), communications using wired or wireless technologies.

We also add to the cyber vulnerabilities the software that manages the different operations and interactions between the components of the CPS. Such software includes operating systems, database management software, etc. Each insecure software is considered a cyber vulnerability.

Regarding cyber-physical vulnerabilities, we summarize them in the protocols and software that act on physical components such as DNP3 [39] and Modbus [40]; insecure protocol of them is considered a vulnerability in CPS. In addition, we include the cyber-physical vulnerabilities: the insecure communications between the CPS components and the absence of access control rules that manage access to devices.

The physical vulnerabilities are generally limited to aging, stealing, and sabotaging the used physical components. Table 2 summarizes the several vulnerabilities of CPS.

The various vulnerabilities threaten the CPS by a set of voluntary (hacker) or involuntary (natural disasters) attacks.

4.2 *Security Attacks in CPS*

Several types of attacks threaten the CPS; some attacks target the three elements of the CPS, namely cyber, physical and cyber-physical entities. Other attacks target the layers of the OSI model. In addition, some attacks prevent satisfying the security aspects of a system, such as: confidentiality, integrity, and availability. Thus, we aim through this sub-section to describe these attacks.

Table 3 Security Attacks in CPS parts

Cyber-physical system	Attacks
Cyber	IP Spoofing Espionage (Flame Attack [41]) Denial of Service (DoS) SQL Injection False Data Injection
Cyber-physical	Legacy communication channels Web-based attacks unauthorized commands injection Malware injection
Physical	Natural and environmental incidents Theft/Stealing/sabotage Terrorist attacks

Table 4 Security attacks in CPS layers

CPS layers	Attacks
Application	Distributed denial-of-service (DDoS) HTTP floods SQL injections Phishing
Presentation	SSL Hijacking DDoS
Session	MITM (Man in The Middle) Poodle [42] SSH Sniffing
Transport	Denial of Service TCP Session hijacking
Network	IP Spoofing Espionage Packet sniffing
Data link	MAC disruption ARP Spoofing
Physical	External source attacks (Natural and environmental incidents, stealing, etc.)

Firstly, we distinguish between:

- Cyber-attacks: they affect communications protocols, wired and wireless networks;
- Cyber-Physical attacks: affect cyber components which act on the physical components.
- Physical attacks: they only affect physical components.

Table 3 summarizes these attacks classification:

Moreover, Table 4 presents the several attacks that affected the OSI model layers in CPS.

Tables 3 and 4 deduce that CPSs encounter several attacks that are the subject of various research to find adequate security solutions.

Another security problem that can be generated in CPSs, particularly during the establishment of interactions and collaborations between the components of a CPS and those of another CPS, is the lack of trust.

Indeed, trust plays a significant role in the realization of collaborations and interactions between different entities, particularly when sharing resources and services [34]. In this context, we devote the following section to presenting the trust concept, its various classifications, and its characteristics.

5 Trust Management in CPS

Regarding all the attacks and vulnerabilities mentioned in this chapter, we must protect and secure the different interactions between CPS entities. In this context, trust management facilitates interactions, especially those that can produce malicious activities. For this, we aim to study trust management in cyber-physical systems.

5.1 *Trust: Definitions and Classifications*

A unified definition of trust does not exist; definitions differ depending on the use of the trust concept and its application areas.

Focusing on the definitions of trust in the computer systems field, it designates *"the trust is a commitment to believe in the well running of the future actions of another entity"* [43]. Whereas [44] defines trust as *"a quantified belief in the abilities of the entity that is believed, so the act of trusting takes place in a specific context"*. Furthermore, Trust can be attached to the risk [45]; a trust level is given according to the risk that can be produced. Other researchers see that the situation (the context that requires the establishment of trust) is the most important in the definition of trust [46]. In addition, other researchers link the concept of trust to the concept of reputation [47], [48]: trusting someone means that he is reputed within his environment. Thus, the trust level of an entity is based on its reputation level.

Although several research works rely on reputation [48] to establish and build trust, the difference between the two concepts should be clear, as presented in [49].

- Trust means the belief that an entity builds towards the honesty of another entity based on its own experiences [49].
- Reputation means the belief that an entity builds towards the honesty of another entity based on recommendations [50] received from other entities. Reputation can be centralized [51], calculated by a trusted third party, or it can be decentralized.

Trust and reputation differ in how they are developed, but they are closely related; both are used to evaluate the trust of an entity.

Each achievement of trust is conditioned by the existence of a set of actors [52]; everyone has a role in the trust system. These actors are as follows:

- Trustor: An entity/organization/agent that makes/establishes trust towards another entity;
- Trustee: an entity/organization/agent which is trusted;
- Situation: the context in which trust is established.

But, the question is at the heart of the interaction and establishing trust between these actors. For this, we aim to study the methodology of trust systems.

5.2 Trust Management: How Does It Work?

To establish trust between different actors, the 'Trustor' entity relies on security policies to manage authorizations between actors (policy-based trust), or it depends on history, experiences, and the interactions of the 'Trustee' entity with other entities (reputation-based trust). According to this description, trust systems are classified into two categories [52]:

- **Policy-based trust systems**: These systems use policies to define whether or not an entity is granted access authorization. The management of certificates and the interpretation of relationships between actors are considered among the most used policies in this system.
- **Reputation-based trust systems**. These systems are based on the interactions, the direct and indirect experiences of the entities, their history, as well as the recommendations of the other entities to trust a particular entity. Generally, this system is based on mathematical models for calculating the trust scores assigned to an entity. The different interactions between entities (Trustor and Trustee), especially in reputation-based trust systems, fall into a centralized and decentralized system.
- **Centralized trust systems:** Within centralized trust systems, each 'Trustor' entity that wishes to establish trust towards another entity, must interrogate a central trusted authority [53] to know the trust score of the Trustee entity in particular. The principle of these systems is that the trust scores of the entities are stored at the central trusted entity. Thus, each entity directly contacts this entity to establish its trust towards another without going through the other entities.
- **Decentralized trust systems:** These systems don't have a central component that manages trust between entities [54], and therefore there is no storage component to record the trust scores of the entities. The principle of these systems is that each entity contacts the other entities to collect the various trust scores and calculate the overall score of an entity. In this sense, an entity that wishes to establish trust must create links with all the entities to retrieve the scores calculated towards a particular entity.

5.3 Trust Criteria

Several criteria have been identified in the literature to evaluate the trust between different entities. Indeed, reputation [47], recommendation [50], and satisfaction [55] present the most important criteria, given their usefulness in judging the behavior and reliability of an entity. Moreover, some other criteria may coexist, such as a number of interactions [47], popularity [50]; updating after the collaboration [28].

5.4 Trust Management in CPS

CPSs play a vital role in the evolution of the industry. Their applications have spread to several fields. In this sense, their security presents a great challenge for the actors involved in their deployment. Several research works and projects have been conducted to study the various vulnerabilities and attacks that threaten the CPS components. Based on these vulnerabilities and attacks, a set of security solutions have been proposed, among them, there are general solutions (authentication, identity management, anti-virus, firewall, etc.), but they are adapted or improved depending on the CPS characteristics.

In our research, we are particularly interested in the security of interactions and collaborations between several entities belonging to the different CPSs. To carry out an operation or action within the CPS, the latter may need help from another CPS (request for a resource, service, information, etc.). Therefore, before establishing an interaction between the entities, a trust evaluation is essential to prevent malicious actions. This evaluation is based on criteria to calculate a trust score (previous experiences, reputation, recommendation, number of successful interactions, etc.).

In this sense, we propose to apply our established mathematical model [34] for the trust evaluation between collaborating CPSs. We have already demonstrated the application of our model in the electrical network [34], particularly in the collaboration of Microgrids.

In addition to the techniques presented above for trust management (security policies, reputation and trust score calculation). Another technique that is considered reliable for managing trust between different actors, particularly in a decentralized environment, is the Blockchain [56].

6 Blockchain-Based Trust Management in CPS

6.1 Blockchain: Definition, Architecture, and Characteristics

Blockchain technology [56] was developed in 2008. It is a technology for storing and transmitting the information. It allows its users to share data with a very high level of security and reliability since it is a decentralized architecture without the intervention of a central (intermediate) entity.

From a technical and computer point of view, the blockchain is considered as a database that can be shared simultaneously with all its users. They can securely store data using cryptography [57]. The blockchain is also called the ledger/register.

A set of aspects characterizes the blockchain:

- **Security and reliability**: the blockchain is based on a consensus mechanism of all nodes participating in the network each time information is added. Indeed, the data is decrypted and authenticated by data centers. Thus, the validated transaction is added to the database as encrypted data block.
- **Decentralization:** the blockchain is protected from any falsification of transactions. Each new block added to the blockchain is linked to the previous one and a copy is transmitted to all nodes in the network.
- **The speed of transactions**: thanks to the fact that the validation of a block takes only a few seconds to a few minutes.
- **Autonomy:** Blockchain works without a control entity: users are authorized to modify this database at any time (add information, verify a transaction, etc.)
- **Non-falsification:** it is impossible to modify a transaction after it has been integrated into the Blockchain. In case of error, a second transaction canceling the first must appear. Both will then be visible.

Blockchain can be used in several areas: banking systems, agriculture-food, energy, logistics, transportation, health care, aeronautics, housing, etc.

The blockchain consists of 3 key elements:

- Blocks: they are defined as a group of transactions. More or less important, depending on the amount of data they contain, they are distinguished from each other by an identifier, a unique code called "hash".
- Nodes: they are connected to the blockchain. Each of them hosts a copy of the database. This copy is downloaded automatically when connecting to the network and contains all the exchanges between users.
- Miners are responsible for checking whether the newly created blocks correspond to security standards. They have an essential role within the blockchain, since they guarantee the authenticity of the blocks and the entire chain.

6.2 *Principle of Operation of Blockchain*

When a transaction is issued, it is immediately combined with other transactions released simultaneously: these transactions are grouped into a block. In this sense, the blockchain does not rely on an external authority to establish block verification. The users (minors) are responsible for guaranteeing the authenticity and security of the blocks. This analysis process is called mining [58]. It consists, of using calculations and algorithms, to certify or not a new block.

Once verified, the block is added to the chain, locked to the previous block, and becomes available to all users. Each additional block strengthens the verification of the previous block and, therefore, the entire blockchain. The other party then receives the transaction.

6.3 *Blockchain-Based Trust Management in CPS*

Blockchain is used to create reliable systems without a third party or intermediate intervention. This is made possible by using an algorithm known as the consensus algorithm [59]. Consensus is the process of elaborating an agreement between users who generally do not trust each other. There are different types of models to create consensus, for example: Proof of Work (PoW) [60], Proof of Stake (PoS) [61], etc.

In addition, the blockchain has shown its efficiency and reliability in several areas [62]. Also, it is considered relevant in CPS security for several reasons:

- The principle of CPS is based on the interaction between the different entities (cyber and physical);
- The various interactions are based on decentralized approaches;
- Sometimes, these entities do not know each other, hence the need for trust management between these entities;

7 Conclusion and Future Works

Through this paper, we have widened our research domain from the collaboration of information systems within Critical Infrastructures to the collaboration within the CPS. We presented the CPS concept, its definitions, and its different characteristics. Focusing on security, we have summarized the various vulnerabilities and attacks in the literature that threaten CPS.

We later presented the notion of trust, and its different classifications. Given the interest brought by the trust in the security of exchanges between systems, it is essential to describe its need in the CPS. Then, we presented the Blockchain as a relevant security technology in CPS.

In future works, we will deepen our research into CPSs in general and their security in particular. Next, we will focus on the Smart Grid as an application of CPS. We aim to apply our proposed approach by studying a real case of an attack threatening the CPS part of the Smart Grid to demonstrate its feasibility.

References

1. Monostori, L., Kádár, B., Bauernhansl, T., Kondoh, S., Kumara, S., Reinhart, G., . . . & Ueda, K. (2016). Cyber-physical systems in manufacturing. Cirp Annals, 65(2), 621-641. https://doi.org/10.1016/j.cirp.2016.06.005.
2. Candanedo, I. S., Nieves, E. H., González, S. R., Martín, M., & Briones, A. G. (2018, August). Machine learning predictive model for industry 4.0. In International Conference on Knowledge Management in Organizations (pp. 501–510). Springer, Cham.
3. Jiang, J., Xiong, Y., Zhang, Z., & Rosen, D. W. (2020). Machine learning integrated design for additive manufacturing. Journal of Intelligent Manufacturing, 1–14.
4. Lee, E. A. (2002). Embedded software. In Advances in computers (Vol. 56, pp. 55–95). Elsevier. https://doi.org/10.1016/S0065-2458(02)80004-3.
5. Strang, D., & Anderl, R. (2014). Assembly process driven component data model in cyber-physical production systems. In Proceedings of the World Congress on Engineering and Computer Science (Vol. 2).
6. Humayed, A., Lin, J., Li, F., & Luo, B. (2017). Cyber-physical systems security—A survey. IEEE Internet of Things Journal, 4(6), 1802–1831. https://doi.org/10.1109/JIOT.2017.2703172.
7. Junejo, A. K., Komninos, N., Sathiyanarayanan, M., & Chowdhry, B. S. (2019). Trustee: A trust management system for fog-enabled cyber physical systems. IEEE transactions on emerging topics in computing, 9(4), 2030–2041. https://doi.org/10.1109/TETC.2019.2957394.
8. Hermann, M., Pentek, T., & Otto, B. (2015). Design principles for Industrie 4.0 scenarios: a literature review. Technische Universität Dortmund, Dortmund, 45. https://doi.org/10.13140/RG.2.2.29269.22248.
9. Lee, E. A. (2006, October). Cyber-physical systems-are computing foundations adequate. In Position paper for NSF workshop on cyber-physical systems: research motivation, techniques and roadmap (Vol. 2, pp. 1–9).
10. Rajkumar, R., Lee, I., Sha, L., & Stankovic, J. (2010, June). Cyber-physical systems: the next computing revolution. In Design automation conference (pp. 731–736). IEEE. https://doi.org/10.1145/1837274.1837461.
11. Monostori, L. (2014). Cyber-physical production systems: Roots, expectations and R&D challenges. Procedia Cirp, 17, 9–13. https://doi.org/10.1016/j.procir.2014.03.115.
12. Alguliyev, R., Imamverdiyev, Y., & Sukhostat, L. (2018). Cyber-physical systems and their security issues. Computers in Industry, 100, 212–223. https://doi.org/10.1016/j.compind.2018.04.017.
13. Lewis, F. L. (2004). Wireless sensor networks. Smart environments: technologies, protocols, and applications, 11–46.
14. Ardelet, C., Veg-Sala, N., Goudey, A., & Haikel-Elsabeh, M. (2017). Entre crainte et désir pour les objets connectés: comprendre l'ambivalence des consommateurs. Décisions Marketing, 86(2), 31–46. https://doi.org/10.7193/DM.086.31.48.
15. Cardin, O. (2016). Contribution à la conception, l'évaluation et l'implémentation de systèmes de production cyber-physiques (Doctoral dissertation, Université de nantes).
16. Alavian, P., Eun, Y., Meerkov, S. M., & Zhang, L. (2020). Smart production systems: automating decision-making in manufacturing environment. International Journal of Production Research, 58(3), 828–845. https://doi.org/10.1080/00207543.2019.1600765.

17. Ahmed, S. H., Kim, G., & Kim, D. (2013). Cyber Physical System: Architecture, applications and research challenges. 2013 IFIP Wireless Days (WD), 1–5. https://doi.org/10.1109/WD.2013.6686528.
18. Ahmadi, A., Cherifi, C., Cheutet, V., & Ouzrout, Y. (2017, December). A review of CPS 5 components architecture for manufacturing based on standards. In 2017 11th International Conference on Software, Knowledge, Information Management and Applications (SKIMA) (pp. 1–6). IEEE. https://doi.org/10.1109/SKIMA.2017.8294091.
19. Tan, Y., Goddard, S., & Perez, L. C. (2008). A prototype architecture for cyber-physical systems. ACM Sigbed Review, 5(1), 1–2. https://doi.org/10.1145/1366283.1366309.
20. Alani, M. M. (2014). OSI model. In Guide to OSI and TCP/IP Models (pp. 5–17). Springer, Cham.
21. Ali, Z. H., Ali, H. A., & Badawy, M. M. (2015). Internet of Things (IoT): definitions, challenges and recent research directions. International Journal of Computer Applications, 128(1), 37–47.
22. Wortmann, F., & Flüchter, K. (2015). Internet of things. Business & Information Systems Engineering, 57(3), 221–224.
23. Li, S., Xu, L. D., & Zhao, S. (2015). The internet of things: a survey. Information systems frontiers, 17(2), 243–259. https://doi.org/10.1007/s10796-014-9492-7.
24. Atzori, L., Iera, A., & Morabito, G. (2017). Understanding the Internet of Things: definition, potentials, and societal role of a fast evolving paradigm. Ad Hoc Networks, 56, 122–140.
25. Sundmaeker, H., Guillemin, P., Friess, P., & Woelfflé, S. (2010). Vision and challenges for realising the Internet of Things. Cluster of European research projects on the internet of things, European Commision, 3(3), 34–36.
26. Tubaishat, M., & Madria, S. (2003). Sensor networks: an overview. IEEE potentials, 22(2), 20–23. https://doi.org/10.1109/MP.2003.1197877.
27. Chen, F., & Li, R. (2013). Sink node placement strategies for wireless sensor networks. Wireless personal communications, 68(2), 303–319.
28. Yick, J., Mukherjee, B., & Ghosal, D. (2008). Wireless sensor network survey. Computer networks, 52(12), 2292–2330. https://doi.org/10.1016/j.comnet.2008.04.002.
29. Malinowski, A., & Yu, H. (2011). Comparison of embedded system design for industrial applications. IEEE transactions on industrial informatics, 7(2), 244–254. https://doi.org/10.1109/TII.2011.2124466.
30. Sadou, N. (2007). Aide à la conception des systèmes embarqués sûrs de fonctionnement (Doctoral dissertation, INSA de Toulouse).
31. Wu, L., Chuang, C. H., & Hsu, C. H. (2014). Information sharing and collaborative behaviors in enabling supply chain performance: A social exchange perspective. International Journal of Production Economics, 148, 122–132. https://doi.org/10.1016/j.ijpe.2013.09.016.
32. Bernard, F. (2017). Un système d'information collaboratif en appui à la gouvernance des territoires d'action agro-environnementale à enjeu eau-pesticides (Doctoral dissertation, Paris, Institut agronomique, vétérinaire et forestier de France).
33. Autefage, V., Chaumette, S., & Magoni, D. (2015, June). Influence des modèles de mobilité sur un système collaboratif pour flottes autonomes hétérogènes. In ALGOTEL 2015-17èmes Rencontres Francophones sur les Aspects Algorithmiques des Télécommunications.
34. Ait Aali, N., Baina, A., & Echabbi, L. (2018). Trust management in collaborative systems for critical infrastructure protection. Security and Communication Networks, 2018. https://doi.org/10.1155/2018/7938727.
35. S. Bortzmeyer, « Centralisé, décentralisé, pair à pair, quels mots pour l'architecture des systèmes répartis ? », p. 6, 2015.
36. Kumar, R., & Verma, R. (2012). Classification algorithms for data mining: A survey. International Journal of Innovations in Engineering and Technology (IJIET), 1(2), 7–14.
37. Obermeyer, Z., & Emanuel, E. J. (2016). Predicting the future—big data, machine learning, and clinical medicine. The New England journal of medicine, 375(13), 1216.
38. Likas, A., Vlassis, N., & Verbeek, J. J. (2003). The global k-means clustering algorithm. Pattern recognition, 36(2), 451–461. https://doi.org/10.1016/S0031-3203(02)00060-2.

39. East, S., Butts, J., Papa, M., & Shenoi, S. (2009, March). A Taxonomy of Attacks on the DNP3 Protocol. In International Conference on Critical Infrastructure Protection (pp. 67–81). Springer, Berlin, Heidelberg.
40. Huitsing, P., Chandia, R., Papa, M., & Shenoi, S. (2008). Attack taxonomies for the Modbus protocols. International Journal of Critical Infrastructure Protection, 1, 37–44. https://doi.org/10.1016/j.ijcip.2008.08.003.
41. K. Munro, « Deconstructing Flame: the limitations of traditional defences », Comput. Fraud Secur., vol. 2012, no 10, p. 8–11, oct. 2012, https://doi.org/10.1016/S1361-3723(12)70102-1.
42. Fogel, B., Farmer, S., Alkofahi, H., Skjellum, A., & Hafiz, M. (2016, April). POODLEs, more POODLEs, FREAK attacks too: how server administrators responded to three serious web vulnerabilities. In International Symposium on Engineering Secure Software and Systems (pp. 122–137). Springer, Cham.
43. Golbeck, J. A. (2005). Computing and applying trust in web-based social networks. University of Maryland, College Park.
44. Grandison, T., & Sloman, M. (2003, May). Trust management tools for internet applications. In International Conference on Trust Management (pp. 91–107). Springer, Berlin, Heidelberg.
45. Pappas, N. (2016). Marketing strategies, perceived risks, and consumer trust in online buying behaviour. Journal of retailing and consumer services, 29, 92–103. https://doi.org/10.1016/j.jretconser.2015.11.007.
46. Riker, W. H. (2017). The nature of trust. In Social power and political influence (pp. 63–81). Routledge.
47. Hawa, M., As-Sayid-Ahmad, L., & Khalaf, L. D. (2013). On enhancing reputation management using Peer-to-Peer interaction history. Peer-to-Peer Networking and Applications, 6(1), 101–113. https://doi.org/10.1007/s12083-012-0142-x.
48. Reddy, T. C., & Seshadri, R. (2014). Reputation-Based Dynamic Trust Evaluation Model for multi-agent Systems based on service satisfaction. International Journal of Emerging Technologies and Advanced Engineering, ISSN, 2250-2459.
49. Wang, Y., & Vassileva, J. (2003, September). Trust and reputation model in peer-to-peer networks. In Proceedings Third International Conference on Peer-to-Peer Computing (P2P2003) (pp. 150–157). IEEE. https://doi.org/10.1109/PTP.2003.1231515.
50. Can, A. B., & Bhargava, B. (2012). Sort: A self-organizing trust model for peer-to-peer systems. IEEE transactions on dependable and secure computing, 10(1), 14–27. https://doi.org/10.1109/TDSC.2012.74.
51. Marti, S., & Garcia-Molina, H. (2006). Taxonomy of trust: Categorizing P2P reputation systems. Computer Networks, 50(4), 472–484.
52. Gaillard, E. (2011). Les systèmes informatiques fondés sur la confiance: un état de l'art (Doctoral dissertation, Loria & Inria Grand Est).
53. Hughes, R. J., Nordholt, J. E., McCabe, K. P., Newell, R. T., Peterson, C. G., & Somma, R. D. (2013). Network-centric quantum communications with application to critical infrastructure protection. arXiv preprint arXiv:1305.0305.
54. Bearly, T., & Kumar, V. (2007). Building trust and security in peer-to-peer systems. In Secure Data Management in Decentralized Systems (pp. 259–287). Springer, Boston, MA. https://doi.org/10.1007/978-0-387-27696-0_8.
55. Zhao, H., & Li, X. (2013). VectorTrust: trust vector aggregation scheme for trust management in peer-to-peer networks. The Journal of Supercomputing, 64(3), 805–829. https://doi.org/10.1007/s11227-011-0576-6.
56. Yli-Huumo, J., Ko, D., Choi, S., Park, S., & Smolander, K. (2016). Where is current research on blockchain technology?—a systematic review. PloS one, 11(10), e0163477. https://doi.org/10.1371/journal.pone.0163477.
57. Buchmann, J. A. (2004). Cryptographic hash functions. In Introduction to Cryptography (pp. 235–248). Springer, New York, NY. https://doi.org/10.1007/978-1-4419-9003-7_11.
58. Chung, K., Yoo, H., Choe, D., & Jung, H. (2019). Blockchain network based topic mining process for cognitive manufacturing. Wireless Personal Communications, 105(2), 583–597. https://doi.org/10.1007/s11277-018-5979-8.

59. Mingxiao, D., Xiaofeng, M., Zhe, Z., Xiangwei, W., & Qijun, C. (2017, October). A review on consensus algorithm of blockchain. In 2017 IEEE international conference on systems, man, and cybernetics (SMC) (pp. 2567–2572). IEEE. https://doi.org/10.1109/SMC.2017.8123011.
60. Gervais, A., Karame, G. O., Wüst, K., Glykantzis, V., Ritzdorf, H., & Capkun, S. (2016, October). On the security and performance of proof of work blockchains. In Proceedings of the 2016 ACM SIGSAC conference on computer and communications security (pp. 3–16). https://doi.org/10.1145/2976749.2978341.
61. Li, W., Andreina, S., Bohli, J. M., & Karame, G. (2017). Securing proof-of-stake blockchain protocols. In Data privacy management, cryptocurrencies and blockchain technology (pp. 297–315). Springer, Cham.
62. Xu, Q., Su, Z., & Yang, Q. (2019). Blockchain-based trustworthy edge caching scheme for mobile cyber-physical system. IEEE Internet of Things Journal, 7(2), 1098–1110. https://doi.org/10.1109/JIOT.2019.2951007.

Blockchain-Based Authentication in IoT Environments: A Survey

Mohammed M. Alani

1 Introduction

As Internet-of-Things (IoT) devices grow to be more ubiquitous, the world is witnessing unprecedented growth in IoT adoption.

According to Statista [6], the number of devices is expected to double within the coming 6 years. This rapid growth, combined with lack of interest in security from the vendors' side, creates a very attractive play field for malicious actors to perform their attacks.

According to the research published in [4], 98% of all IoT traffic is unencrypted, and 57% of IoT devices are have medium to high severity vulnerabilities that can be exploited to attacks. As most operating systems and applications in IoT are smaller in size, in comparison to other sophisticated devices such as personal computers or mobile devices, most IoT vulnerabilities are simple to patch. However, the focus of most vendors is on providing commercial low-cost solutions, and ignore security concerns even for well-known security threats. In addition, as explained in [3], most of these vulnerable IoT devices are insecure by design or by deployment.

Most malicious actors take control over IoT devices to use them in performing attacks on other targets [11].

A large amount of IoT risks come from botnets. Botnets, which can be defined as networks formed by compromised machines, spread the infection to other devices and recruit them to be part of the botnets [31]. Botnets are controlled by malicious actors who utilize them to attack different targets. These orchestrated large-scale attacks can cause much larger damage to the targets as these botnets grow in size.

M. M. Alani (✉)
Computer Science Department, Toronto Metropolitan University, Toronto, ON, Canada
e-mail: m@alani.me

Y. Maleh et al. (eds.), *Blockchain for Cybersecurity in Cyber-Physical Systems*,
Advances in Information Security 102, https://doi.org/10.1007/978-3-031-25506-9_5

In the past few years, a specific infamous botnet, named Mirai, gained popularity. Mirai botnet was built using Mirai malware that was first known in 2016, and soon was utilized in a massive orchestrated Distributed Denial-of-Service (DDoS) attack. In September 2016, Mirai temporarily brought down many high-profile online services such as OVH, Dyn, and Krebs on Security with a massive DDoS that exceeded 1 Tbps rate. This DDoS attack was the highest in history at its time. The infrastructure behind this enormous attack was nothing but small IoT devices such as home routers, air-quality monitors, and personal surveillance cameras. At its peak, Mirai infected over 600,000 vulnerable IoT devices [19].

In 2017, the creator of Mirai, released its source code to the public. This led to a large surge in variant botnets developed using Mirai's code base. These variants, such as Katana, Boonet Botnet, Mukashi, among over 60 variants, were utilized in many large-scale attacks over the years [2, 27, 29].

Another large-scale DDoS attack hit Cloudflare, a cloud computing service provider, that was based on UDP protocol took place in September, 2020. The attack peaked at 654 Gbps and lasted for about 2 min [3]. This massive-scale attack was initiated by the Moobot botnet built from infected IoT devices. Analysis have shown that the attack originated from 18,705 unique IP addresses located in 100 countries. The attack was successful in exploiting two vulnerabilities that were know for a long time, but not patched in most vulnerable IoT devices.

A botnet attack starts with active IP scanning, as the infected device tryies to find other IoT devices to infect. As the malware finds other active IP addresses, it starts port scanning to find open ports and active services. The malware then tries to perform Operating System (OS) fingerprinting to find whether the IP address belongs to another IoT device or not [13]. Once the malware finds another IoT device, it tries to bypass the device's authentication mechanism through brute-force or by exploiting other authentication vulnerabilities. If the malware succeeds, it will infect the other IoT devices and recruit them to join the botnet [15].

Many research publications tried to address IoT security challenges through machine learning, and big data [12]. However, the limited memory and processing power in IoT devices makes many of these proposed solutions unrealistic.

1.1 Problem Definition

Having limited processing power is an inherent disadvantage to IoT devices in relevance to security. Many security-related operations such as authentication and encryption are resource intensive, at least the complex and secure algorithms. Resource limitations, in addition to other limitations relevant to identity and access management makes IoT devices in need of drastic solutions to strengthen them against authentication attacks [14].

Password-based authentication is considered one of the weakest authentication factors. Within the IoT context, password-based authentication is even a bigger problem because most IoT devices lack simple protections against brute-force attacks [13]. This has led to many research directions to strengthen the security of the authentication process in IoT devices. Many research papers addressed the authentication problem by proposing a blockchain-based authentication system. These research papers approached the problem in different ways and proposed different solutions relying on the concepts of block chain. This diversity in the proposed solutions makes it difficult to choose a suitable solution for a particular implementation.

1.2 Research Contribution

In this chapter, we survey research in the area of blockchain applications in authentication within the IoT environment. We review various proposed solutions presented in state-of-the-art and produce a comparative study that would help practitioners in selecting a suitable blockchain-based authentication mechanisms for their IoT applications.

In addition, we introduce a detailed analysis of advantages and challenges facing the implementation of blockchain-based IoT authentication solutions.

1.3 Chapter Layout

The next section of the chapter provides preliminary information about blockchain, authentication, and authentication attacks. Section 3 explains the methodology followed in conducting this survey, while Sect. 4 presents the findings of the survey along with the details of the different authentication schemes reviewed. Discussion of the findings is presented in Sect. 5. Finally, Sect. 6 provides the conclusions of our study.

2 Preliminaries

After providing a short introduction about IoT and IoT security in Sect. 1, in this section, we will provide a brief introduction into the two remaining topics of this survey; blockchain, and authentication.

2.1 Blockchain

A blockchains, or distributed ledgers, are defined in [36] as "tamper evident and tamper resistant digital ledgers implemented in a distributed fashion". these distributed ledgers operate without the need of a central authority.

Blockchains can be categorized into permisionless, and permissioned. Permissionless blockchains are decentralized ledger platforms where anyone can publish blocks without the need of a permission from any authority. In a permissioned blockchain, the process of publishing blocks requires permissions by some authority, whether the blockchain was centralized or not. Based on these permissions, it is also possible to control which entities has read access on the blocks as well [36].

Figure 1 shows an overview of the operation of blockchain. As shown in Fig. 1, each block consists of a block header, and block data. The block data would contain any data relevant to the transaction that is to be maintained by the blockchain. The block header consists of the following parts:

- A hash of the header of t he previous block. This help maintain the chaining of the blocks together such that whenever a block is edited or removed, it can directly be detected, because it would generate a different hash.
- A timestamp that would mark the exact time of the transaction. This helps in maintaining the transaction history in order.
- A random number used only once (nonce) that helps in maintaining the integrity of the block and counters replay attacks.
- A hash of the block's data. This helps in maintaining the integrity of the data contained in the block.

As shown in Fig. 1, the blocks are chained together, and hence the name blockchain. If an attacker manipulates the content of one block within the chain, it can be easily detected.

The adoption of blockchains have been rapidly expanding in various area of our lives, such as electronic currencies, banking and financial transactions, health care records, records management, voting, and cybersecurity. The main features of blockchain that makes it suitable for these various applications are:

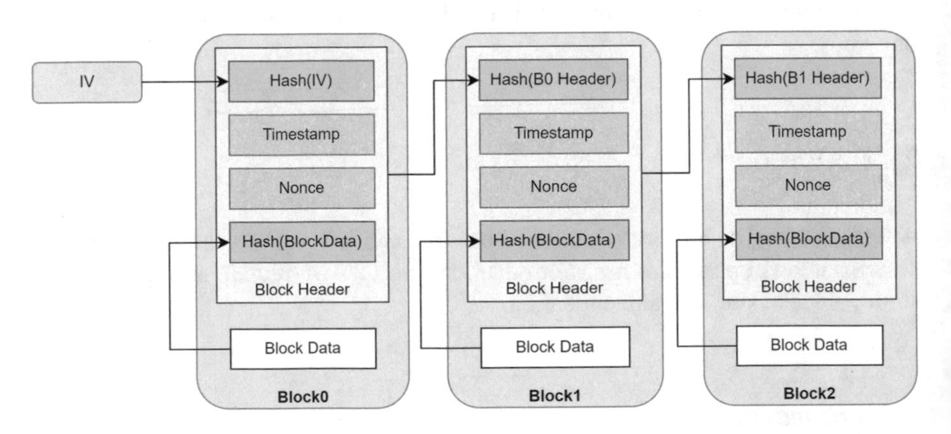

Fig. 1 An overview of blockchain operation

- Distributed: In most blockchain implementations, the ledger of transactions is not stored in one place.
- Tamper Resistant: Tampering can be easily detected in blockchains.
- Immutable: The data stored in the blockchain cannot be edited. There are specific types of blockchains where editing is available with certain limitations.

The aforementioned features make it suitable for many scenarios where there is no centralized entity that manages the communications between peers. Scenarios where there is no inherent trust relation between peers can be a suitable place to implement blockchains.

The specific use case that this chapter is focused on is the case of authentication of IoT devices. When IoT devices need to communicate with each other, mutual authentication plays an important role in securing the communication and preventing an abundance of attack types. Most of such scenarios require authentication without a third entity managing the process. This is especially evident in context such as self-driving cars and other intelligent types of transportation.

2.2 *Authentication*

Authentication can be defined as the process of validating the identity of a user or process. The interaction starts with identification, where a user confesses an identity. The next step would for that user to validate that they are who they claim they are, through authentication. Once authenticated, the resources that this user has access to, and the specific type of access are defined in the process of authorization [10].

Authentication is considered an important security control that supports many security services such as identity management and access control. Authentication can be achieved using different factors. The following list shows the different authentication factors [9]:

- Something you know: Such as a password, Personal Identification Number (PIN), or an answer to a security question.
- Something you have: Such as a physical token, a USB device, or an authenticator application on your phone.
- Something you are: Such as a fingerprint, voice print, retina scan, or face recognition.
- Something you do: Such as a gesture performed on a touch screen.
- Somewhere you are: Such as geolocation authentication.

When multiple factors are combined together, it is referred to as Multi-Factor Authentication (MFA). MFA is generally considered much more secure in comparison to single factor authentication.

Most weak authentication used in IoT relies significantly on the use of passwords. With the rapid developments in processing power in the recent years, passwords are becoming the weakest authentication technique used. These advances in processing power has made it possible to break short passwords in a few seconds [5].

Authentication mechanisms popular in IoT environment can be categorized into the following three categories [30]:

- Symmetric-Key Based: Such as Kerberos, and Hash-based Message Authentication Code (HMAC).
- Certificate-Based: Such as Transport-Layer Security (TLS), Elliptic-Curve Cryptography (ECC), and Elliptic-Curve Digital Signature Algorithm (ECDSA).
- Identity-Based: Such as Identity-Based Signatures (IBS), and Identity-Based Encryption (IBE).

2.3 Authentication Attacks

A successful attack on an authentication mechanism employed in a system would probably lead to a complete system compromise. If authentication is successfully bypassed by a malicious actor, they can pursue further damage through privilege escalation or lateral movement in the system [24].

Authentication attacks can be divided into the following categories:

1. Eavesdropping attacks: In eavesdropping attacks, the attacker tries to obtain the authentication credentials by capturing unencrypted traffic, looking over the person's shoulder while they're typing their credentials, or through the use of keyloggers. A keylogger, is a software or a hardware device that captures are records all keystrokes on a user's keyboard.
2. Brute-force attacks: The older attack in the books. The attack s based on trying all the possible passwords until finding the right one. A special type of this attack is called "dictionary attack" where the attacker uses commonly used words from a dictionary to brute-force a user's password.
3. Social Engineering: Social engineering refers to the non-technical attacks that try to convince the user to give up their credentials. Phishing and spoofing are two commonly used social engineering techniques where a user is tricked into giving their credentials through well crafted emails, web pages, or applications.
4. Insider Attacks: Insider attacks are a wide range of attacks where threat actors are from within the organization or connected to the organization's network. In the context of authentication, an insider attack could be originating from an administrator with high privileges, or from a regular user account trying to perform privilege-escalation.
5. Session Hijacking: In this attack, the attacker takes over the session after authentication. After authentication, an authentication token is stored at the victim's browser cookies. The attacker steals this token using malware or other techniques, and sends it to the server to hijack the session that has already been authenticated.

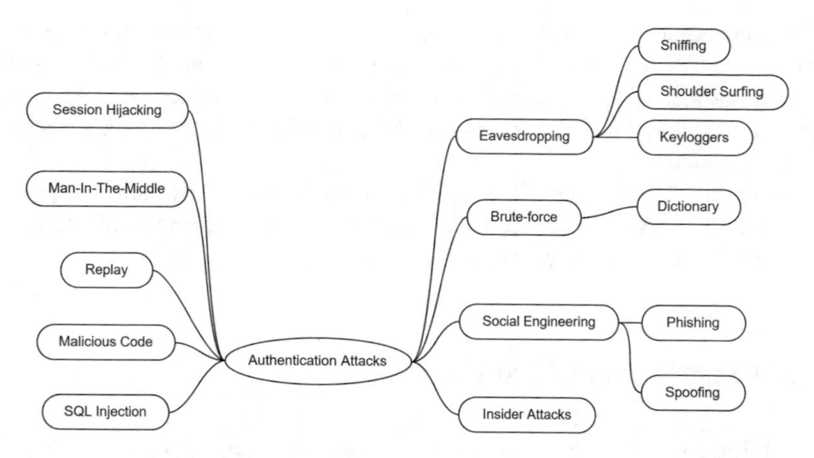

Fig. 2 Taxonomy of authentication attacks

6. Man-In-The-Middle: In this attack, the attacker operates in the middle between the victim and the server. On the victim's side, the attacker impersonates the server, while on the server's side, the attacker impersonates the client. When successful, the attacker can read all the exchanged information including credentials.
7. Replay Attacks: In replay attacks, the attacker captures the network traffic carrying the credentials as it is sent to the server. These traffic might be encrypted. However, instead of reading the credentials, the attacker re-sends the encrypted traffic to the server to impersonate the victim.
8. Malicious Code: Malicious code can be in different forms. The code can be a Trojan horse that implants a keylogger, a virus that steals session information, or other types of code that can help the attacker gain access to the authentication credentials or bypass the authentication process completely.
9. SQL Injection: Structured Query Language (SQL) injection is an attack where the attacker injects SQL commands within an input field on a webpage. If the webpage is not built properly, this input can execute commands directly on the database where data can be deleted, or authentication process bypassed.

The aforementioned list is not a comprehensive list of all the possible attacks on authentication. The list explains the most common attacks. Figure 2 shows simplified taxonomy of authentication attacks.

3 Methodology

Most related works, such as [7, 20, 34], are focused on surveying general applications of blockchain in IoT, including authentication. However, our work focuses specifically on blockchain-based authentication techniques in IoT environments.

The methodology adopted in our research was forward snowballing [35]. In this technique, the survey starts with highly cited well-known works, and then starts crawling the research that have cited these highly-cited publications. This step is followed by proper filtering to remove publications that are not directly relevant to the research area.

In our final step, we filtered the resulting references based on their relevance and selected the 12 papers that we considered providing most effective techniques with proper testing and proof of concept.

4 Blockchain and IoT Authentication

IoT environments are known to generate large amounts of data. Especially in industrial contexts, more than other contexts. IoT devices exchange information usually without human intervention. Often in IoT environments, there is a lack of a centralized entity that can handle the identification and authentication processes. Due to the autonomy needed in this information exchange, there is an inherent need for these devices to be able to authenticate and trust each other.

In 2018, Hammi et al. published one of the most influential papers in the area of blockchain-based authentication for IoT that presented a robust decentralized identification and authentication system named "Bubbles of Trust" [23]. The proposed approach is based on creating secure virtual zones (bubbles) within the IoT environment where each device is allowed to communicate with devices within its own bubbles. Devices from outside of the bubble are considered malicious. The proposed system is built using public blockchain and smart contracts. The choice of public over private blockchain was to ensure the system's flexibility after deployment. The process of communication between the two IoT devices takes place on the blockchain where each message sent is considered a blockchain transaction. when device A wants to send a message to device B, the message would be sent to the blockchain. Once the blockchain authenticates A, the transaction is validated and device B can read the message. The proposed system was implemented and tested on Raspberry Pi device and proved resistance to many attacks such as Sybil attacks, spoofing attacks, replay attacks, and impersonation attacks.

Li et al. presented, in 2018, a blockchain based authentication mechanisms for IoT device [26]. The proposed system was based on creating a unique identity for each blockchain device and record that on the blockchain along with a public key allocated to that IoT device. When device A wants to communicate with device B, it sends a message with its identity, message content, and a signature. Once device B receives the message, it contact the blockchain to validate the signature using the public key of A that is stored in the trusted blockchain. The proposed model divides the nodes int he network to two types: consensus nodes, that generate and verify blocks, and participate in the consensus, and non-consensus nodes that are used to transfer data. The proposed system was implemented on Raspberry Pi IoT cluster, based on Hyperledger Fabric [1]. Testing showed that the proposed system was capable of preventing malicious nodes from intruding, preventing firmware backdoors, and resist DDoS attacks.

In 2019, Lin et al. proposed a blockchain based mutual authentication system for smart homes [28]. The proposed approach utilizes blockchain, message authentication codes, and group signatures to provide a secure authentication method that records access history, and anonymously authenticate group members. The proposed approach maintains confidentiality as well through employing encryption. The proposed system was implemented in a virtualized environment. Testing results showed that the proposed system, in addition to providing secure mutual authentication, provides confidentiality as well.

Inter-device communication intensifies in industrial contexts. In 2020, Shen et al. presented a blockchain-assisted device authentication for cross-domain Industrial IoT (IIoT) environments, named BASA [30]. The proposed system introduces consortium blockchain to construct trust among different domains. IBSes are used in the proposed authentication process. To address privacy concerns, the BASA utilizes a mechanism that keeps the identity of the authenticated device anonymous. In addition, the process includes the negotiation of session keys to maintain the communication's confidentiality. Thorough testing was conducted on the proposed system that yielded resistance to authentication attacks. Testing also included measuring computational overhead, communication overhead, writing and query latency in a virtualized environment.

Khalid et al. presented, in 2020, a lightweight blockchain-based authentication mechanism based on blockchain [25]. The proposed system provides access control, in addition to decentralized authentication. The proposed system utilizes fog computing to host the blockchain used for mutual authentication. An IoT node would first perform registration at the nearest fog node. This registration is recorded in the blockchain. After registration, two IoT devices can only communicate with each other after mutual authentication. The mutual authentication processes, each device authenticates using the information and keys stored on the blockchain. The proposed system was tested in a combination of hardware and virtualized devices. Testing results showed that the proposed system counters many authentication attacks, and its timing parameters were suitable for a smart city environment.

Al-Naji and Zagrouba presented, in 2020, a continuous authentication architecture based on blockchain for IoT, named CAB-IoT [8]. The proposed solution utilizes fog nodes to provide the processing power needed to host and run the blockchain, and to handle the continuous-authentication activities. The proposed solution also utilizes face-recognition based on machine learning to eliminate the possibility of unauthorized physical access to the devices. The proposed system was tested through simulation and provided acceptable performance measures. It also succeeded in countering eavesdropping, DoS, and MITM among other attacks.

In 2020, Ali et al. presented a decentralized blockchain based access control and permission delegation framework for IoT devices, named xDBAuth [16]. The proposed system includes a hierarchy of local and global smart contracts that perform permission delegation and access control for both internal and external IoT devices. The proposed framework also considers the privacy of external devices connecting to the internal devices. Once a device is authenticated, the framework authorizes the device based on the policies stored on the blockchain. The proposed framework was implemented using Node.js [28] in a virtualized environment. Testing showed good performance in terms of timing and load, and the proposed framework was capable of countering different authentication attacks.

Goyat et al. presented, in 2020, a decentralized blockchain-based framework for data storage, authentication, and privacy preservation for Wireless Sensor NEtworks (WSN)s [22]. In this proposed work, a cloud-based base station handles the device registration, certification, and revocation. In addition, the base station handles all collected information sent from cluster heads. This means that the base station would record all the key parameters on the distributed blockchain and the large bulk of data to be stored on the cloud. The proposed system was tested through simulation. Testing results showed good timing parameters, and resistance to authentication and confidentiality attacks.

In 2020, Tahir et al. presented a lightweight blockchain-based authentication and authorization mechanisms for Health-IoT (HIoT) [32]. The proposed framework uses a probabilistic model combined with blockchain approach to authenticate and authorize access to devices. It utilizes random numbers in the authentication process which is further connected through joint conditional probability. The proposed framework was tested through simulation using AVISPA tool [18]. Simulation results showed good resistance to authentication attack, and reasonable timing overhead.

Gong et al. presented, in 2021, Blockchain of Things (BCoT) Gateway [21]. A gateway that records authentication transactions in a blockchain network without the need to modify the IoT device hardware or applications. The proposed system is said to be lightweight identity authentication system based on recording all authentication activities in a blockchain. In this method, the blockchain is not stored at the IoT devices, but on independent nodes. The proposed method utilizes a machine-learning based model to detect device fingerprint and be able to identify the device. Testing showed processing costs, and acceptable resistance to authentication attacks.

Alzubi presented, in 2021, a blockchain-based authentication tool for IoT in healthcare that utilizes Lamport-Merkle Digital Signature(LMDS) [17]. At the initialization phase, LMDS model generation takes place through constructing a tree in which the leaves symbolize sensitive patient medical data's hashes. The verification process happens through a centralized controlled that determines the root of LMDS using the has of the public-key. The proposed method performance was measured through simulation to calculate the computational overhead and computation time. The proposed system was proven to counter most authentication attacks.

Wang et al. proposed, in 2021, an IIoT authentication mechanism based on Transfer Learning empowered Blockchain, named ATLB [33]. The proposed mechanism employs blockchains to achieve privacy preservation for industrial applications. Transfer learning is utilized in building the authentication mechanism. At the start, ATLB uses a guiding deep deterministic policy gradient algorithm to train the device authentication model of a specific region. The trained model is transferred locally for outsider device authentication. The proposed system was implemented in a simulation environment, and was able to achieve high throughput and low latency, in addition to the secure authentication.

Table 1 shows a summary of features for the reviewed research. While all proposed systems achieved the goal of countering authentication attacks and provide secure authentication, as shown in Table 1, only two of the proposed solutions had a holistic take on security to provide secure authentication, in addition to confidentiality and integrity.

5 Discussions

Authentication in IoT, as discussed in Sect. 2, has a broader attack surface in comparison to other authentication contexts. This is due to many factors such as inherent insecurity of many IoT devices' firmware, low processing power, continuous connectivity, and begin ubiquitous in many applications in our daily lives.

As we explored how previous research explored different ways of utilizing blockchain technologies to replace vulnerable pieces within the authentication process. In the following subsections, we will discuss the advantages, and challenges of utilizing blockchain in IoT authentication.

5.1 Advantages of Using Blockchain in IoT Authentication

All of the solutions reviewed in this chapter provided secure authentication that can counter most authentication attacks mentioned in Sect. 2. This comes from several advantages the blockchain presents in comparison to classical authentication solutions.

Being a decentralized solution, blockchain-based authentication provides higher reliability in comparison to classical systems that have a signal point of failure. In a classical solution, the authentication process is centralized because there is only on node that has the information necessary to make an authentication decision. However, the decentralized nature f blockchain makes it more resilient to attack-caused, or non-deliberate failures in authentication nodes.

Another advantage is being tamper-proof. Blockchains, with the way each block is chained to other blocks, makes it impossible for someone to alter the data without being detected. Although the specific use of the blockchain within the authentication system varied, as we have seen in Sect. 4, this feature is a big advantage. Blockchains can be used to hold public keys used for authentication, hold device identities, or hold authentication activity logs, or to hold other authentication data.

Table 1 Summary of reviewed research

Research	Blockchain type	Context	Testing type	Secure authentication	Other attacks	Confidentiality	Integrity
Al-Naji et al. [8]	Private	IoT	Simulated	✓	✓	✓	✓
Lin et al. [28]	Private	Smart home	Virtualized	✓	✓	✓	✓
Shen et al. [30]	Private	IIoT	Virtualized	✓	✗	✓	✓
Goyat et al. [22]	Private	IoT	Simulated	✓	✗	✓	✓
Hammi et al. [23]	Public	IoT	Implementated	✓	✓	✗	✓
Khalid et al. [25]	Public	Smart city	Hybrid	✓	✓	✗	✓
Want et al. [33]	Private	IIoT	Simulated	✓	✓	✗	✓
Ali et al. [16]	Public	IoT	Virtualized	✓	✓	✗	✗
Li et al. [26]	Private	IoT	Implementated	✓	✓	✗	✗
Gong et al. [21]	Private	IoT	Virtualized	✓	✗	✗	✓
Alzubi [17]	Private	HIoT	Simulated	✓	✗	✗	✓
Tahir et al. [32]	Private	HIoT	Simulated	✓	✗	✗	✓

5.2 *Challenges in Using Blockchain in IoT Authentication*

A major challenge in IoT authentication is device identification. Older systems used hardware-based features for identification, such as Media Access-Control (MAC) addresses. MAC addresses are supposed to be unique. However, many attackers have the capacity to spoof MAC addresses using cheap and widely available hardware and software. In other scenarios, the identity is chosen by the user. This case is difficult to configure because there needs to be a centralized database to avoid identity re-use by other users on other devices.

Different publications addressed the identification issue in different ways. In [25], the unique system identity was created using the system name in addition to the last 5 hashed digits of the fog node's MAC address. In [23], the identity is obtained through the first 20 digits of the SHA-3 hash of the device's public key. Although the probability is low, it is possible for identities to be duplicated using this method.

Another, rather obvious, challenge is processing overhead. IoT devices, by design, have lower processing power compared to personal computers. Although many of the solutions discussed in Sect. 4 were "lightweight" by design, the computational load remains higher in blockchain-based authentication solutions, in comparison to weaker classical methods. This remains an interesting area of research in the future.

Another challenge to consider is the lack of support and adoption by vendors. Most IoT vendors look for commercial solutions that are cheaper and can sell more. Hence, making drastic changes to the insecure firmware they use is not always favorable. With the exception of large vendors who are security-aware, most vendors will prefer older solutions that are proven to work previously.

In general, the biggest challenge that blockchain adoption in IoT authentication faces is lack of standardization. Many reasonably secure solutions were presented in the literature in the past few years. However, there is general hesitance in adopting these directions as the vendors want to be "compatible" with the rest of the market. Interoperability is an important aspect that encourages adoption of technologies manufactured by different vendors. Once a standardized authentication protocol is created, vendors will start following the standard.

6 Conclusions

In this chapter, we presented a survey of recent research utilizing blockchain in authentication for IoT environments. The chapter explored different methods in which blockchains were employed in IoT authentication, and at different stages, or roles within the authentication process. Blockchains, being decentralized, tamper-proof method of holding authentication information, has a great potential in serving the purpose of providing secure authentication.

We also explored the advantages and challenges that blockchain utilization in IoT authentication faces. As we examine these challenges and advantages in Sect. 5, we realize that there is a lot more to explore in this area and further intensive research is required in the future to help the community harness the full potential of blockchains in IoT authentication.

References

1. Hyperledger Fabric – Hyperledger Foundation (2020). URL https://www.hyperledger.org/use/fabric. [Online; accessed 18. Apr. 2022]
2. Mirai Variant Targeting New IoT Vulnerabilities, Network Security Devices (2021). URL https://unit42.paloaltonetworks.com/mirai-variant-iot-vulnerabilities. [Online; accessed 1. May 2021]
3. OT/IoT Security Report February 2021 | Nozomi Networks (2021). [Online; accessed 1. May 2021], https://www.nozominetworks.com/ot-iot-security-report
4. OT/IoT Security Report: Rising IoT Botnets and Shifting Ransomware Escalate Enterprise Risk (2021). [Online; accessed 1. May 2021], https://www.nozominetworks.com/blog/what-it-needs-to-know-about-ot-io-security-threats-in-2020
5. Use this chart to see how long it'll take hackers to crack your passwords (2021). URL https://www.komando.com/security-privacy/check-your-password-strength/783192. [Online; accessed 12. Apr. 2022]
6. IoT connected devices worldwide 2019–2030 | Statista (2022). URL https://www.statista.com/statistics/1183457/iot-connected-devices-worldwide. [Online; accessed 10. Apr. 2022]
7. Akram, S.V., Malik, P.K., Singh, R., Anita, G., Tanwar, S.: Adoption of blockchain technology in various realms: Opportunities and challenges. Security and Privacy **3**(5), e109 (2020)
8. Al-Naji, F.H., Zagrouba, R.: Cab-iot: continuous authentication architecture based on blockchain for internet of things. Journal of King Saud University-Computer and Information Sciences (2020)
9. Alani, M.M.: Elements of cloud computing security: A survey of key practicalities. Springer (2016)
10. Alani, M.M.: Security Threats in Cloud Computing, pp. 25–39. Springer International Publishing, Cham (2016). URL https://doi.org/10.1007/978-3-319-41411-9_3
11. Alani, M.M.: Iot lotto: Utilizing iot devices in brute-force attacks. In: Proceedings of the 6th International Conference on Information Technology: IoT and Smart City, pp. 140–144 (2018)
12. Alani, M.M.: Big data in cybersecurity: a survey of applications and future trends. Journal of Reliable Intelligent Environments **7**(2), 85–114 (2021)
13. Alani, M.M.: Botstop : Packet-based efficient and explainable iot botnet detection using machine learning. Computer Communications (2022). DOI https://doi.org/https://doi.org/10.1016/j.comcom.2022.06.039. URL https://www.sciencedirect.com/science/article/pii/S0140366422002419
14. Alani, M.M.: Iotprotect: A machine-learning based iot intrusion detection system. In: Proceedings of the 6th International Conference on cryptography, security and privacy. IEEE (2022)
15. Alani, M.M., Alloghani, M.: Security challenges in the industry 4.0 era. In: Industry 4.0 and engineering for a sustainable future, pp. 117–136. Springer (2019)
16. Ali, G., Ahmad, N., Cao, Y., Khan, S., Cruickshank, H., Qazi, E.A., Ali, A.: xdbauth: Blockchain based cross domain authentication and authorization framework for internet of things. IEEE Access **8**, 58800–58816 (2020)
17. Alzubi, J.A.: Blockchain-based lamport merkle digital signature: authentication tool in iot healthcare. Computer Communications **170**, 200–208 (2021)

18. Armando, A., Basin, D., Boichut, Y., Chevalier, Y., Compagna, L., Cuellar, J., Drielsma, P.H., Heám, P.C., Kouchnarenko, O., Mantovani, J., Mödersheim, S., von Oheimb, D., Rusinowitch, M., Santiago, J., Turuani, M., Viganò, L., Vigneron, L.: The avispa tool for the automated validation of internet security protocols and applications. In: K. Etessami, S.K. Rajamani (eds.) Computer Aided Verification, pp. 281–285. Springer Berlin Heidelberg, Berlin, Heidelberg (2005)
19. Bursztein, E.: Inside the infamous Mirai IoT Botnet: A Retrospective Analysis. Cloudflare Blog (2020). URL https://blog.cloudflare.com/inside-mirai-the-infamous-iot-botnet-a-retrospective-analysis
20. Ferrag, M.A., Derdour, M., Mukherjee, M., Derhab, A., Maglaras, L., Janicke, H.: Blockchain technologies for the internet of things: Research issues and challenges. IEEE Internet of Things Journal **6**(2), 2188–2204 (2019). DOI https://doi.org/10.1109/JIOT.2018.2882794
21. Gong, L., Alghazzawi, D.M., Cheng, L.: Bcot sentry: A blockchain-based identity authentication framework for iot devices. Information **12**(5), 203 (2021)
22. Goyat, R., Kumar, G., Saha, R., Conti, M., Rai, M.K., Thomas, R., Alazab, M., Hoon-Kim, T.: Blockchain-based data storage with privacy and authentication in internet-of-things. IEEE Internet of Things Journal (2020)
23. Hammi, M.T., Hammi, B., Bellot, P., Serhrouchni, A.: Bubbles of trust: A decentralized blockchain-based authentication system for iot. Computers & Security **78**, 126–142 (2018)
24. Jesudoss, A., Subramaniam, N.: A survey on authentication attacks and countermeasures in a distributed environment. Indian Journal of Computer Science and Engineering (IJCSE) **5**(2), 71–77 (2014)
25. Khalid, U., Asim, M., Baker, T., Hung, P.C., Tariq, M.A., Rafferty, L.: A decentralized lightweight blockchain-based authentication mechanism for iot systems. Cluster Computing **23**(3), 2067–2087 (2020)
26. Li, D., Peng, W., Deng, W., Gai, F.: A blockchain-based authentication and security mechanism for iot. In: 2018 27th International Conference on Computer Communication and Networks (ICCCN), pp. 1–6. IEEE (2018)
27. Montalbano, E.: New Mirai Variant 'Mukashi' Targets Zyxel NAS Devices. Threatpost (2020). URL https://threatpost.com/new-mirai-variant-mukashi-targets-zyxel-nas-devices/153982
28. Node.js: Node.js (2022). URL https://nodejs.org/en. [Online; accessed 18. Apr. 2022]
29. O'Donnell, L.: Latest Mirai Variant Targets SonicWall, D-Link and IoT Devices. Threatpost (2021). URL https://threatpost.com/mirai-variant-sonicwall-d-link-iot/164811
30. Shen, M., Liu, H., Zhu, L., Xu, K., Yu, H., Du, X., Guizani, M.: Blockchain-assisted secure device authentication for cross-domain industrial iot. IEEE Journal on Selected Areas in Communications **38**(5), 942–954 (2020)
31. Silva, S.S., Silva, R.M., Pinto, R.C., Salles, R.M.: Botnets: A survey. Computer Networks **57**(2), 378–403 (2013)
32. Tahir, M., Sardaraz, M., Muhammad, S., Saud Khan, M.: A lightweight authentication and authorization framework for blockchain-enabled iot network in health-informatics. Sustainability **12**(17), 6960 (2020)
33. Wang, X., Garg, S., Lin, H., Piran, M.J., Hu, J., Hossain, M.S.: Enabling secure authentication in industrial iot with transfer learning empowered blockchain. IEEE Transactions on Industrial Informatics **17**(11), 7725–7733 (2021)
34. Wang, X., Zha, X., Ni, W., Liu, R.P., Guo, Y.J., Niu, X., Zheng, K.: Survey on blockchain for internet of things. Computer Communications **136**, 10–29 (2019). DOI https://doi.org/10.1016/j.comcom.2019.01.006. URL https://www.sciencedirect.com/science/article/pii/S0140366418306881
35. Wohlin, C.: Guidelines for snowballing in systematic literature studies and a replication in software engineering. In: Proceedings of the 18th international conference on evaluation and assessment in software engineering, pp. 1–10 (2014)
36. Yaga, D., Mell, P., Roby, N., Scarfone, K.: Block chain technology overview. (2018). National Institute of Standards and Technology. https://doi.org/10.6028%2Fnist.ir.8202

Blockchain Technology-Based Smart Cities: A Privacy-Preservation Review

Yeray Mezquita, Ana-Belén Gil-González, Javier Prieto, and Juan-Manuel Corchado

1 Introduction

Devices and sensors make up the smart city's sophisticated infrastructure. There is a continuous flow of data because of the automated communication between components in those systems. The integrity of communications and sensor data is crucial to the smooth operation of these urban areas. Blockchain technology, by definition, may be used to safeguard networks in this setting, where ever-increasing volumes of data are generated via direct interactions between devices. The availability of such technology has made it feasible to build a trustworthy, fully automated platform that eliminates the need for human mediators.

Smart cities can leverage the massive amounts of data they create, providing residents with a plethora of individualized offerings. However, to give these kinds of services, collecting and analyzing massive amounts of multidimensional data is necessary, which might compromise users' privacy [10]. The need to ensure the confidentiality of sensitive data has never been greater, especially in light of recent legislation like the General Data Protection Regulation (GDPR) [39].

Data transfer can also be protected via asymmetric encryption and access controls instead of blockchain technology; but, these methods have their own set of problems, including a single point of failure [21]. Data privacy and security in smart cities built on the blockchain is a popular topic for academic studies. However, none of the recommendations work with the current crop of Internet of Things devices [12, 35, 36]. In the context of smart cities, where data must be processed and analyzed in near real-time, the technological limitations of IoT devices make it challenging to deploy advanced cryptographic algorithms and protocols [12, 20, 22].

Y. Mezquita (✉) · A.-B. Gil-González · J. Prieto · J.-M. Corchado
BISITE Research Group, University of Salamanca, Salamanca, Spain
e-mail: yeraymm*@usal.es; abg@usal.es; javierp@usal.es; corchado@usal.es

Y. Maleh et al. (eds.), *Blockchain for Cybersecurity in Cyber-Physical Systems*, Advances in Information Security 102, https://doi.org/10.1007/978-3-031-25506-9_6

This chapter's significant contribution is a theoretical analysis of existing research and published literature on blockchain-based smart city platforms and preserving user privacy and anonymity in the context of suggested use cases. Researchers and practitioners in the field of urban planning may both benefit from and build upon the study's results. Furthermore, we offer helpful resources to a wide variety of stakeholders interested in learning about the disruptive dynamics of blockchain technology. Finally, the study sheds light on the potential remedies to users' concerns about privacy in blockchain-powered smart cities.

The chapter is organized as follows. This introduction is followed by a theoretical background, Sect. 2, where several interesting ideas related to blockchain technology and smart cities are laid forth, afterward, we'll check out the published works in the literature for further development, Sect. 3, and the detailed presentation of the studied solutions, Sect. 4. In Sect. 5 a discussion of the information found in this work is carried out. Finally, Sect. 6 concludes this chapter.

2 Theoretical Background

The term "smart city" was introduced with the creation of pervasive Internet of Things platforms meant to enhance users' lives. The Internet of Things (IoT) devices that make up these platforms collect and analyze data in real time to power various helpful citizen apps. Further, a smart city optimizes urban planning, accessibility, transportation, sustainability and energy savings, people's health, and the environment to promote more efficient and sustainable growth [10].

The massive exchange of data between the IoT devices of any smart city could be exploited by attackers, disrupting the smooth operation of the system. With the implementation of blockchain technology, the control of smart city's processes is no longer centralized, being possible to achieve a secure environment for authentication and transmission of devices communications [28]. Blockchain technology is a distributed digital ledger that cannot be altered, making it ideal for recording many types of valuable transactions, not just monetary ones [38]. A blockchain network stores the information redundantly in each of its peer-to-peer (P2P) nodes. Implementing that network in any IoT system, it is possible to get rid of centralized controllers and human intermediaries, by enforcing traceability and provenance of data [27].

The public key signature process allows a blockchain network to confirm the legitimacy of data easily and ensures that it was not tampered with during its creation. Blockchain networks have open, decentralized, and cryptographic features, its main benefits could be listed as follows [29]:

- A public-key system signs data for authenticity. Because of this, you can confidently determine the credibility of any information you see on the blockchain.
- Using blockchain technology permits getting rid of some intermediaries that drive up the cost of the system and leave it open to human mistakes. This function

arises because blocks of immutable code, whose execution can be easily verified, may be stored in the blockchain.
- Protect sensitive information by using consensus techniques and protocols to ensure that all network nodes use the same version of the blockchain. This technique ensures that the ledger's data cannot be altered once it has been recorded.

A consensus algorithm is a primary rule in any blockchain network. Consensus algorithms, along with their variants, continue to increase. Although the most popular and successful algorithms may be broken down into three distinct types, this is not always the case [42]:

1. In the **Proof-of-Work** (PoW) algorithm, a node must first solve a cryptographic puzzle to add a new block of data to the blockchain. Nodes that add new blocks (the so-called miners) are dissuaded from conducting illicit transactions due to the computational cost and difficulty of solving the problem, the energy consumed in discovering its solution (work), and the ease with which its verification may be performed. This approach uses much computing power to generate new blocks by solving cryptographic puzzles. Furthermore, its scalability is restricted [7].
2. **Proof-of-Stake** (PoS) is a consensus algorithm in which miners add new blocks to the network alternate. The quantity of bitcoin held in escrow for a miner determines how likely he is to obtain his turn to contribute a block (Stake). The technique relies on the honesty of a node when it generates a block so that escrow funds are not compromised. The delegated Proof-of-Stake (dPoS) method is one variant of this system; in this system, users vote on which nodes are allowed to add new data to the blockchain depending on their stake. The nothing at stake hypothesis is a theoretical flaw in this algorithm that increases the likelihood of blockchain splits compared to other consensus methods [24].
3. Within the **Practical Byzantine Fault Tolerance** (PBFT) protocol, a round is an act of adding a new block to the ledger. In each round, a node is chosen randomly to propose a new block, which is subsequently sent to the network to verify. Every node in the network checks the block's data and casts a vote if the data is correct. The broadcasted block is considered genuine and added to the blockchain if it receives 2/3 of the votes from all nodes in the network. This consensus algorithm's biggest flaw is that it isn't completely decentralized like others [40].

The ability to record and execute smart contracts is yet another benefit of blockchain implementation. Contracts are the stipulations of a deal, written in code and saved in the blockchain. The network nodes then execute the programs and agree on the output. These applications help negotiate, confirm, and enforce terms of a deal that have already been settled upon [13]. In any scenario where parties with divergent interests must interact, "smart contracts" can be used to improve processes, ensure all parties' requirements are met, and reduce the likelihood of conflicts [30].

Although smart contracts may be deployed on first-generation blockchains like Bitcoin [18], it wasn't feasible to create Turing-complete Smart Contracts until the advent of Ethereum in the second generation [9]. The inability to develop Turing-complete smart contracts means that this type of blockchain has the same scalability issues as the first-generation blockchain systems. The emergence of third-generation blockchain platforms has facilitated the proliferation of blockchain-based use cases. These blockchain networks are designed to address the scalability issues of their forerunners and pave the way for the widespread adoption of decentralized applications (DApps). Tron, Cardano, and EOS are all excellent examples of third-generation blockchains. If we talk tech, it's because of blockchains of the third generation that we can construct a blockchain-based system where lots of data can be traded easily. Nonetheless, smart contracts may be executed on any IoT platform, including a smart city.

3 Methodology

In this chapter, it is studied the state of the art on data privacy and user anonymity in smart city platforms based on blockchain technology. We study the key aspects needed to protect the citizens' privacy in any implementation of a blockchain based-smart city. To achieve our goals, we have proposed the following questions to be answered during our study:

- **Q1**: What has been proposed, in the literature, regarding the privacy preservation of citizens in smart cities based on blockchain technology?
- **Q2**: Are there viable solutions for actual real implementation?

Related to this research study, the following terms and regular expressions were found to be useful:

- **Blockchain**: ("distributed ledger technolog" OR blockchain*).
- **Smart city**: ("smart city").
- **Privacy**: (privacy).

To answer the defined research questions, a search for papers was carried out in the ScienceDirect[1] and Springer-Link[2] databases, obtaining 630 and 257 papers respectively, of which 17 have been used in this study. Besides, from those 17 articles selected, it has been found within them more useful papers for this work, the snowballing technique, see Tables 1, 2, 3, 4, and 5. The papers studied a specific proposed solution to achieve privacy in either the data generated and/or the user anonymity in any blockchain based-smart city. Besides, the study covers some relevant review works to improve it while snowballing other manuscripts from the

[1] https://www.sciencedirect.com/.

[2] https://link.springer.com/.

Table 1 Summary of science direct works (I)

Title	Summary	Classification. RQ
TP2SF: a Trustworthy Privacy-Preserving Secured Framework for sustainable smart cities by leveraging blockchain and machine learning [20]	A three-module Trustworthy Privacy-Preserving Secure Framework (TP2SF) for smart cities is proposed. In the two-level privacy module, a blockchain-based enhanced Proof of Work (ePoW) technique is simultaneously applied with Principal Component Analysis (PCA) to transform data for preventing inference and poisoning attacks.	Generated data protection. Q1, Q2
The Blockchain Random Neural Network for cybersecure IoT and 5G infrastructure in Smart Cities [35]	This article presents a Blockchain Random Neural Network for Cybersecurity applications. Authors propose to use the weight of a neural network to keep the identity of the users secret. While the information is codified, it is also possible to restore it again in case of a security breach.	Generated data protection, User identity privacy. Q1, Q2
Blockchain for IoT-based smart cities: recent advances, requirements, and future challenges [22]	In this paper, it is stated that security is one of the key challenges of a smart city. It is discussed and critically evaluated various smart applications based on blockchain technology. Besides, the authors present real-world blockchain implementation in smart cities as case studies and the key requirements to integrate both technologies.	Review. Q1
Blockchain-empowered cloud architecture based on secret sharing for smart city [10]	In this manuscript, it is described a distributed system based on blockchain technology and a Secret Sharing algorithm. Thanks to which it is possible to improve the integrity and security of the data in external cloud services.	User identity privacy. Q1, Q2
Blockchain: the operating system of smart cities [6]	This paper presents the challenges that must be overcome to enable widespread adoption and deployment of blockchain technology within a smart city context.	Review. Q1
Blockchain for smart cities: a review of architectures, integration trends and future research directions [8]	This paper presents the state-of-the-art of blockchain technology to solve the security issues of smart cities. Besides, some future research directions are identified through an extensive literature survey on blockchain-based smart city systems.	Review. Q1
Quantum-inspired blockchain-based cybersecurity: securing smart edge utilities in IoT-based smart cities [2]	It is proposed a protocol that is employed to design a quantum-inspired blockchain for the secure transmission of data among IoT devices for smart water utilities.	Generated data protection, User identity privacy. Q1, Q2

Table 2 Summary of science direct works (II)

Title	Summary	Classification. RQ
PrivySharing: a blockchain-based framework for privacy-preserving and secure data sharing in smart cities [23]	It is proposed a blockchain-based innovative framework for privacy-preserving and secure IoT data sharing in a smart city environment. Data privacy is preserved by dividing the blockchain network into various channels, where every channel comprises a finite number of authorized organizations and processes a specific type of data. The users' data is stored in private databases, while in the blockchain channels are stored the hashes of these data. Also, the data is stored encrypted in the databases.	Generated data protection, User identity privacy. Q1, Q2
Blockchain-based authentication and authorization for smart city applications [12]	This work proposes a solution for distributed management of identity and authorization policies by leveraging blockchain technology to hold a global view of the security policies within the system. This solution has the issue that it is possible to relate the data stored to a user, it doesn't bring anonymity to the stored data.	User identity privacy. Q1, Q2
Convergence of blockchain and artificial intelligence in IoT network for the sustainable smart city [36]	In this paper, it is stated that the smart city context is being revolutionized by the convergence of Artificial Intelligence (AI) and blockchain technology. Besides, it is presented a detailed discussion of several key factors for the convergence of Blockchain and AI technologies that will help form a sustainable smart society.	Review. Q1
A deep learning-based IoT-oriented infrastructure for secure smart City [37]	Authors propose a Deep Learning-based IoT-oriented infrastructure for a secure smart city where Blockchain provides a distributed environment at the communication phase of CPS, and Software-Defined Networking (SDN) establishes the protocols for data forwarding in the network. This framework doesn't provide real anonymity, because it manages the access of the users through a smart contract, being easily related to any user with their corresponding addresses.	User identity privacy. Q1, Q2
Achieving efficient and Privacy-preserving energy trading based on blockchain and ABE in smart grid [14]	Ciphertext-Policy Attribute-Based Encryption (CP-ABE) is introduced as the core algorithm to reconstruct the transaction model of a blockchain-based smart city. To log and store produced data, it is needed a complex communication protocol, impeding this algorithm to be used in a constrained real-time environment.	Generated data protection. Q1, Q2

Table 3 Summary of science direct works (III)

Title	Summary	Classification. RQ
A survey of privacy enhancing technologies for smart cities [11]	It is provided a review of the state of Smart Cities around the world, some examples of implemented solutions, and explored how the privacy of individuals could be exposed. Authors claim that there is a great risk of violating citizens' privacy, due to gathering a mixture of data and analyzing them with AI techniques.	Review. Q1
A survey on cybersecurity, data privacy, and policy issues in cyber-physical system deployments in smart cities [15]	Here it is described that, although challenges include important technical questions, it is equally important to address policy and organizational questions. Policy and technical implementation hurdles are perhaps equally likely to slow or disable smart city implementation efforts.	Review. Q1
Privacy-preserving blockchain-based federated learning for traffic flow prediction [34]	It is proposed a consortium blockchain-based federated learning framework to enable decentralized, reliable, and secure federated learning without a centralized model coordinator as an efficient solution for achieving privacy protection. Sensible data is stored locally, each party performs the training locally and updates it for each global iteration. It is still possible to steal that information by inferring the parameters of a local model stored in the blockchain.	Generated data protection. Q1, Q2
Privacy protected blockchain based architecture and implementation for sharing of students' credentials [31]	To protect the sharing of students' credentials, authors propose the use of a distributed system based on blockchain technology. Although it is not possible to know who is the intended student for any deployed credential, it is still possible to know how many times a student is offering their credentials to get a job.	Generated data protection. Q1, Q2
Privacy preservation in blockchain based IoT systems: integration issues, prospects, challenges, and future research directions [16]	The authors discuss the privacy issues caused due to the integration of blockchain in IoT applications by focusing on the applications of our daily use. Besides, it has been described the implementation of five privacy preservation strategies in blockchain-based IoT systems named anonymization, encryption, private contract, mixing, and differential privacy.	Review. Q1
Privacy preservation in permissionless blockchain: a survey [32]	In this paper, it is stated that, though numerous surveys reviewed the privacy preservation in blockchain, they failed to reveal the latest advances, nor have they been able to conduct a unified standard comprehensive classification of the privacy protection of permissionless blockchain.	Review. Q1

Table 4 Summary of Springer-link works (I)

Title	Summary	Classification. RQ
Distributed ledger technology for securities clearing and settlement: benefits, risks, and regulatory implications [33]	It is stated that a lot of currently repetitive business processes could be eliminated because there are fewer intermediaries involved in a DLT system. Besides, Although the industry and scholars are attempting to solve the technological and operational issues that DLT systems still face, outstanding legal risks are such that the financial industry is asking for more regulatory guidance and intervention.	Review. Q1
Blockchain technology in IoT systems: current trends, methodology, problems, applications, and future directions [4]	In this paper, it is researched the security and privacy concerns of IoT from the lens of current trends, pertinent challenges, security methodologies, applications, and gaps for future research directions. Because of that, high-performance and scalable cryptographic schemes (that is, those in the class of lightweight approach) are suggested to deal with the privacy and security of data in a Blockchain-based IoT system.	Review. Q1
A survey on boosting IoT security and privacy through blockchain [5]	The authors conduct a comprehensive literature review to address recent security and privacy challenges related to IoT where they are categorized according to IoT layered architecture: perception, network, and application layer. It is stated that blockchain's secure decentralization can overcome the security, authentication, and maintenance limitations of the current IoT ecosystem.	Review. Q1
Cloud-based vs. blockchain-based IoT: a comparative survey and way forward [26]	In this article, it is provided a taxonomy of the challenges in the current IoT infrastructure, and a literature survey with a taxonomy of the issues to expect in the future of the IoT after adopting blockchain as an infrastructure. Authors found that there are some good features in both Cloud Based-IoT and Blockchain Based-IoT, so instead of shifting to an entirely new infrastructure, they propose that it would be better to come up with something in between rather than a full migration.	Review. Q1

studied ones. Papers that do not propose a solution in this specific topic have been excluded.

To provide customized services, a smart city has to digitalize all kinds of data provided by the citizens, link and analyze those data, which increases the risk of

Table 5 Summary of Springer-link works (II)

Title	Summary	Classification. RQ
Privacy protection for fog computing and the internet of things data based on blockchain [21]	This paper proposes a distributed access control system based on blockchain technology to secure IoT data. The proposed mechanism is based on fog computing and the concept of the alliance chain. This method uses mixed linear and nonlinear spatiotemporal chaotic systems and the least significant bit to encrypt the IoT data on an edge node and then upload the encrypted data to the cloud.	Generated data protection. Q1, Q2
Security, privacy and risks within smart cities: literature review and development of a smart city interaction framework [17]	The study carried in this paper is organized around a number of key themes within smart cities research like privacy and security of mobile devices and services, and algorithms and protocols to improve security and privacy. This comprehensive review provides a useful perspective on many of the key issues and offers key direction for future studies.	Review. Q1

privacy violations [10, 11, 17]. To ensure citizens' privacy, the data generated in a smart city and its level of access should be controlled by its owner [22].

4 Proposed Solutions

Some research have offered different frameworks, models, and algorithms to deal with the security and privacy challenges that smart cities confront. In this section, we will describe and compare the studied solutions found and proposed in the literature for smart cities based on blockchain technology. Authors in [11] claim that to employ technologies and protocols to enhance the security of these platforms, it is needed, to address the anonymity issue of the citizens whose data is recorded. In addition, each person's unique, personal attributes must be preserved [25], Wherein the authors propose a model of citizen privacy consisting of five dimensions (identification, query, location, footprint, and owner).

In [36] it is stated that Artificial intelligence (AI) and blockchain technologies are allegedly revolutionizing the smart city scenario. Numerous critical reasons for the confluence of Blockchain and AI technologies that will assist in forming a sustainable smart society are discussed in length. In addition, the fog computing paradigm should be leveraged in constructing security algorithms to help overcome the constraints of IoT devices [5].

4.1 User Identity Privacy

In the paper [10], a Secret Sharing is an algorithm designed to enhance the confidentiality and authenticity of data stored in cloud services outside blockchain-based distributed ledgers. The blockchain in this work aims to verify user data's authenticity and make it available and easy to access across several cloud service providers. Users' data is recorded on the blockchain using the suggested distributed paradigm, which improves security compared to existing centralized systems.

Authors suggest in [12] using blockchain technology to manage citizens' identities and access policies, with a comprehensive overview of all system security measures. The fundamental issue with this technique is that it does not guarantee the anonymity of the stored data; it is still possible to link the data to a specific user.

Singh et al. [37] proposes a solution that combines AI with blockchain technology. It is demonstrated that a Deep Learning-based IoT-oriented infrastructure is provided for a safe, smart city. Blockchain technology offers a decentralized setting throughout the Cyber-Physical System's communication phase. Whereas, Software-Defined Networking (SDN) uses a three-layer architecture consisting of (i) connection, (ii) conversion and (iii) application to set the protocols for data flow in the network. However, as the proposed framework governs user access via a smart contract, any user's corresponding addresses may be readily determined. Hence it does not offer true anonymity to its user.

4.2 Generated Data Protection

To protect the generated data, in [14], a general distributed transaction model called PP-BCETS (Privacy-preserving Blockchain Energy Trading Scheme) is constructed. In addition, Ciphertext-Policy Attribute-Based Encryption (CP-ABE) has been introduced in this paper as a core algorithm to reconstruct the smart city transaction model stored in the blockchain. Because of the complex communication protocol used to store data, this method is not suitable in real-time systems.

Another privacy protection method for the generated data has been proposed in [34]. This method makes use of federated learning without exchanging raw data. To achieve privacy of the generated data, a differential privacy method with a noise addition mechanism is applied. In the framework of the proposed model, a permissioned blockchain is made use of to enable decentralized, reliable, and secure federated learning without a centralized model coordinator.

In [31], to protect the sharing of students' credentials, authors propose the use of a distributed system based on blockchain technology. Although it is not possible to know who is the intended student for any deployed credential, it is still possible to know how many times a student is offering their credentials to get a job.

Authors in [20], present a framework based on blockchain technology to preserve the privacy of the generated data within a smart city. The framework transform the data generated with Principal Component Analysis (PCA) algorithm, thanks to which the data is protected against inference and poisoning attacks. To detect intrusions to the platform, it is used, also, an optimized gradient tree boosting system (XGBoost). Finally, blockchain technology is used along with an InterPlanetary File System (IPFS) architecture, to deploy the framework within a smart city. Thanks to the IPFS system it is possible to store data off-chain while using the blockchain as a way to store data hashes in an effective way of maintaining the integrity of the data. Blockchain technology helps in data authentication and is used for maintaining the integrity of the data.

In [2] it is proposed the use of an encryption algorithm that takes advantage of a p2p verification protocol. This means that, when node A sends a transaction to node B, the sender encrypts the data with the key parameters of node B. Then, node B receives the notification by the blockchain network and is asked to validate it by using the key parameters of node A. Finally, that transaction is added to the blockchain if it is validated.

Finally, in [21] coupled nonlinear spatiotemporal chaotic systems with Least Significant Bit (LSB) data encryption technique Lattice of Coupled Maps with Linear and Nonlinear Effects (MLNCML). Following are the steps taken by the suggested method, which is based on fog computing and the idea of the alliance blockchain: To acquire encrypted data, I plaintext information is transformed to Unicode encoding; (ii) chaotic series is used to encrypt the Unicode encoding to gain encrypted information; and (iii) encrypted information is rendered concealed in the original data using the LSB algorithm. (iii) The encrypted data is subjected to a secondary encryption procedure using the MLNCML series to produce the final encrypted data. The alliance blockchain is responsible for storing the access control model strategy that restricts access to the strategic data to just the alliance blockchain members.

4.3 Protection to the Data and to the Privacy of the Users

To enhance the privacy preservation of the citizens, along with securing IoT communications and data generated, authors in [23] propose a framework based on a multi-channel blockchain network. Each channel handles a unique category of information, such as medical, automobile, utility, or banking records, and comprises a limited number of permitted businesses. Smart contracts may be programmed with rules governing who can access user data inside a certain channel. Data from experiments indicate that using several channels inside a blockchain system improves performance over using only one. Committing peers in a large-scale smart city network may need a lot of resources because of the several ledgers they need to keep up with, thanks to the multi-channel design. Additionally, user information is encrypted in private databases, with only hashes of that information saved in the blockchain channels.

Authors in paper [1] propose the use of different channels in a multi-channel blockchain network to obtain a certain level of privacy to the users' generated data and identity. Although this kind of solution does not protect from all the involved parties, it is possible to get some level of privacy with this kind of architecture. In [3] it is proposed the use of middleware to protect the data generated, although the paper does not elaborate on this topic.

The last studied work of this category, [35], presents a Blockchain Random Neural Network for Cybersecurity applications. The authors propose hiding users' identities in the neural network's weight. Because the data is encoded, it can be recovered in the event of a hack, making this approach to authentication viable on a decentralized site.

5 Discussion

In this chapter, we have proposed two research questions to conduct the study. By answering Q1, we seek to understand the current context of the state of the art in solving the privacy-preservation issue within a smart city based on blockchain technology. This question has been addressed in the Sects. 3 and 4. The first conclusion from the research is that the present limits of the IoT ecosystem in terms of security, authentication, and maintenance may be facilitated through the adoption of blockchain technology's safe decentralization [5].

The literature has made patent the potential of blockchain-based IoT systems, in particular smart cities. It allows the development of urban applications in the supply chain, transportation, logistics, and governance domains [6, 27]. However, to allow real adoption of the technology within an IoT system in general, and a smart city in particular, it is needed to face several challenges [6, 8, 22, 26]: (i) **Security and privacy**. Blockchain technology could be used as a defensive framework to protect the interconnection and data transmission of IoT devices, guaranteeing the provenance of data. Nevertheless, because of the transparent nature of this technology, a new challenge in the privacy preservation of the users arises [30]. (ii) **Storage**. Storing the citizen's data in a cloud system could compromise its integrity and the citizens' privacy. It has been proposed frameworks based on blockchain technology, although the issues of privacy are still patent [19, 26]. (iii) **Energy efficiency and scalability**. In an environment where the energy demand is always increasing like in a smart city, consensus algorithms like PoW cannot be used. Other alternatives like PoS or PBFT, in case of a permissioned network, are energy-efficient and scalable, besides, there are being proposed new ones like Proof of Trust (PoT) [43]. In the case of using a cloud-based system, these costs would not be lower, due to large cloud data centers used to maintain the services up [26]. (iv) **Incentive mechanism**. In case that blockchain technology is used to enhance a smart city system, incentives must be used to reward nodes that contribute to the maintenance of the network and the validation and verification of the data. Furthermore, a punishment mechanism is needed to avoid nodes with malicious intentions operating in the network. More research should be carried out in this regard because the solutions proposed are not enough [8]. (v) **Interoperability**. No work in the literature has addressed the challenge of creating an interoperable blockchain-based system. (vi) **Regulation**. In the literature, it has been highlighted the regulatory problems that distributed platforms like those enhanced with blockchain technology face [41]. It is needed a proper regulatory framework to solve conflicts between parties [33].

From the works studied, it can be concluded that the future of any encryption protocol to ensure the privacy of information within an IoT system, involves the use of the fog computing paradigm. Apart from this precept, authors have opted for different encryption techniques, highlighting the use of machine learning in these techniques to store the data generated and ensure privacy, while also allowing the identification of malicious users. Also, it is needed any way to ensure the anonymity of the users, to preserve the privacy of their identity, while avoiding the possibility of making profiles from their use of the platforms. The privacy preservation of the data generated along with the anonymity of the users that interact with the system must be preserved in conjunction, like some of the works have proposed in their solutions [1, 3, 23, 35]. However, and answering Q2, the solutions studied do not go beyond the conceptual level, and other authors claim that more work is needed in this area, to achieve a lightweight solution that can be implemented in a real use case scenario [32].

Distributed ledger technology (DLT) uses a fundamentally different mechanism than traditional systems, necessitating a new approach to regulation. Existing standards were designed with a view to what financial markets were like today, and lawmakers could not have foreseen the potential role that DLT will play in those markets. However Authors like [33] argue that the current rules are a hindrance since they do not adequately account for the hazards that this new technology poses to market infrastructures, their users, and society at large.

Summarizing the discussion, it should be emphasized that it is possible to solve some of the smart city problems such as security, maintenance, and authentication with the use of blockchain technology. However, there are still some challenges like the preservation of citizens' privacy and anonymity. Lightweight approaches to cryptographic schemes on privacy preservation remain an unresolved issue, with the only viable option being the use of approaches that rely on fog computing, limiting the implementation of the proposed proofs-of-concept in the literature in a real smart city [4, 32]. In addition, it should be noted that while the improvement of advanced cryptography to provide real anonymity to the citizens, it also hinders the resolution of legal disputes [15].

6 Conclusion

In the presented chapter, we have studied the issues and challenges on the privacy of communications, along with user-generated data and anonymity of the citizens in smart cities based on blockchain technology. The study is limited to the use of two databases, Springer-Link and ScienceDirect, along with the snowball technique, which has helped to find more valuable works in other reviews of the literature. It is being motivated, in most of the studied works, more research regarding the topic. Because the prototypes proposed in the literature are not sufficient for the implementation of a resource-constrained real case study. Besides most of the solutions are not detailed enough.

From our research, it has been found that in any smart city environment, enhanced with blockchain technology, the ledger maintained by the network is used as a mechanism for generated data verification, not as a database to store them. Moreover, it has become clear that the only way to obtain minimally scalable solutions is thanks to the fog computing paradigm.

When exchanging information within a smart city, there is a tendency in the literature to use machine learning techniques to mask the identity of the user who generates the data. Others are more orthodox and bet on frameworks in which they make use of encryption algorithms, with lightweight ones being those that scale best. In some works, it is being used a multi-chain based system to protect the data generated and the anonymity of the users.

Finally, it is found that the proposed prototypes of the study obtain good results in data management and protection of anonymity of the users. Although most of the prototypes are far from implementable in a scalable and resource-constrained environment like a big smart city. In addition, it has been found that the possibility of citizens' anonymity in these platforms, comes with the challenge of appearing conflicts between citizens, ending up generating disputes that would be difficult to resolve without an adequate legal framework.

Acknowledgments Yeray Mezquita has been awarded a pre-doctoral scholarship by the University of Salamanca, with funding assistance from Banco Santander, to further her study. In addition, funding for this study came in part from the project "Computación cuántica, virtualización de red, edge computing, and registro distribuido para la inteligencia artificial del futuro," Reference: CCTT3/20/SA/0001, which was supported by the European Regional Development Fund and the Institute for Business Competitiveness of Castilla y León (FEDER).

References

1. Abbas, K., Tawalbeh, L.A., Rafiq, A., Muthanna, A., Elgendy, I.A., El-Latif, A., Ahmed, A.: Convergence of blockchain and iot for secure transportation systems in smart cities. Security and Communication Networks 2021 (2021)
2. Abd El-Latif, A.A., Abd-El-Atty, B., Mehmood, I., Muhammad, K., Venegas-Andraca, S.E., Peng, J.: Quantum-inspired blockchain-based cybersecurity: securing smart edge utilities in iot-based smart cities. Information Processing & Management 58(4), 102549 (2021)
3. Abou-Nassar, E.M., Iliyasu, A.M., El-Kafrawy, P.M., Song, O.Y., Bashir, A.K., Abd El-Latif, A.A.: Ditrust chain: towards blockchain-based trust models for sustainable healthcare iot systems. IEEE Access 8, 111223–111238 (2020)
4. Alfa, A.A., Alhassan, J.K., Olaniyi, O.M., Olalere, M.: Blockchain technology in iot systems: current trends, methodology, problems, applications, and future directions. Journal of Reliable Intelligent Environments pp. 1–29 (2020)

5. Alfandi, O., Khanji, S., Ahmad, L., Khattak, A.: A survey on boosting iot security and privacy through blockchain. Cluster Computing pp. 1–19 (2020)
6. Bagloee, S.A., Heshmati, M., Dia, H., Ghaderi, H., Pettit, C., Asadi, M.: Blockchain: The operating system of smart cities. Cities 112, 103104 (2021)
7. Beikverdi, A., Song, J.: Trend of centralization in bitcoin's distributed network. In: 2015 IEEE/ACIS 16th International Conference on Software Engineering, Artificial Intelligence, Networking and Parallel/Distributed Computing (SNPD). pp. 1–6. IEEE (2015)
8. Bhushan, B., Khamparia, A., Sagayam, K.M., Sharma, S.K., Ahad, M.A., Debnath, N.C.: Blockchain for smart cities: A review of architectures, integration trends and future research directions. Sustainable Cities and Society 61, 102360 (2020)
9. Buterin, V., et al.: A next-generation smart contract and decentralized application platform. white paper 3, 37 (2014)
10. Cha, J., Singh, S.K., Kim, T.W., Park, J.H.: Blockchain-empowered cloud architecture based on secret sharing for smart city. Journal of Information Security and Applications 57, 102686 (2021)
11. Curzon, J., Almehmadi, A., El-Khatib, K.: A survey of privacy enhancing technologies for smart cities. Pervasive and Mobile Computing 55, 76–95 (2019)
12. Esposito, C., Ficco, M., Gupta, B.B.: Blockchain-based authentication and authorization for smart city applications. Information Processing & Management 58(2), 102468 (2021)
13. Gazafroudi, A.S., Mezquita, Y., Shafie-khah, M., Prieto, J., Corchado, J.M.: Islanded microgrid management based on blockchain communication. In: Blockchain-based Smart Grids, pp. 181–193. Elsevier (2020)
14. Guan, Z., Lu, X., Yang, W., Wu, L., Wang, N., Zhang, Z.: Achieving efficient and privacy-preserving energy trading based on blockchain and abe in smart grid. Journal of Parallel and Distributed Computing 147, 34–45 (2021)
15. Habibzadeh, H., Nussbaum, B.H., Anjomshoa, F., Kantarci, B., Soyata, T.: A survey on cybersecurity, data privacy, and policy issues in cyber-physical system deployments in smart cities. Sustainable Cities and Society 50, 101660 (2019)
16. Hassan, M.U., Rehmani, M.H., Chen, J.: Privacy preservation in blockchain based iot systems: Integration issues, prospects, challenges, and future research directions. Future Generation Computer Systems 97, 512–529 (2019)
17. Ismagilova, E., Hughes, L., Rana, N.P., Dwivedi, Y.K.: Security, privacy and risks within smart cities: Literature review and development of a smart city interaction framework. Information Systems Frontiers pp. 1–22 (2020)
18. Kaiser, I.: A decentralized private marketplace: Draft 0.1
19. Kopp, H., Bösch, C., Kargl, F.: Koppercoin–a distributed file storage with financial incentives. In: International Conference on Information Security Practice and Experience. pp. 79–93. Springer (2016)
20. Kumar, P., Gupta, G.P., Tripathi, R.: Tp2sf: A trustworthy privacy-preserving secured framework for sustainable smart cities by leveraging blockchain and machine learning. Journal of Systems Architecture p. 101954 (2020)
21. Liu, Y., Zhang, J., Zhan, J.: Privacy protection for fog computing and the internet of things data based on blockchain. Cluster Computing pp. 1–15 (2020)
22. Majeed, U., Khan, L.U., Yaqoob, I., Kazmi, S.A., Salah, K., Hong, C.S.: Blockchain for iot-based smart cities: Recent advances, requirements, and future challenges. Journal of Network and Computer Applications p. 103007 (2021)
23. Makhdoom, I., Zhou, I., Abolhasan, M., Lipman, J., Ni, W.: Privysharing: A blockchain-based framework for privacy-preserving and secure data sharing in smart cities. Computers & Security 88, 101653 (2020)
24. Martinez, J.: Understanding proof of stake: The nothing at stake theory (June 2018), https://medium.com/coinmonks/understanding-proof-of-stake-the-nothing-at-stake-theory-1f0d71bc027, [Accessed; 09/10/2019]
25. Martínez-Ballesté, A., Pérez-Martínez, P.A., Solanas, A.: The pursuit of citizens' privacy: a privacy-aware smart city is possible. IEEE Communications Magazine 51(6), 136–141 (2013)

26. Memon, R.A., Li, J.P., Ahmed, J., Nazeer, M.I., Ismail, M., Ali, K.: Cloud-based vs. blockchain-based iot: a comparative survey and way forward. Frontiers of Information Technology & Electronic Engineering 21, 563–586 (2020)
27. Mezquita, Y., Casado-Vara, R., GonzÁlez Briones, A., Prieto, J., Corchado, J.M.: Blockchain-based architecture for the control of logistics activities: Pharmaceutical utilities case study. Logic Journal of the IGPL (2020)
28. Mezquita, Y., Gazafroudi, A.S., Corchado, J., Shafie-Khah, M., Laaksonen, H., Kamišalić, A.: Multi-agent architecture for peer-to-peer electricity trading based on blockchain technology. In: 2019 XXVII International Conference on Information, Communication and Automation Technologies (ICAT). pp. 1–6. IEEE (2019)
29. Mezquita, Y., González-Briones, A., Casado-Vara, R., Chamoso, P., Prieto, J., Corchado, J.M.: Blockchain-based architecture: A mas proposal for efficient agri-food supply chains. In: International Symposium on Ambient Intelligence. pp. 89–96. Springer (2019)
30. Mezquita, Y., Valdeolmillos, D., González-Briones, A., Prieto, J., Corchado, J.M.: Legal aspects and emerging risks in the use of smart contracts based on blockchain. In: International Conference on Knowledge Management in Organizations. pp. 525–535. Springer (2019)
31. Mishra, R.A., Kalla, A., Braeken, A., Liyanage, M.: Privacy protected blockchain based architecture and implementation for sharing of students' credentials. Information Processing & Management 58(3), 102512 (2021)
32. Peng, L., Feng, W., Yan, Z., Li, Y., Zhou, X., Shimizu, S.: Privacy preservation in permissionless blockchain: A survey. Digital Communications and Networks (2020)
33. Priem, R.: Distributed ledger technology for securities clearing and settlement: benefits, risks, and regulatory implications. Financial Innovation 6(1), 1–25 (2020)
34. Qi, Y., Hossain, M.S., Nie, J., Li, X.: Privacy-preserving blockchain-based federated learning for traffic flow prediction. Future Generation Computer Systems 117, 328–337 (2021)
35. Serrano, W.: The blockchain random neural network for cybersecure iot and 5g infrastructure in smart cities. Journal of Network and Computer Applications 175, 102909 (2021)
36. Singh, S., Sharma, P.K., Yoon, B., Shojafar, M., Cho, G.H., Ra, I.H.: Convergence of blockchain and artificial intelligence in iot network for the sustainable smart city. Sustainable Cities and Society 63, 102364 (2020)
37. Singh, S.K., Jeong, Y.S., Park, J.H.: A deep learning-based iot-oriented infrastructure for secure smart city. Sustainable Cities and Society 60, 102252 (2020)
38. Valdeolmillos, D., Mezquita, Y., González-Briones, A., Prieto, J., Corchado, J.M.: Blockchain technology: a review of the current challenges of cryptocurrency. In: International Congress on Blockchain and Applications. pp. 153–160. Springer (2019)
39. Voigt, P., Von dem Bussche, A.: The eu general data protection regulation (gdpr). A Practical Guide, 1st Ed., Cham: Springer International Publishing 10, 3152676 (2017)
40. Witherspoon, Z.: A hitchhiker's guide to consensus algorithms (November 2017), https://hackernoon.com/a-hitchhikers-guide-to-consensus-algorithms-d81aae3eb0e3, [Accessed; 09/10/2019]
41. Yeoh, P.: Regulatory issues in blockchain technology. Journal of Financial Regulation and Compliance (2017)
42. Zheng, Z., Xie, S., Dai, H., Chen, X., Wang, H.: An overview of blockchain technology: Architecture, consensus, and future trends. In: 2017 IEEE International Congress on Big Data (BigData Congress). pp. 557–564. IEEE (2017)
43. Zou, J., Ye, B., Qu, L., Wang, Y., Orgun, M.A., Li, L.: A proof-of-trust consensus protocol for enhancing accountability in crowdsourcing services. IEEE Transactions on Services Computing 12(3), 429–445 (2018)

Security in Electronic Health Records System: Blockchain-Based Framework to Protect Data Integrity

Md Jobair Hossain Faruk, Hossain Shahriar, Bilash Saha, and Abdul Barek

1 Introduction

Electronic Health Records (EHRs) are intended to improve the quality, efficiency, and safety of patient care. The HITECH Act mandated that the Office of the National Coordinator (ONC) for Health Information Technology create an infrastructure for a nationwide health information technology exchange that would allow the use and exchange of health records and information electronically [1–5]. When a patient visits a different provider or an external hospital system, record transmittal to the next point of care is not always guaranteed unless a reciprocity agreement is already in place [6]. A survey of Health Information Exchange (HIE) organizations revealed that more than 170 regional health information organizations did not meet the criteria for comprehensive health information exchange [5, 7]. Based on ONC's interoperability road map, a functional interoperable system should include the following: a secure network infrastructure, verification of identity, and authentication of all participants as well as consistent proof of authorization of electronic health records [8]. If existing EHR systems implemented these and allowed data sharing among providers, it would bring major benefits such as decreased cost of accessing healthcare [9], timely access to patient data by providers, improving clinical workflow, and quality of care [10, 11]. A survey [12] found that only 6% of clinicians reported it was easy for them to obtain information from other organizations without interruptions to their workflow, and fewer than one-third said they could easily access data from other EHRs [13, 14]. Legal mandates also prohibit data sharing for secondary use [15].

M. J. H. Faruk (✉) · H. Shahriar · B. Saha · A. Barek
College of Computing and Software Engineering, Kennesaw State University, Marietta, GA, USA
e-mail: mhossa21@students.kennesaw.edu; hshahria@kennesaw.edu; bsaha@students.kennesaw.edu; mbarek@students.kennesaw.edu

Y. Maleh et al. (eds.), *Blockchain for Cybersecurity in Cyber-Physical Systems*, Advances in Information Security 102, https://doi.org/10.1007/978-3-031-25506-9_7

Blockchain systems have been demonstrated to address these issues by allowing healthcare providers, through prior agreement, to share and receive data in real-time without the risk of data corruption [16], [41], [42]. A blockchain system provides a decentralized, tamper-proof data repository across a shared platform, on which all parties can maintain security through audit trails of access authorization [10]. Developing a blockchain-based application for patients and providers will able to overcome the interoperability barrier between commercial EHR systems and allow patients to take full control of their health records by granting healthcare providers permission to share data seamlessly based on need. The primary contribution of this chapter is the following:

- We study the HITECH act and ONC's requirements to provide an understanding of health information technology for secure electronic health records (EHR) network.
- We introduce a novel blockchain-based EHR database framework for patients and providers that allows data sharing among providers seamlessly.
- We evaluate the security risk of existing applications and provide a comparison with the proposed framework.

The rest of this paper will be organized as follows: In Sect. 2, we define blockchain technology and hyperledger fabric followed by a discussion on the recent work about electronic health record (EHR) and provide comparison between existing and proposed system in Sect. 3. Section 4 presents the proposed framework and the system architecture while Sect. 5 discusses the evaluation of the framework, challenges, limitations, and future work. Section 6 concludes the paper.

2 Blockchain Technology

A blockchain system offers an alternative to the traditional model of data sharing by providing a complete and clear source of data. By embracing blockchain technology, healthcare providers would have access to coherent, collaborative records without the cost of reconciling different EHR interfaces [17, 23]. In addition, blockchain systems could address the issues surrounding continuity of care by granting patients greater access to their own health records [3, 14]. The introduced system shall demonstrate the feasibility of integrating a blockchain-based EHR data-sharing prototype that provides patients with more freedom to grant data access to healthcare providers based on need. Seamless integration of mobile app-based point-of-care system with EHRs. When patients provide data-sharing permission to healthcare providers, they often have no option but to revoke this permission at a later time. Our prototype tool will be capable of connecting patients with clinics in which they have existing EHRs and allow them to communicate any changes needed to their data-sharing agreements.

3 Related Work

MedRec is a Hyperledger Fabric-based private blockchain system that was developed to store patient metadata using smart contracts through which providers could retrieve and authenticate records [17]. While it is a viable proof of concept, the system queries for or requests files, rather than receiving updates or files pushed to the blockchain [24]. Hong et al. [18] described a platform and a health level 7 (HL7) fast healthcare interoperability resource (FHIR) system. Here, data was anonymized through pseudonyms, but not encrypted that raises security concerns. 43% of security breaches are related to health data in the USA [19].

Use of FHIRChain architecture has been proposed for patient identity management in addition to the MedRec proof of concept, leveraging token-based access exchanges that were FHIR-compliant based on user authorization [20]. This approach did not address the storage concern. To focus more on patients and less on data management, most facilities store patient health information in the cloud rather than in house data servers [10]. Dalianis et al. [21] presented an idea for data-sharing infrastructure based on the Stockholm EPR Corpus, which is a repository of 2 million+ patients.

Although a number of additional studies and research projects have explored the use of blockchain [18, 19], data security [19, 23], token-based access exchanges [21, 22], and processing engines, among other concepts, mobile device and app-based identity management systems that are capable of overcoming data interoperability issues have not yet been developed [25]. We provide a high-level comparison Fig. 1 between the proposed platform and other existing platforms in terms of some essential blockchain-based healthcare properties including Network Permission. According to the findings, the proposed model achieves unique and novel properties of a blockchain-based healthcare platform.

Blockchain Applications	Network Permission	Approach	Unique ID	HIE	Data Backup	FHIR	Credential Management	Mobile App	Cloud Network	Data verifiability
Gem Health Net.	Public	Ethereum	✓	✓	x	x	x	X	x	x
MedRec	Public	Ethereum	x	✓	x	x	x	X	x	x
Carechain	Public	Ethereum	✓	✓	x	x	x	X	x	x
Dovetail	Permissioned	H. Fabric	x	✓	x	x	x	X	x	x
Axuall	permissioned	H. Indy	x	✓	x	x	x	X	✓	x
MedHypChain	Permissioned	H. Fabric	x	✓	x	x	x	X	x	x
Proposed Model	Permissioned	H. Fabric	✓	✓	✓	✓	✓	✓	✓	✓

Fig. 1 Comparison of the proposed platform with other existing applications

4 Proposed Framework

4.1 Preliminary Studies

We conducted a preliminary study on the feasibility of a prototype architecture framework [26]. The architecture was developed by examining the industry and academic knowledge as well as public data from the GitHub data repository of the Centers for Disease Control [27].

Figure 2 illustrates the workflow of proposed system. For this architecture, each EHR system is set up as a node on a Hyperledger Fabric blockchain, which is a global, enterprise-based platform for decentralized applications [28].

Since current Health Information Exchange (HIE) is solely known as a way to exchange clinical information between professionals and organizations electronically and primary barriers include privacy and security, lack of data standards that permit the exchange of clinical data, and complex systems are top concerns within the conventional database-based application [29]. In accordance with various studies within this domain [4], researchers reported that blockchain-based HIE and EHR applications could be a solution and it has been proven by several blockchain-based systems already [30–32]. Based on the preliminary study, we design a low-level data sharing architecture within a blockchain network.

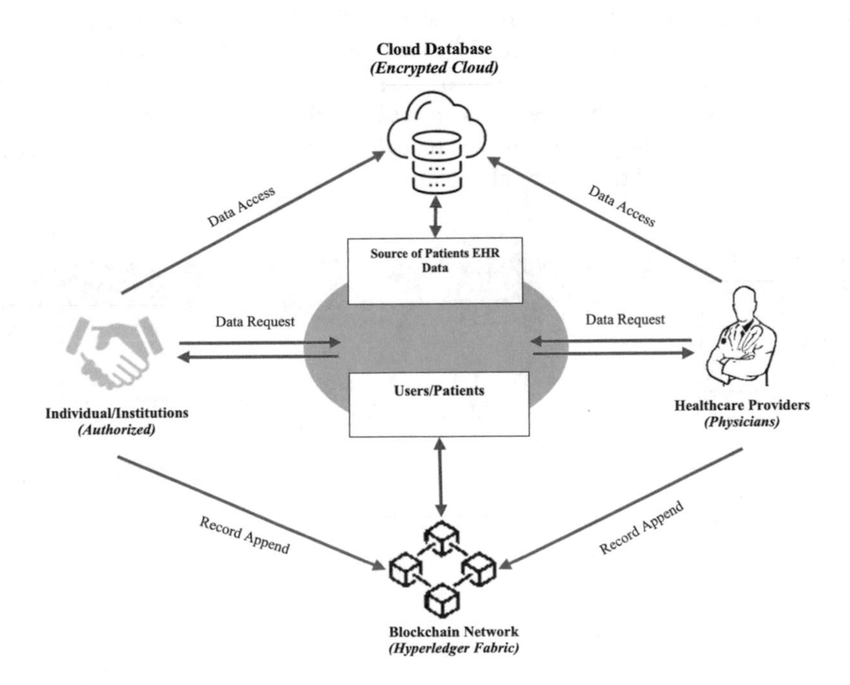

Fig. 2 Workflow of the EHR framework

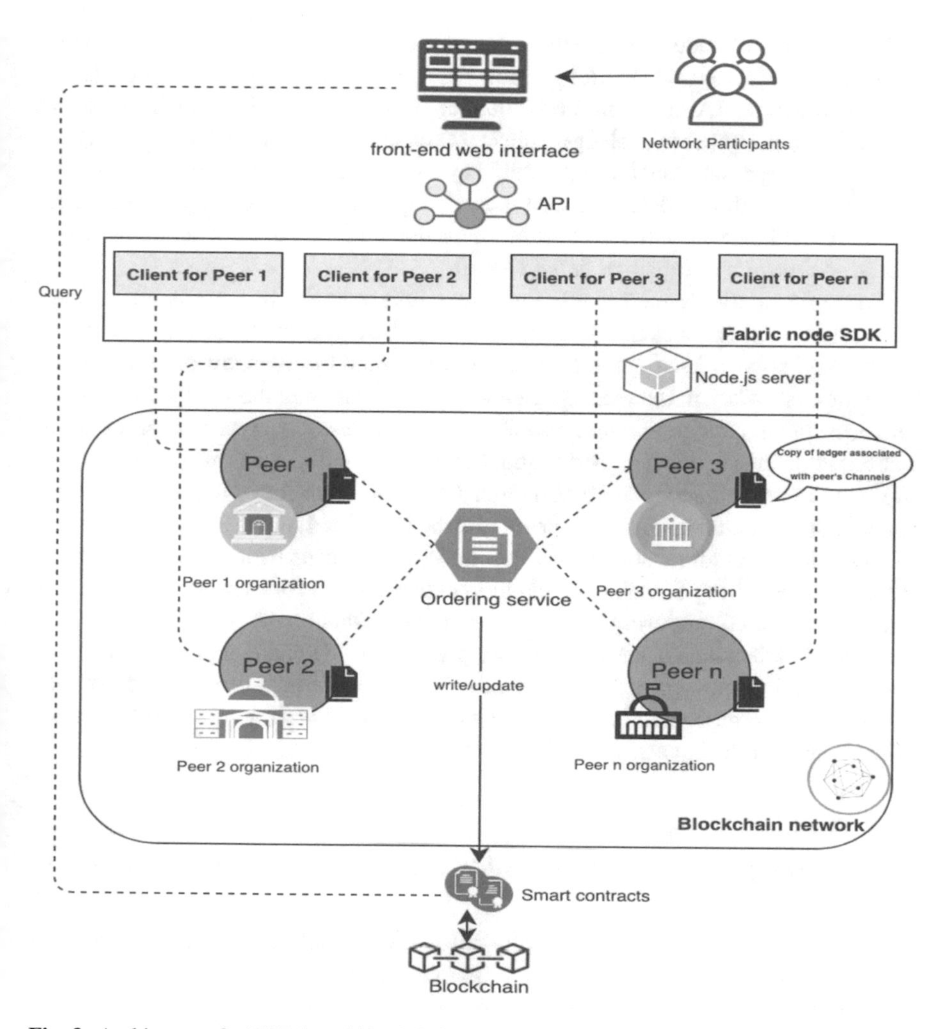

Fig. 3 Architecture for EHR based blockchain application

The network will allow healthcare providers, individuals, and institutions to access the previous medical record with proper authorization. The user is the owner of personal health data and is responsible for granting, denying, and revoking data access from any other parties, such as healthcare providers with a decentralized permission management protocol. Users can also share medical records with the desired doctors or institutes for a specific time-frame. Due to security and transparency concerns, every corresponding data access will be recorded on the blockchain. We will develop a blockchain-based EHR data-sharing platform. Figure 3 provides an overview of the EHR system.

The proposed application will be smart contracts based that shall implement business logic specifically oriented towards a blockchain's distributed, transparent ledger. Smart contracts service two classes of the message: transactions, which change ledger state, and queries, which read ledger state. The chaincode (smart contracts) will be developed in the Go language. We use multilayer encryption as a modular component of hyperledger fabric that enables to mediate secure interactions including encryption at rest, unique data keys, private keys, and signed transactions as well as separation of data to ensure security. By virtual private computers for each node, and API gateways, the EHR database is not exposed publicly, reducing risk of attack or exposure of sensitive data [33–36].

To begin, a file is encrypted, then sent to a storage bucket. A message is transmitted via the blockchain to all listening EHR nodes; this message conveys those changes have been made to a file or that a new file is ready through the use of a unique data key in the message header. A listening node then pulls the message and decrypts it using public and private keys. After the first layer is decrypted, the pulling node then sends a signed request to the originating node to receive the file. We will deploy chaincode to Hyperledger Fabric by executing a deploy transaction for enabling interactions between peers and the shared ledger consists of three types of operations including deploy, invoke, and query. To provide isolation between different data sharing domains, blockchain network peers for transaction validation, and the orderers (for ordering service) to the Fabric client on the cloud server. For the purpose of patient data access seamlessly, we established channels for patients with synchronization on their patient's mobile platforms using the introduced model. After verifying the credential, the framework will send web requests to the Fabric cloud server for data synchronization, query, access or update data.

4.2 Develop a blockchain-Based EHR Data-Sharing Prototype

A smart contract executes automatically based on a predefined trigger; thus, a smart contract can execute and run without human action or oversight, adding to its reliability and security [37]. The platform will be implemented as a user-friendly application in conventional mobile smartphones. All records will be encrypted and will require matching digital key pairs to access data. Special permission scenarios must be established using key pairings. Every participant in the blockchain will have a private and a public key that will be cryptographically connected. The model will generate a data key that can be shared freely with healthcare providers because it is encrypted.

- Expected results. The use of blockchain technologies will allow us to share data across various EHRs across multiple providers.

4.3 Implement a Web Application for Blockchain-Enabled EHR Data Sharing with Authentication

Because patient data is valuable and often the target of attacks, we focused on redundant data encryption for our app. The workflow is as follows: (1) a master key is created, and the proposed application provides decryption permissions for all authorized providers; and (2) an additional data key is also created and used to encrypt all information stored on the blockchain. The creation of a data key is a sensible measure; however, loss of this key would allow anyone to read the data stored on the blockchain. So, we encrypt the data key with the master key. Following three phases of User-Centered Design (UCD), we have first reviewed existing blockchain-based platforms related to EHR. We first conduct two focus groups with the stakeholders of the system and devise additional requirements for the application. Based on the review of the literature and other platforms and early user feedback, we implement a high-fidelity prototype of the app that we evaluate with the stakeholders (evaluation phase) through 1:1 usability study session. The use of UCD ensures that the tool reflects the needs of the system, providers, and patients.

4.4 A Case Study: Blockchain-Based Medical Image Sharing and Critical-Result Notification

To help with prompt diagnosis and effective patient care, teleradiology involves sending medical pictures to radiologists who are located off-site for interpretation. The radiologists then send back the dictation report to the primary site. One of the primary causes of longer radiology turnaround time is image unavailability and delayed critical result communication that resulted by a lack of integrated systems between teleradiology practice and healthcare organizations where lack of system integration is a primary issue [38]. In this chapter, we address the current limitation and demonstrate a Hyperledger Fabric-based teleradiology application for secure and efficient medical image sharing. The system also sends critical-result notifications automatically which shall help physicians and radiologists to improve patient care and reduce turnaround time.

Figure 4 presents the application flow when teleradiologists request images or critical connections. Radiologists can request images by submitting an image request and are instantly provided with a link to access the requested images. In addition, radiologists can check the Critical Notification box to communicate critical findings or the need for a critical connection with the patient's care team. When the box is checked, the patient's physician and care team will immediately get a notification on their user application.

Figure 5 demonstrates the application flow regarding critical notifications. The Critical tab on the physician and the care team members' home screen is in green

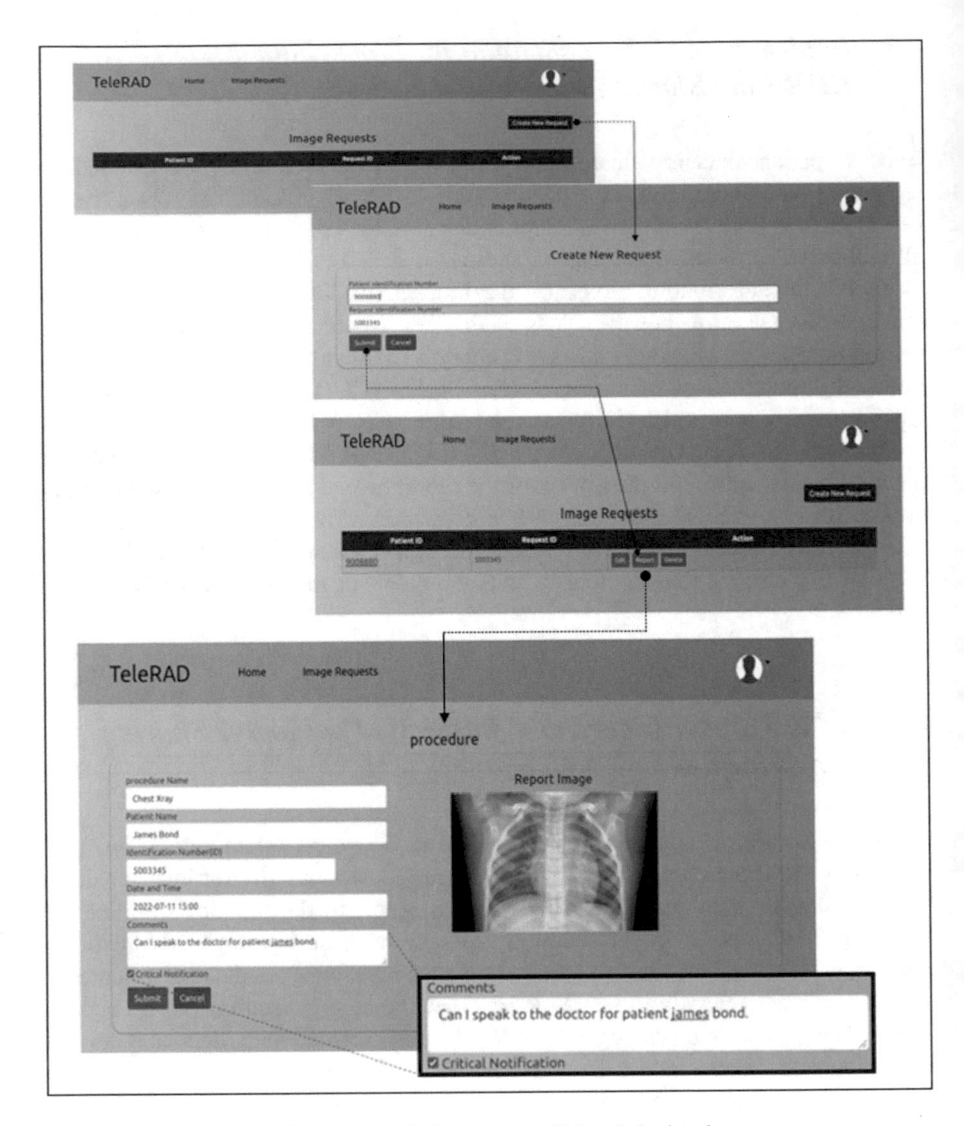

Fig. 4 illustrates the interface of teleradiology system: Teleradiologist view

by default. Upon receiving new critical notifications, the tab is highlighted red with a number of new critical notifications inside a bracket. When clicking on the tab, they can check the detail of the critical notifications. Once resolved, the Critical tab turns back to green.

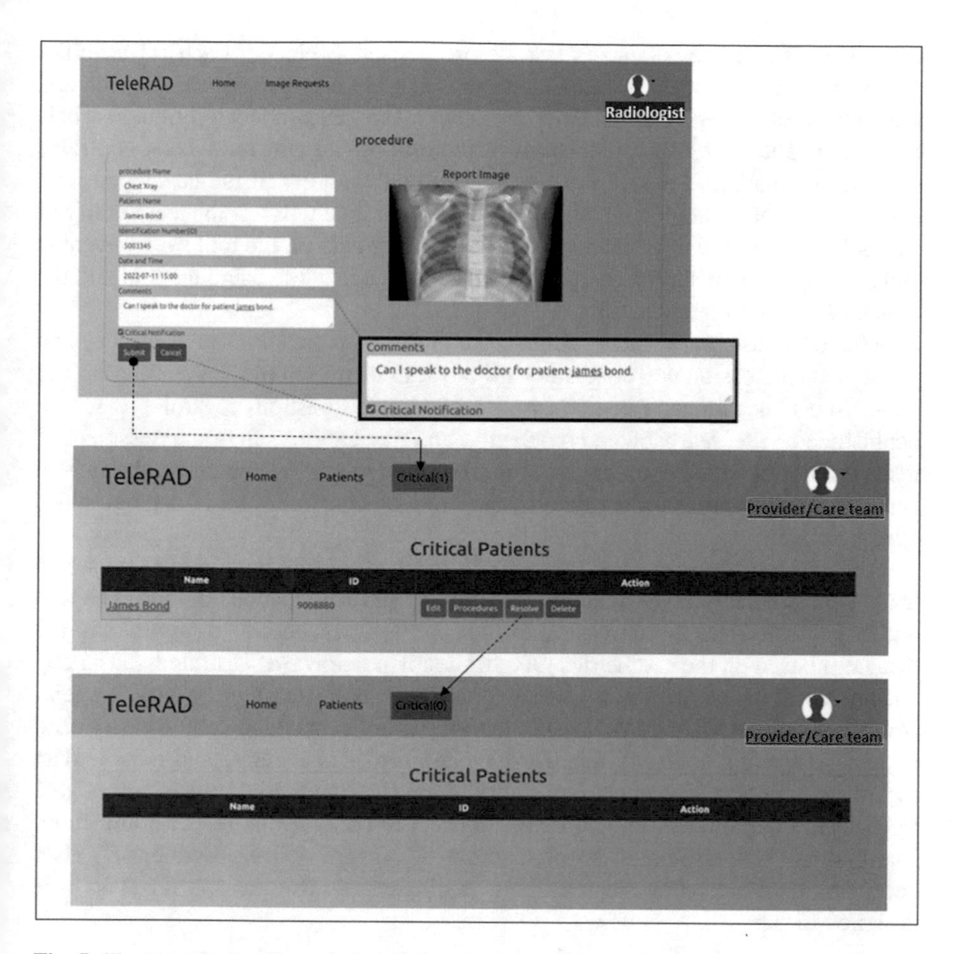

Fig. 5 illustrates the interface of teleradiology system: Critical notifications

5 Discussion and Challenges

After implementation, we will perform application testing including (i) functional, (ii) component, (iii) system, and (iv) regression testing. We design evaluation based on the evaluation matrix including reliability, acceptance, accessibility, and product quality metrics. The developed software will be evaluated based on application error, fault, and failure (both time and failure based). We will conduct a study during the beta release to find out the problem per user month (PUM) as $PUM = X/(Y * Z)$ where $X =$ # of problems reported by users; $Y =$ # of installations of the software; $Z =$ # of months in the calculation period.

Our aim is to evaluate the performance characteristics in a real-world clinical setting. We will demonstrate that by using the proposed application, patients are able to access and share their data with healthcare providers. We will recruit five professionals and five patients to test the prototypes developed. Each 1:1 usability session will be conducted by one facilitator trained in user experience research and one to two note-takers in a standard office room. We will conduct a formal heuristic assessment and cognitive walkthrough of the proposed application using Nielsen's heuristic evaluation to ensure the usability and navigability of the developed tool. A cognitive walkthrough with the think-aloud technique will test the learnability of the tool and inform design changes. We will collect data on the tool's effectiveness and performance, including the number of errors made while using the tool and the time taken to complete the tasks.

After each task, participants will be asked follow-up questions on what they had just done. The session will conclude with the administration of the system usability scale (SUS) questionnaire that consists of 10 simple questions regarding ease and confidence of use, learnability, complexity, and navigability, an oral assessment of their familiarity with technology, a semi-structured interview about how the system can potentially minimize their data sharing roadblocks, and a basic demographics questionnaire.

For primary analyses, we will solicit feedback from both participants and healthcare providers by conducting technology and satisfaction surveys. The app will be evaluated on a 5-point scale (strongly agree, agree, neutral, disagree, strongly disagree) based on the six metrics [39, 40] detailed below: (a) Usefulness: The app enables users to achieve their specific goals and motivates them to use it again; (b) Effectiveness: The app behaves in the way users expect and can be easily used to achieve specific goal; (c) Veracity: The app provides accurate and reliable data regarding patient management (e.g., reminders); (d) Interactivity: The app is able to engage, entertain, satisfy, and motivate users to use it again; (e) Customization: The systems supports users in one or more healthcare delivery domain; (f) User acceptability: Users show demonstrable willingness to use the app for tasks it is designed to support.

6 Conclusion

Emerging blockchain technology has the potential to address the security concern of electronic health record (EHR) storing and sharing. In this chapter, we proposed a novel blockchain-based EHR database framework for patients and providers that allows data storing and sharing among providers transparently. The introduced model shall allow patients to take full control of their health records by granting healthcare providers permission to share data seamlessly by overcoming the interoperability challenges. In order to improve the security concern, the proposed framework eliminates the threats of malicious attacks by adopting permission-based Hyperledger fabric and implementing features including secure network infrastructure, verification of identity, and authentication of all participants as well as consistent proof of authorization of electronic health records. We also demonstrate a case study using Hyperledger fabric for teleradiology application for secure and efficient medical image sharing. The application shows the capability to share medical images securely.

Acknowledgments The work is partially supported by the U.S. National Science Foundation Awards #2100115 and #1723578. Any opinions, findings, and conclusions or recommendations expressed in this material are those of the authors and do not necessarily reflect the views of the National Science Foundation.

References

1. D. Kim, J. H. Kagel, N. Tayal, S. Bose-Brill, and A. Lai, "The Effects of Doctor-Patient Portal Use on Health Care Utilization Rates and Cost Savings," SSRN Electron. J., 2017, doi: https://doi.org/10.2139/ssrn.2775261.
2. L. A. Linn and M. B. Koo, "Blockchain For Health Data and Its Potential Use in Health IT and Health Care Related Research," ONC/NIST Use Blockchain Healthc. Res. Work., pp. 1–10, 2016.
3. M. S.T., M. N., R. S., and F. E.W., "Impact of the HITECH act on physicians' adoption of electronic health records," J. Am. Med. Informatics Assoc., vol. 23, no. 2, pp. 375–379, 2016, [Online]. doi: http://dx.doi.org/10.1093/jamia/ocv103.
4. D. Blumenthal and M. Tavenner, "The 'Meaningful Use' Regulation for Electronic Health Records," N. Engl. J. Med., vol. 363, no. 6, pp. 501–504, 2010, doi: https://doi.org/10.1056/nejmp1006114.
5. V. Lapsia, K. Lamb, and W. A. Yasnoff, "Where should electronic records for patients be stored?," Int. J. Med. Inform., vol. 81, no. 12, pp. 821–827, 2012, doi: https://doi.org/10.1016/j.ijmedinf.2012.08.008.
6. H. B. Bosworth et al., "Health information technology: Meaningful use and next steps to improving electronic facilitation of medication adherence," JMIR Med. Informatics, vol. 4, no. 1, 2016, doi: https://doi.org/10.2196/medinform.4326.
7. M. L. Braunstein, "Healthcare in the age of interoperability: The promise of fast healthcare interoperability resources," IEEE Pulse, vol. 9, no. 6, pp. 24–27, 2018, doi: https://doi.org/10.1109/MPUL.2018.2869317.
8. "Why blockchain offers a fresh approach to interoperability — Health Data Management," [Online]. Available: https://www.healthdatamanagement.com/opinion/why-blockchain-offers-a-fresh-approach-to-interoperability.
9. J. Studeny and A. Coustasse, "Personal health records: is rapid adoption hindering interoperability?," Perspect. Health Inf. Manag., vol. 11, 2014.
10. M. Mettler, "Blockchain technology in healthcare: The revolution starts here," 2016 IEEE 18th Int. Conf. e-Health Networking, Appl. Serv. Heal. 2016, 2016, doi: https://doi.org/10.1109/HealthCom.2016.7749510.
11. L. Winfield, "A look at the Trump administration's approach to HIT," Healthc. Financ. Manag., 2018, [Online]. Available: https://www.hfma.org/topics/hfm/2018/february/59347.html.
12. "Challenges and opportunities for blockchain powered healthcare systems: a review."
13. by M. Ariel C Ekblaw, "Blockchain for Medical Data Access, Permission Management and Trend Analysis Signature redacted Signature redacted," 2017, [Online]. Available: https://dspace.mit.edu/handle/1721.1/109658.
14. K. Peterson, R. Deeduvanu, P. Kanjamala, and K. Boles, "A Blockchain-Based Approach to Health Information Exchange Networks.(2017)," NIST Work. Blockchain Healthc., no. 1, pp. 1–10, 2016, [Online]. Available: http://kddlab.zjgsu.edu.cn:7200/research/blockchain/huyiyang-reference/ABlockchain-BasedApproachtoHealthInformationExchange.pdf.

15. B. Shen, J. Guo, and Y. Yang, "MedChain: Efficient healthcare data sharing via blockchain," Appl. Sci., vol. 9, no. 6, 2019, doi: https://doi.org/10.3390/app9061207.
16. Y. Zhuang, L. Sheets, Z. Shae, J. J. P. Tsai, and C. R. Shyu, "Applying Blockchain Technology for Health Information Exchange and Persistent Monitoring for Clinical Trials," AMIA ... Annu. Symp. proceedings. AMIA Symp., vol. 2018, pp. 1167–1175, 2018.
17. M. A. Engelhardt, "Hitching Healthcare to the Chain: An Introduction to Blockchain Technology in the Healthcare Sector," Technol. Innov. Manag. Rev., vol. 7, no. 10, pp. 22–34, 2017, doi: https://doi.org/10.22215/timreview/1111.
18. J. Hong, P. Morris, and J. Seo, "Interconnected Personal Health Record Ecosystem Using IoT Cloud Platform and HL7 FHIR," Proc. - 2017 IEEE Int. Conf. Healthc. Informatics, ICHI 2017, pp. 362–367, 2017, doi: https://doi.org/10.1109/ICHI.2017.82.
19. J. L. T. Chang, "The Dark Cloud of Convenience: How the New HIPAA Omnibus Rules Fail to Protect Electronic Personal Health Information," Loyola Los Angeles Entertain. Law Rev., vol. 34, p. 119, 2014.
20. G. Slabodkin, "Blockchain remains a work in progress for use in healthcare," Heal. Data Manag., vol. 25, no. 3, pp. 37–39, 2017.
21. H. Dalianis, A. Henriksson, M. Kvist, S. Velupillai, and R. Weegar, "HEALTH BANK - A workbench for data science applications in healthcare," CEUR Workshop Proc., vol. 1381, no. January, pp. 1–18, 2015.
22. H. Ulrich, A. K. Kock, P. Duhm-Harbeck, J. K. Habermann, and J. Ingenerf, "Metadata repository for improved data sharing and reuse based on HL7 FHIR," Stud. Health Technol. Inform., vol. 228, pp. 162–166, 2017, doi: https://doi.org/10.3233/978-1-61499-678-1-162.
23. K. Rabah, M. Research, and N. Kenya, "Challenges & Opportunities for Blockchain Powered Healthcare Systems: A Review," Mara Res. J. Med. Heal. Sci., vol. 1, no. 1, pp. 45–52, 2017, [Online]. Available: www.mrjournals.org.
24. A. Azaria, A. Ekblaw, T. Vieira, and A. Lippman, "MedRec: Using blockchain for medical data access and permission management," Proc. - 2016 2nd Int. Conf. Open Big Data, OBD 2016, pp. 25–30, 2016, doi: https://doi.org/10.1109/OBD.2016.11.
25. "One giant leap for Mexico," Economist, vol. 373, no. 8395, p. 66, 2004, [Online]. Available: https://www.saavha.com.
26. G. Carter, H. Shahriar, and S. Sneha, "Blockchain-based interoperable electronic health record sharing framework," Proc. - Int. Comput. Softw. Appl. Conf., vol. 2, pp. 452–457, 2019, doi: https://doi.org/10.1109/COMPSAC.2019.10248.
27. G. Website, "CDC collaboration on blockchain prototypes," GitHub website, [Online]. Available: https://github.com/CDCgov/blockchain-collab.
28. M. J. Hossain Faruk, H. Shahriar, M. Valero, S. Sneha, S. Ahamed, and M. Rahman, "Towards Blockchain-Based Secure Data Management for Remote Patient Monitoring," 2021, [Online]. Available: https://www.researchgate.net/publication/353588418
29. H. Wu and E. Larue, "Barriers and facilitators of health information exchange (HIE) adoption in the United States," Proc. Annu. Hawaii Int. Conf. Syst. Sci., vol. 2015-March, pp. 2942–2949, 2015, doi: https://doi.org/10.1109/HICSS.2015.356.
30. A. Dubovitskaya, Z. Xu, S. Ryu, M. Schumacher, and F. Wang, "Secure and Trustable Electronic Medical Records Sharing using Blockchain," AMIA ... Annu. Symp. proceedings. AMIA Symp., vol. 2017, pp. 650–659, 2017.
31. A. Dubovitskaya et al., "ACTION-EHR: Patient-centric blockchain-based electronic health record data management for cancer care," J. Med. Internet Res., vol. 22, no. 8, 2020, doi: https://doi.org/10.2196/13598.
32. R. Coelho, R. Braga, J. M. N. David, M. Dantas, V. Ströele, and F. Campos, "Integrating blockchain for data sharing and collaboration support in scientific ecosystem platform," Proc. Annu. Hawaii Int. Conf. Syst. Sci., vol. 2020-Janua, pp. 264–273, 2021, doi: https://doi.org/10.24251/hicss.2021.031.
33. Hyperledger Project, "Fabric CA User's Guide," 2017, [Online]. Available: http://hyperledger-fabric-ca.readthedocs.io/en/latest/users-guide.html.

34. X. Liang, J. Zhao, S. Shetty, J. Liu, and D. Li, "Integrating blockchain for data sharing and collaboration in mobile healthcare applications," IEEE Int. Symp. Pers. Indoor Mob. Radio Commun. PIMRC, vol. 2017-Octob, pp. 1–5, 2018, doi: https://doi.org/10.1109/PIMRC.2017.8292361.
35. "MediaRecorder overview," Android Dev. website, [Online]. Available: https://developer.android.com/guide/topics/media/mediarecorder.
36. Samsung, "Galaxy Fold," Samsung, 2019, [Online]. Available: https://www.samsung.com/global/galaxy/galaxy-fold/.
37. "HTC phones," Amaz. website, [Online]. Available: https://www.amazon.com/slp/htc-phones/9d4mevkzaqe7zvs.
38. "Audio quality when recording in a loud setting," Reddit website, 2016, [Online]. Available:
39. G. Wenger, "EBmH evidence based mHealth app development platform," Prezi website, 2013, [Online]. Available: https://prezi.com/tziutsdhsvwv/ebmh-evidence-based-mhealth-app-development-platform.
40. L. Zhou, J. Bao, I. M. A. Setiawan, A. Saptono, and B. Parmanto, "The mhealth app usability questionnaire (MAUQ): Development and validation study," JMIR mHealth uHealth, vol. 7, no. 4, 2019, doi: https://doi.org/10.2196/11500.
41. Randolph, J., Faruk, M. J. H., Saha, B., Shahriar, H., Valero, M., Zhao, L., & Sakib, N. (2022, June). "Blockchain-based Medical Image Sharing and Automated Critical-results Notification: A Novel Framework". In 2022 IEEE 46th Annual Computers, Software, and Applications Conference (COMPSAC) (pp. 1756–1761). IEEE.
42. Faruk, MJ Hossain, S. Hossain, and M. Valero. "Ehr data management: Hyperledger fabric-based health data storing and sharing." The Fall 2021 Symposium of Student Scholars. 2021.

A Secure Data-Sharing Framework Based on Blockchain: Teleconsultation Use-Case

Hossain Kordestani, Roghayeh Mojarad, Abdelghani Chibani, Kamel Barkaoui, and Wagdy Zahran

1 Introduction

The emergence of technology has made IoT an integral parts of modern day. IoT networks play an important roles in several application domains, including ambient assisted living [19, 21, 24], rehabilitation [20], and healthcare [16, 17, 22, 23]. The data and service must be secure and robust in such applications. For instance, in e-health, the data at stake include sensitive patients' medical information and priceless medical knowledge. The former is considered one the most private information, while the latter should hold guaranteed integrity. Moreover, any failure in service could result in irrecoverable damages. Therefore, it is critical to have a secure and robust communication and storage platform for IoT networks.

H. Kordestani (✉)
Department of Research and Innovation, Maidis SAS, Chatou, France
Centre d'études et de recherche en informatique et communications, Conservatoire National des Arts et Métiers, Paris, France
e-mail: hossain.kordestani@maidis.fr

R. Mojarad · A. Chibani
Laboratoire Images, Signaux et Systmes Intelligents (LISSI), University Paris-Est Créteil, Créteil, France
e-mail: roghayeh.mojarad@u-pec.fr; chibani@u-pec.fr

K. Barkaoui
Centre d'études et de recherche en informatique et communications, Conservatoire National des Arts et Métiers, Paris, France
e-mail: kamel.barkaoui@cnam.fr

W. Zahran
Department of Research and Innovation, Maidis SAS, Chatou, France
e-mail: wagdy.zahran@maidis.fr

Y. Maleh et al. (eds.), *Blockchain for Cybersecurity in Cyber-Physical Systems*, Advances in Information Security 102, https://doi.org/10.1007/978-3-031-25506-9_8

Robustness refers to tolerance of handling perturbations without affecting the functional body of a system. In IoT applications, e.g., in e-health, it is vital to keep the communication and storage available. Current e-health solutions face a significant impediment in centralization, increasing the possibility of a single point of failure and prone to attacks against reliability and availability [15]. The decentralization improves the overall robustness of current healthcare systems, ensuring that medical data are protected from malicious attacks or accidental data loss [2].

Blockchain, a secure decentralized data structure, provides interoperability, integrity, and availability; many studies integrate blockchain technology in IoT applications. Although blockchain provides a distributed platform for storing and sharing data, it lacks enough measures to guarantee the confidentiality of the data. One possible approach is using permissioned blockchains, e.g., hyperledger fabric, which has embedded access control. In a permissioned blockchain, access to some or all blockchain nodes is restricted using access control. This approach is suitable for closed ecosystems, e.g., internal organization networks. However, access control management requires a centralized authority, limiting the benefits of using blockchain. Another group of approaches is extending public permissionless blockchain with a cryptosystem as an additional layer of security. Blockchain technology itself uses secure hash algorithms to guarantee immutability and integrity. Moreover, the additional layer uses encryption algorithms for privacy and confidentiality.

Some of existing solutions adopt the well-established approach in centralized applications, i.e., asymmetric encryption, for use in the blockchain [7, 9]. In centralized applications, e.g., an online web application, using asymmetric encryption is convenient; because each communication channel is dedicated to only one application. In other words, the encrypted data is meant for only one recipient. On the other hand, in a distributed system, various services might exist on the communication channel, requiring the same data. In such cases, using asymmetric encryption creates multiple cryptograms of a single data, causing redundancy of communication channels.

In this paper, we have proposed a secure data-sharing framework based on blockchain. In our framework, we use broadcasting encryption, which allows secure and efficient data sharing. The main contributions of the proposed framework are the followings:

- Defining the security concerns for data sharing in IoT applications
- Proposing a secure framework for data sharing using broadcast encryption
- Avoiding unnecessary redundancy because of broadcast encryption
- Enabling secure data query using homomorphic encryption
- Ensuring the integrity of data using hash algorithms in blockchain technology

The rest of this paper is structured as follows: in Sect. 2, the preliminaries of this framework are explained. The related works are then reviewed in Sect. 3. The security concerns of a data sharing framework is presented in Sect. 4. The details of our proposed framework are detailed in Sect. 5. The security evaluation of the proposed framework using formal approaches is presented in Sect. 6. A use case of teleconsultation is described in Sect. 7 to evaluate the proposed framework. Lastly, the conclusion and future works are provided in Sect. 8.

2 Preliminaries

2.1 Blockchain

Blockchain is a data structure formed by chains of blocks linked together using one-way cryptographic algorithms called hash algorithms. Hence, any block change requires modification of all blocks after that one in the blockchain, which is computationally impossible. Therefore, blockchain can be seen as an immutable data structure.

In [33], it has been demonstrated that the read availability of blockchain is typically high. On the other hand, blockchain guarantees data integrity by leveraging cryptographic techniques, notably, two following mechanisms [6]:

- A linked list of blocks: this structure enforces that the latest appended block should include the hash value of the proceeding block. Hence, any modification of the previous blocks invalidates all the subsequent blocks.
- Merkle tree structure: in this structure, each block holds a root hash of a Merkle tree of all the transactions. In Merkle-tree, each non-leaf node is the hash value of the concatenated values of its children nodes. Therefore, any modification on the transaction logs causes a new hash value in the above layer, leading to a falsified root hash. Ensuring any modification is easily detectable.

  Moreover, incremental construction of the Merkle tree is $O(h)$, where h is the height of the tree; however, reconstructing the root of the tree would require up to $O(2^h)$ time or space complexity [27]. The exponential complexity of modification of the Merkle tree shows that blockchain computationally guarantees integrity.

The confidentiality in the blockchain typically depends on the type of the blockchain. The public blockchains do not aim for confidentiality and focus on transparency, while the permissioned blockchains follow confidentiality requirements using access controls.

2.2 Distributed File Systems

Blockchain is not a general-purpose technology [5]. One of this technology's limitations is scalability in extensive data storage, as storing them on-chain can grow very expensive; one of the well-established solutions is storing data off-chain and managing it on-chain [13]. Blockchain-based distributed file systems, e.g., InterPlanetary File System (IPFS) [1], can be classified into seven layers [14]:

1. Identity layer: this layer allows each node of the distributed file system to have a unique identification information.
2. Data layer: this layer allows organizing the file structure in the distributed file system.
3. Data-swap layer: this layer allows formulating the file-sharing strategy among the nodes.
4. Network layer: this layer allows discovering, establishing connections, and exchanging files among the nodes.
5. Routing layer: this layer allows each piece of files to be found and accessed by the nodes.
6. Consensus layer: this layer ensures the correctness in the ledger recording transactions and maintains the network's consistency.
7. Incentive layer: this layer allows reward/punishment mechanisms to encourage the nodes to be active and honest.

2.3 Homomorphic Encryption

Homomorphic encryption (HE) is a type of cryptographic scheme to address outsourcing computations' privacy issues. It can be traced back to the 1970s when privacy homomorphisms were discussed in [29]. A typical homomorphic encryption scheme consists of four procedures [32]:

- Key generation: in this procedure, the encryption is set up, and the related keys are generated.
- Encryption: in this procedure, the user encrypts his/her data using the generated key.
- Evaluation: in this procedure, the user submits his/her encrypted data and a function to the server and gets an encrypted result.
- Decryption: in this procedure, the user can decrypt the evaluation function's encrypted result.

Homomorphic encryption allows the third party to execute (limited types of) operations on the cryptograms without decrypting them. This definition is presented in Eq. (1), in which m_1 and m_2 are any two messages and E represents encryption function. If an encryption scheme satisfies this equation's condition, it is defined as homomorphic over some operator $\odot$.

$$\forall m_1, m_2 : Enc(m_1) \odot Enc(m_2) = Enc(m_1 \odot m_2) \tag{1}$$

Based on the limitation on the operator, homomorphic encryption schemes can be classified into three categories (generations):

- Partially homomorphic encryption: The first generation of homomorphic encryption handles one type of operation on the ciphertext. For instance, the RSA's asymmetric encryption scheme [30] is homomorphic for multiplication. In other words, the multiplication of two cryptograms yields to a cryptogram equivalent to the cryptogram of the multiplication of their plain text:

$$c_i = Enc_{RSA}(m_i)$$
$$\prod m_i = Dec_{RSA}(\prod c_i)$$

  Similarly, Paillier's encryption scheme [26] is additionally homomorphic.
- Somewhat homomorphic encryption: The homomorphic encryption algorithms allow both addition and multiplication. However, in somewhat homomorphic encryption, only a limited number of computations is allowed in contrast to partially homomorphic encryption, which allows an unlimited number of computations.
- Fully Homomorphic Encryption (FHE): The idea of FHE is to remove any computations' constraints, neither the type nor the times.

2.4 Broadcast Encryption

Broadcast encryption [10] is a type of encryption scheme aiming to deliver encrypted content, e.g., multimedia content, over a broadcast channel with multiple receivers. At the same time, only authorized users, e.g., those who have paid the subscription fee, can decrypt it. Each receiver has a unique private key in this scheme, and the broadcaster has a dedicated key. Assume $R = r_1, r_2, \ldots, r_n$ is the set of receivers, if the broadcaster B decides a subset $S \subseteq R$ of receivers with whom share his/her content M. Hence any user in this subset should be able to decrypt it. In contrast, any member outside this subset should not access information about M. A broadcast encryption (BE) scheme consists of three randomized algorithm [3]:

- $Setup(t, n)$: given $t \in \mathbb{Z}$ is a security parameter and n is the number of receivers, $setup$ function yields n private keys $d_1, \ldots, d_n$ and a broadcaster key T.
- $Encrypt(S, T)$: this function yields a pair of header h and symmetric encryption key k. The message m is encrypted symmetrically using the key k, yielding a ciphertext C_m. The latter is sent in the broadcasting channel alongside h and S.
- $Decrypt(S, d_i, h)$: if the peer is in the list of authorized receivers ($i \in S$), then $decrypt$ function yields the message encryption key k. Given the encryption key, the receiver can decrypt the ciphertext C_m and achieve the message m.

An efficient broadcast encryption solution using n-mulinear maps has been proposed in [3]. The proposed broadcast encryption scheme consists of no header, and the size of private keys is $O((\log t)^2)$.

3 Related Works

The characteristics of blockchain, in particular immutability and decentralization, made it a suitable solution for IoT networks. Blockchain allows a secure sharing of resources immune to malicious tampering. Numerous works have studied blockchain for IoT, some of which are discussed in this section.

A group of studies altered blockchain networks to meet the limitations and requirements of IoT applications. IoT networks' main limitation is the computation power, and its most considered requirement is data security. In [9], blockchain modification has been proposed for the application of IoT devices. They have applied asymmetric encryption suites to design a decentralized platform for secure and efficient medical data transmission. They discuss their proposed architecture to withstand Denial-of-Service attacks, mining attacks, storage attacks and drop attacks. The healthcare data is stored in cloud storage servers connected to an overlay network in their proposed architecture. The latter is a peer-to-peer network of IoT devices. In the overlay network, the nodes are authenticated using valid certificates. In this approach, before sending, the message is signed with the sender's private key and encrypted with the receiver's public key. Upon receipt, the message is decrypted with the receiver's private key and verified with the sender's public key. They have used smart contracts to evaluate whether an IoT reading is normal or abnormal; consequently, to send an alert in the latter case. Similarly, in [7], a Lightweight Scalable Blockchain (LSB) has been proposed, which is based on an overlay network for preserving security in the context of IoT requirements. They have optimized the algorithms for use in the IoT environment; they have opted for a lightweight consensus algorithm. In the overlay network, the nodes are grouped in clusters; and each cluster has an elected cluster head which allows for managing the blockchain network. Transactions in LSB are secure using asymmetric encryption, digital signatures, and cryptographic hash functions.

The LSB concept is explored in smart home [7] and smart vehicle [8] settings to demonstrate its resistance against common security attacks. Block4Forensic [4] is a lightweight blockchain infrastructure proposed for forensics services in smart vehicle environments. Public key infrastructure is used in the proposed blockchain. Moreover, a fragmented ledger is designed to store detailed information about the vehicle. Block4Forensic allows trustless, traceable, privacy-preserving forensics of smart vehicles after accidents. Furthermore, in [28], a privacy-preserving blockchain architecture based on attribute-based encryption techniques has been proposed. Attribute-based encryption scheme allows confidentiality and access control in single encryption [11]. This scheme utilizes distributed management nodes to overcome the need for a centralized access control server.

Another group of studies focuses on the data storage of IoT in the blockchain. A well-established data-sharing architecture using blockchain, including IoT-based applications, uses distributed file systems (see Sect. 2.2). Moreover, an encryption scheme is used on top of the blockchain architecture for data confidentiality. For instance, in [31], a healthcare architecture using IPFS, Ethereum, and AES as distributed file systems, blockchain infrastructure, and encryption schemes has been presented, respectively. Another approach is using asynchronous encryption schemes. For instance, in [18], a scheme for storing and protecting IoT data has been proposed. In this scheme, certificateless cryptography is used for transaction security. This choice is to reduce the redundancies of traditional public key infrastructure. In certificateless cryptography, users establish their private keys using their secrets and partial private keys; the key generation center provides the latter. In the proposed scheme, the IoT data are stored distributedly using blockchain technology.

Another aspect of IoT network which can be handled using blockchain is access management. In [25], a blockchain-based access control system has been proposed, destined for arbitrating roles and permissions in IoT. The proposed architecture eliminates the need for a centralized access management system. It simplifies the process and minimizes communication overheads using a single smart contract. In this proposed approach, the direct integration of blockchain technology into IoT devices is avoided to allow higher usability, given IoT devices' limited capabilities. Furthermore, BCTrust [12] is an authentication mechanism for wireless sensor networks based on blockchain technology. The proposed mechanism is suitable for environments with resource constraints. BCTrust utilizes Ethereum for the blockchain layer. Only a group of trustworthy nodes have access to write on the blockchain of this approach. The proposed approach presents a decentralized authentication system with a global vision as blockchain networks. Hence, the operations are realizable autonomously, transparently, and securely.

The existing studies provide valuable efforts in enhancing the security of IoT applications using blockchain technology. However, in the context of confidentiality, the redundancy of data encryption is often overlooked. In one-to-many data sharing, the data is encrypted multiple times, resulting in extensive computation and storage overhead in a resource-limited IoT network. This paper considers this challenge and proposes a secure data-sharing framework with minimum overhead.

4 Security Concerns

This proposed framework tackles the confidentiality and privacy concerns regarding data sharing in distributed IoT applications. We consider blockchain as the distributed platform of our work, given its security and interoperability features.

The following security concerns are addressed in the proposed framework:

- Transparency: In this framework, the activities are necessarily traceable, such that an auditor can verify all the transactions.
- Integrity: Any data stored in this framework should stay intact, and any data modification should be detectable.
- Fine-grained access control: Owners should be able to adapt their data access in our framework. In our framework, data owners, at any time, may alter the access to their data, i.e., allowing/revoking access to the data.
- Traceable access control: The transactions regarding the access controls should be traceable in our proposed framework. In other words, requests for data access and their response, i.e., granting or denying their requested access, should be available for auditors to verify.

5 A Secure Data-Sharing Framework Based on Blockchain

5.1 Architecture

The proposed framework is designed in two primary layers to take advantage of two technologies: blockchain and broadcast encryption. The former provides a distributed immutable platform for ensuring integrity. The latter is for enabling secure data sharing without unnecessary replications. The general architecture of the proposed framework is depicted in Fig. 1, and the details of each layer and component are presented as follows.

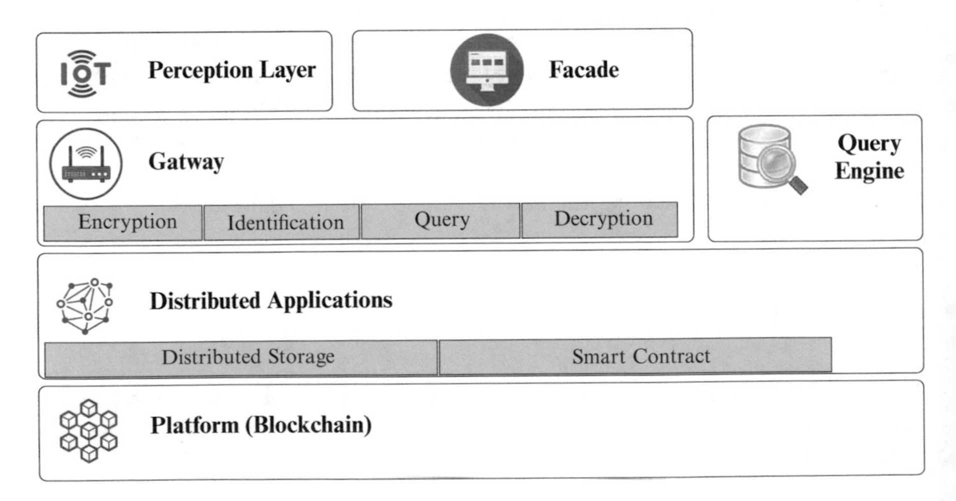

Fig. 1 General architecture of the proposed framework

5.2 *Perception Layer*

The perception layer is the nearest layer to the physical layer in the proposed framework. This layer consists of the IoT devices interacting with the physical world. These IoT devices, such as wearable sensors, environmental sensors, and mobile phones, collect data for IoT applications.

5.3 *Platform Layer*

The blockchain consists of distributed ledger, which can be seen as an immutable decentralized database. Additionally, smart contracts are event-driven scripts that can trigger data manipulation in the blockchain in defined cases. Blockchain is the backbone of the proposed framework, and all the other components of this framework interact using the blockchain network.

5.4 *Decentralized Applications Layer*

A Decentralized Application (DApp) is saved and executed through a decentralized network, e.g., blockchain. In the proposed framework, DApp is the logical layer orchestrating the interactions. Generally, we have two types of DApp in our framework: smart contracts and data storage. The former manages the workflow of the system, while the latter controls the distributed data storage.

5.4.1 Distributed Storage Component

Since storing extensive data in the blockchain is not efficient, in the proposed framework, we store the data off-chain, i.e., outside the blockchain, and only store its address in the blockchain. A well-established solution for distributed storage is the InterPlanetary File System (IPFS) [1], which is a decentralized file system that allows access, storage, and security of data in a distributed network. It combines distributed hash tables, incentivized block exchange, and self-certifying namespaces.

5.4.2 Smart Contract Component

The smart contracts provide event-driven execution of scripts that allows a guaranteed flow of work. The smart contract in the proposed framework manages the underlying transactions, including the following ones:

- Register user: This transaction allows new users to join the network. All users should have been registered prior to any other activities. This transaction allows meaningful traceability.
- Retrieve access control list (ACL): This transaction allows users to get the ACL regarding their data.
- Grant access: This transaction allows data owners to grant access to their data.
- Decline access: This transaction allows data owners to remove existing access to their data.
- Request access: This transaction allows peers to request access to the data; the data owner can use grant/decline transactions to respond to request transactions.

5.5 Gateway Component

Traditional IoT gateways collect and transfer data from IoT devices to external systems. In the proposed framework, the external systems are accessible via the blockchain. The peers in the proposed framework can be data producers and data consumers. Hence, two supplementary processes in the gateway of this proposed framework enable querying and retrieving the data. Moreover, this proposed framework's encryption scheme and identification are slightly different from the gateway mentioned above.

5.5.1 Encryption Process

As we detailed in Sect. 2.4, broadcast encryption is an answer for distributing information among multiple recipients. The idea of such an encryption system is to enable multiple parties to decrypt the ciphertext with only one encryption. Broadcast encryption scheme provides a multi-purpose secure channel without the need for redundant encryption of the same message.

In the proposed framework, we utilize broadcast encryption in the IoT gateways to encrypt the incoming IoT data for the authorized group of recipients. This process exports a single cryptogram that all the peers can decrypt in the authorized group.

5.5.2 Identification Process

In this framework, we need to allow other peers to find their interest data. Hence, in this framework, we encrypt the identification information regarding the data using an homomorphic encryption scheme, see Sect. 2.3. Using homomorphic encryption allows the data owner to keep the confidentiality of their identification information and allows other peers to query the existing data based on their requirements. Since homomorphic encryption allows computation without revealing the data itself, homomorphic encryption of identification information allows querying based on this information without revealing any additional knowledge from the identification information.

5.5.3 Query Process

Since our proposed framework's data are kept private, data consumers can not easily search or select their intended data using traditional approaches. Data consumers can establish some structured query language (SQL)-like queries and encrypt them. The queries are sent to the query execution component to allow its homomorphic execution.

5.5.4 Decryption Process

Data consumers are allowed to decrypt any cryptograms that are destined for them. Since the data are encrypted using broadcast encryption scheme, the data consumers can decrypt the encrypted data using their key without any further required steps.

5.6 Query Engine Component

A traditional query execution component is designed to allow queries on the stored data in a database and retrieve the query's data. In the proposed framework, the database is stored distributively, and the query-able information is encrypted homomorphically.

The encrypted queries from the data consumers are processed on homomorphically encrypted identification data and yield encrypted results. The result of a query consists of the data owners' address, which the data consumer would use to create a request access transaction.

5.7 Facade Layer

In order to facilitate the use of the proposed framework, a facade layer is proposed on top of all the other layers. The facade layer is the abstraction of the underlying complex layers.

5.8 Cryptology Algorithms

Cryptology algorithms are designed to provide confidentiality and integrity of the messages. In the proposed framework, we take advantage of various cryptology algorithms. The use of cryptology algorithms in the proposed framework is summarized as follows:

- Hash before Encryption: we assume the integrity of all messages is essential. Hence, all the data in the proposed framework accompany their secure hash signature prior to their encryption. The secure hash signature allows verification of their integration upon receipt.
- Broadcast encryption on IoT data: we assume all the IoT data are confidential and can be destined to multiple recipients. Therefore, IoT data are encrypted using broadcast encryption scheme respecting the authorized group of access.
- Homomorphic encryption on identification information: the identification information allow data consumers to find their intended data; however, they also contain private information that should be kept confidential. Hence, in the proposed framework, the homomorphic encryption scheme is applied to the identification information to allow querying without revealing the contents.
- Asymmetric encryption on queries/query results: The query and their result might contain sensitive information. However, since it is only destined for one recipient, applying broadcast encryption is not encouraged. Therefore, in the proposed framework, such messages are encrypted with the public key of the recipient.

5.9 Workflow

In the proposed framework, two prominent roles, data owner and data consumer, exist. Moreover, smart contracts, distributed storage, gateways, and query engines are intermediate roles that enable the interactions between the two prominent roles. Three main workflows occur in the proposed framework: data owner registration, request access, and data sharing. A recurrent workflow in the proposed framework is data storage. We describe these workflows in the next sections (Fig. 2).

5.9.1 Storage

The workflow regarding storing any data in the proposed framework is depicted in Fig. 2. For any data storage: upon arrival of new data, regardless of its type, smart contracts verify the data's integrity and its sender; then the contents of data is transmitted to the distributed storage; when the latter has saved the data, it responds with the address of the stored data; the address of the data in the distributed storage is saved in the blockchain.

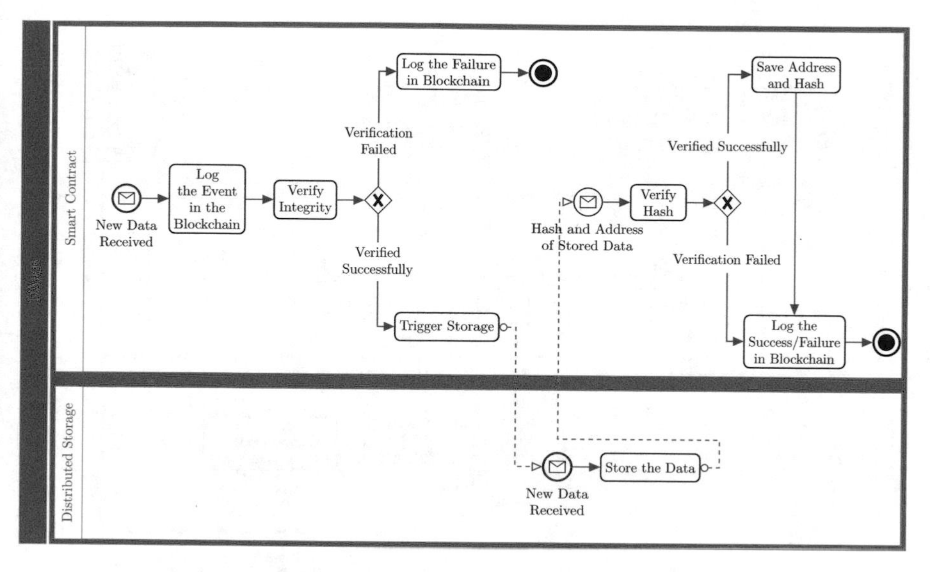

Fig. 2 Workflow of storage of any data in the proposed secure framework based on blockchain

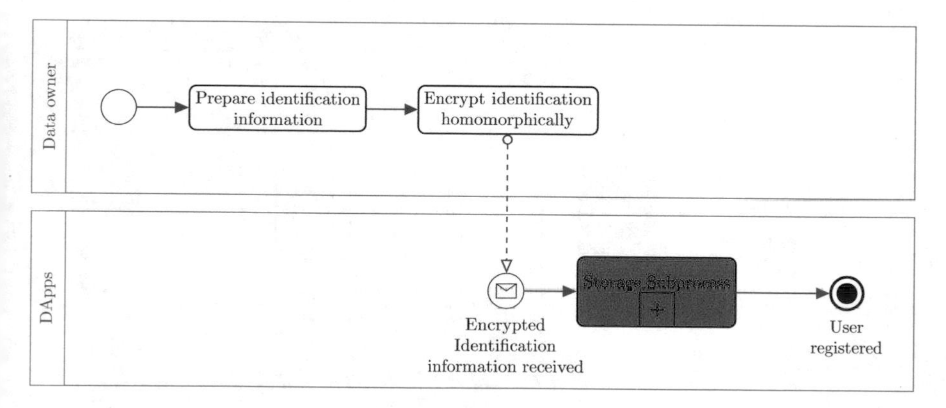

Fig. 3 Workflow of registration in the proposed secure framework based on blockchain

5.9.2 Data Owner Registration

Figure 3 depicts the general workflow of data owner registration. This simple workflow aims to store the data owner's identification information for future queries. The data owner prepares his/her identification information and homomorphically encrypts it. This encrypted information is stored using the storage subprocess, discussed in Sect. 5.9.1. Once the identification information is successfully stored, the registration workflow is terminated.

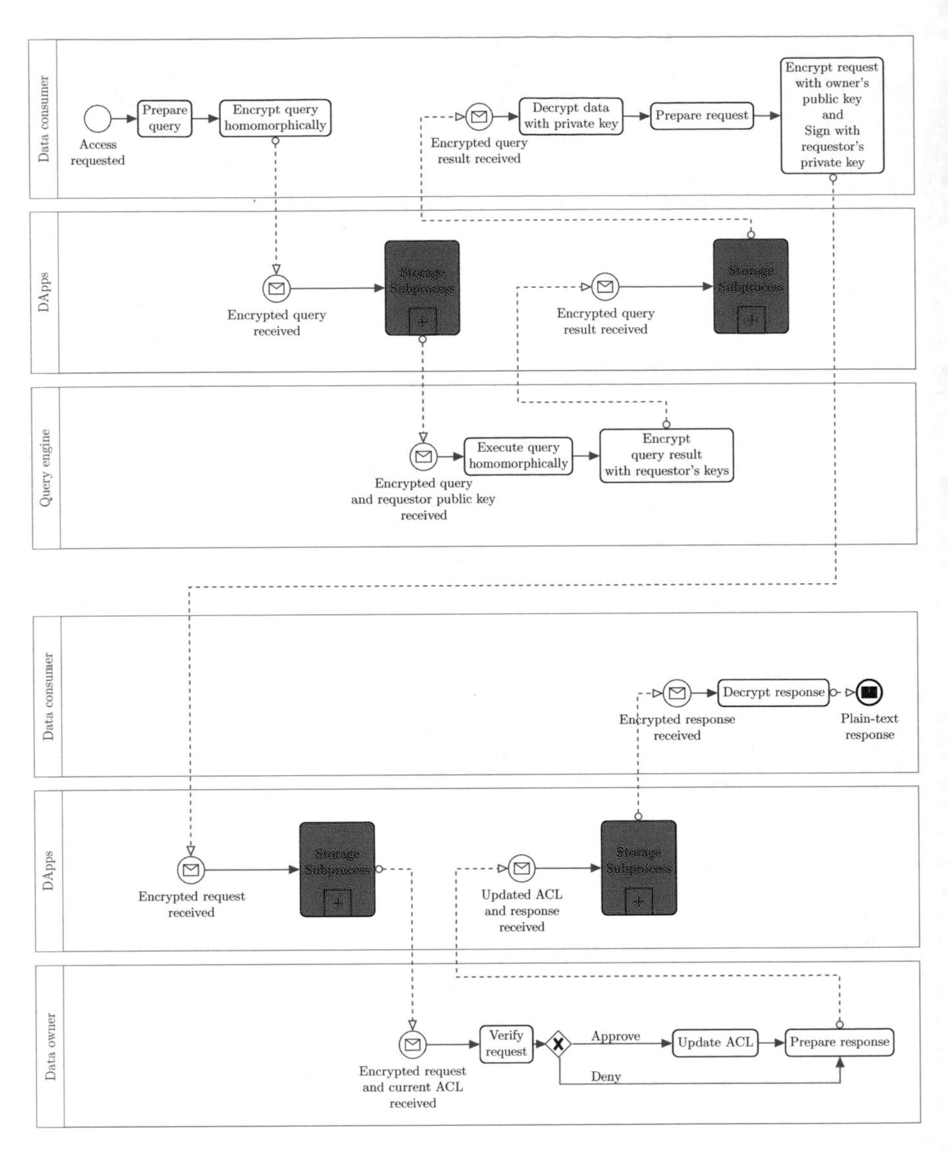

Fig. 4 Workflow of request access in the proposed secure framework based on blockchain

5.9.3 Request Access

The request access workflow involves all the major roles of the proposed framework; this workflow is depicted in Fig. 4. Data consumer prepares a query based on his/her requirement. The query might include intended identification information such as the name of the data owner. This query is encrypted homomorphically and send to the DApps. Before forwarding the query to the query engine, DApps store the query for future references. The query engine executes the query homomorphically and re-encrypts the result for the requestor. The result might include the encryption of access information, e.g., the intended data owner's address and public key. DApps store and forward the encrypted result to the data consumer; then, the latter decrypts and prepares an access request based on the query results. The data consumer signs the access request with his/her key and encrypts it with the data owner's public key. DApps store and shares the access request with the data owner. The latter might decrypt the request with his/her key and verify the signature based on the requestor's public key. Data owner might update his/her ACL if he/she decides to grant permission to the requestor. The request access workflow terminates by informing the data consumer about the data owner's response to the access request.

5.9.4 Data Sharing

Data sharing is arguably the most frequent workflow compared to registration and request access. Figure 5 presents an overview of the data sharing workflow. The workflow initiates with capturing new data in an IoT device on the data owner's side. Data owner might retrieve their ACL from DApps. ACL includes the keys

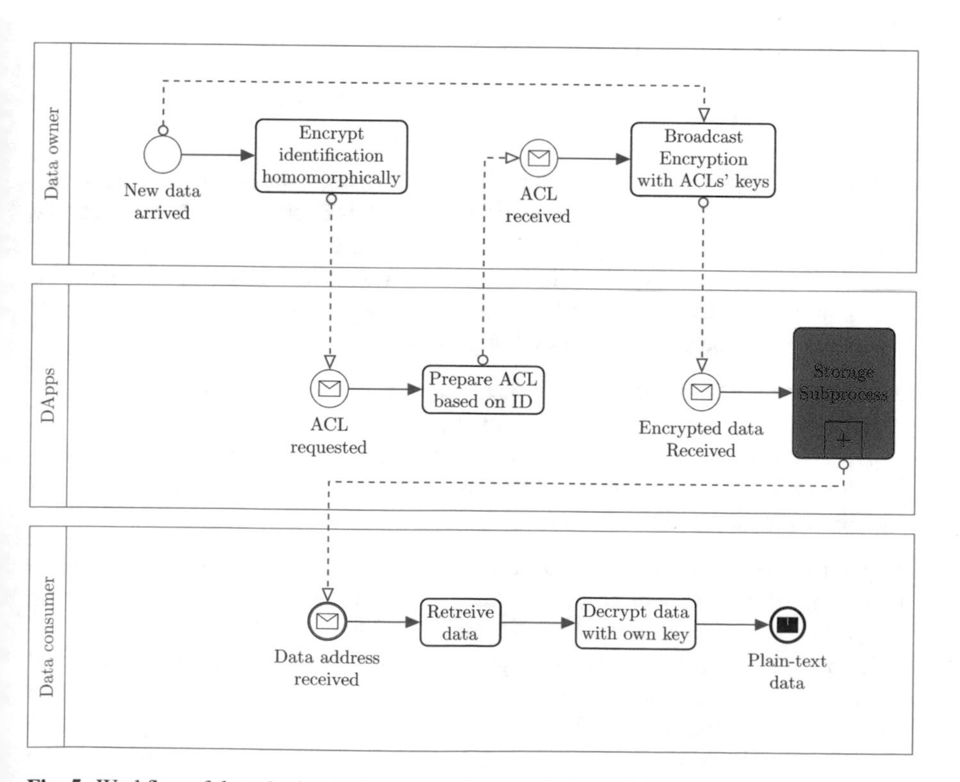

Fig. 5 Workflow of data sharing in the proposed secure framework based on blockchain

of authorized recipients, which is used for the broadcast encryption scheme. Upon receiving the ACL, the data owner's gateway can encrypt the new data using the broadcast encryption scheme and transfer it to DApps for storage. Any peers can retrieve the encrypted data from DApps. While only the authorized recipients defined in the ACL can decrypt and access the new data in plain form. The latter terminates the data sharing workflow.

The required encryption and storage for data sharing in the proposed framework is $O(N_{messages})$; in which $N_{messages}$ represents the number of messages, respectively. In the existing frameworks in this context that use asymmetric encryption for privacy conserving data sharing, the complexity is $O(N_{messages} \times N_{receivers})$, in which $N_{receivers}$ represents the number of receivers.

6 Security Evaluation

In this section, we analyze the security characteristics of our proposed framework. The notations used in this section are summarized in Table 1.

6.1 Formal Description

6.1.1 Proposed Framework

As discussed in Sect. 5.9, there are three main processes in the proposed framework. The logical dataflow regarding these three processes is depicted in Fig. 6, where the processes are segregated using a horizontal dotted separator. The inner processes of the data lanes are omitted for simplicity. All messages sent to DApps are coupled with their hash digest for verification in smart contracts; however, these hash digests are not shown in the data flow for better readability.

Data Owner's Registration For the registration, data owners prepare their identification information internally (id) and encrypt it using homomorphic encryption (C^{π}_{id}). This message, alongside the hash digest value of the identification information (h_{id}) and hash digest of the whole message, is transmitted to DApps. DApps store these data in the blockchain.

Request Access For requesting access, the first step is to find the data owner. To this end, the data consumer prepares a query (qry) and encrypts using homomorphic encryption (C^{π}_{qry}) which are sent to DApps alongside its hash value.

After internal verification and storage, DApps allow the query engine to read and execute the encrypted query. The latter re-encrypts the results using the requestor's public key (C^{Pc}_{lst}) and sends it alongside the hash digest of the results (h_{lst}), and the hash digest of the whole message to DApps. DApps internally verify and store the encrypted results, allowing the data consumer to read the encrypted result. The data

Table 1 Notations used in formal modeling of the proposed blockchain-based secure data-sharing framework

Notation	Description
ID	Identification information
qry	Query
lst	List of matched items, i.e., the result of a query
c	Data consumer
o	Data owner
d	Raw data
acl	ACL, i.e., the keys of authorized recipients
P_u	Public key of user u
req	Access request
π	Key used for homomorphic encryption
S_u	Signature of user u.
$E(m)$	Encryption on message m
$C_m^{P_u}$	Cryptogram on message m with the public key of user u
C_m^{π}	Cryptogram on message m using homomorphic encryption
C_m^{acl}	Cryptogram on message m using broadcast encryption
$D(c)$	Decryption of message m
$Query(qry)$	Classification of message m
$H(m)$	Secure hash digest on message m
h_m^S	Digest value of message m with the signature S
a_m	Address of message m in the distributed storage
u	User of the system
$Own(u, m)$	Is u the owner of message m
$Acc(u, m)$	Has u read access to message or transaction m
$Send(u_1, u_2, m)$	User u_1 has send message m to user u_2
$Allow(u_1, u_2, m)$	User u_1 has granted access for message m to user u_2

consumer, who has issued the query, can decrypt the result of his/her query (lst). He/she may now use this information and form an access request(s) (req); he/she then encrypt using the public key of the intended data owner(s) (C_{req}^{Po}) and sign it using his/her signature (h_{req}^{Sr}). These messages are transmitted to the data owner(s) via DApps. The data owner decides whether to approve or reject the request using a response message res and signs it using his/her signature (h_{res}^{So}). In the case of approval, he/she updates ACL and encrypts it using his/her key (C_{acl}^{Po}); and sends all these messages to DApps for storage. Although, for simplicity, the workflow and dataflow did not include message type for the access revocation, the latter is easily addable to the request and response, allowing a fine-grained access control.

Data Sharing For data sharing, the first step is to retrieve the list of authorized recipients. The data owner can retrieve his/her the encrypted ACL from DApps (C_{acl}^{Po}) and decrypt it locally. Then data owner uses the keys of authorized recipients to encrypt the new data (d) using broadcast encryption, resulting in a single cryptogram (C_d^{acl}). The latter, alongside its hash value, is verified and stored using

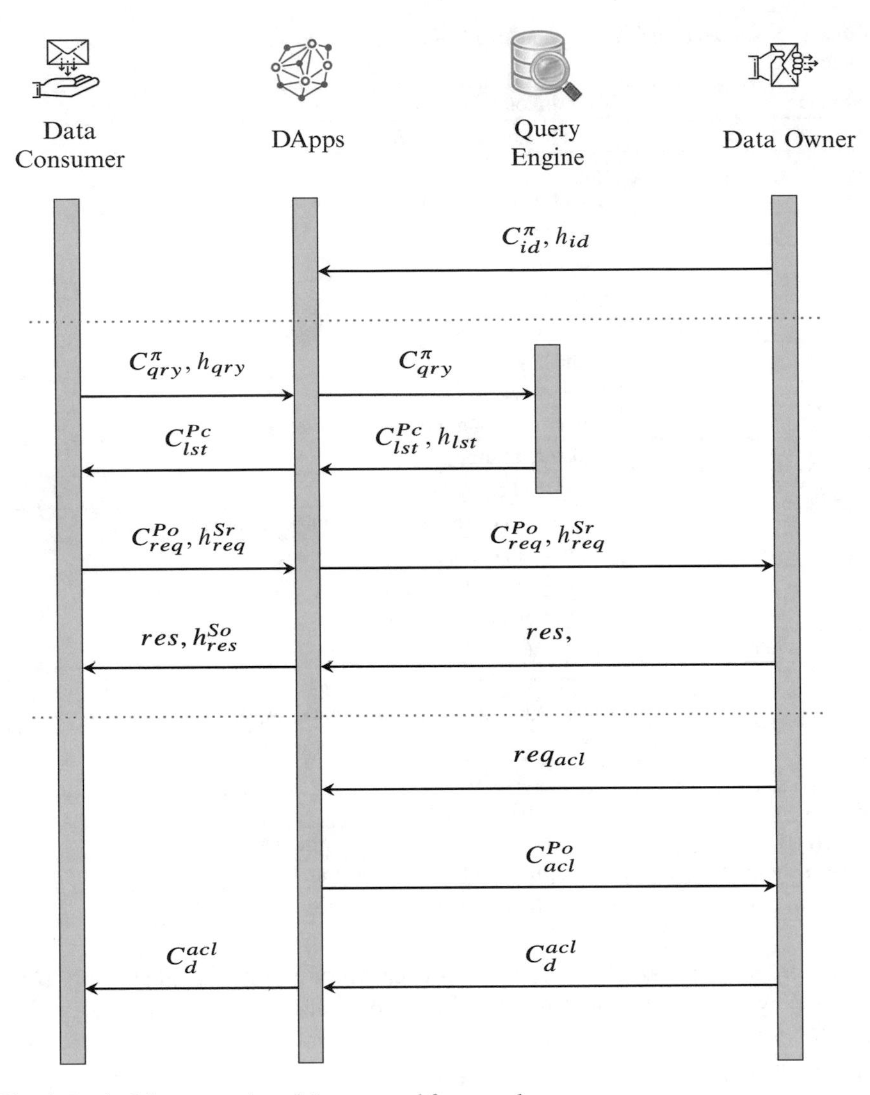

Fig. 6 Logical flow execution of the proposed framework

DApps. Any authorized recipient can now retrieve and decrypt the cryptogram to access the data (d) in plain form.

6.1.2 Requirements

The security concerns are broadly classified into three main categories: (1) Confidentiality: only authorized entities can access data. (2) Integrity: only authorized entities can change data. (3) Availability: access of authorized entities is always possible. The security concerns are focused on the authorized entities, which should be defined. In the proposed framework, only the data owner is the intrinsically authorized entity, and the data owner should authorize any other entity. Moreover, since the transactions are essential to be kept as archives, no entity is allowed to change the data once entered into the framework.

The confidentiality requirement of this proposed framework is categorized into threes forms:

- R1: confidentiality of identification information and query,
- R2: confidentiality of metadata, e.g., request, results, ACL,
- R3: confidentiality of data IoT captured data.

The first two forms can be formalized using Eqs. (2) and (3); while the last one is, in fact, the access control requirement.

$$\mathrm{R1}: \forall ID, \forall u : Acc(u, ID) \Longleftrightarrow Own(u, ID) \vee u = q \tag{2}$$

$$\mathrm{R2}: \forall m, \forall u : Acc(u, m) \Longleftrightarrow Own(u, m) \vee Send(o, u, m) \tag{3}$$

The traceable access control requirement is, in fact, a subset of transparency, Eq. (4); and the access control requirement and integrity requirement can be formalized in Eqs. (5) and (10), respectively:

$$\text{Transparency}: \quad \forall e \in T, \forall u \in \mathcal{U} : Acc(u, e) \tag{4}$$

$$\text{Access control(R3)}: \forall Message\ m, \forall User\ u : Acc(u, m)$$

$$\Longleftrightarrow \quad User\ o : Own(o, m) \wedge Allow(o, u, m) \tag{5}$$

$$\text{Integrity}: \quad \forall m \in \mathcal{M} : m_t \Longrightarrow m_{t+1} \tag{6}$$

6.1.3 Assumptions

In this proposed framework, all users have access to the public keys of all nodes, and only the owner of the key has access to the private key.

Moreover, we also assume the security of the broadcast encryption scheme as well as asymmetric encryption one; both can be similarly formalized as Eq. (7). Moreover, at least one trusted party is required for handling the homomorphic encryption and decrypting the query results; for simplicity, we assumed this is the same user as the query engine; however, this trusted user can be any other node.

$$\Big(Acc(u, c_m) \Rightarrow Acc(u, m)\Big) \Longleftrightarrow Acc(u, \pi_{c_m}) \tag{7}$$

Since finding a hash collision is computationally impossible, we assume that the hash digests are unique; this assumption is formalized in Eq. (8).

$$H(m_1) = H(m_2) \Longleftrightarrow m_1 = m_2 \tag{8}$$

The immutability of the blockchain network can be formalized as in Eq. (9).

$$\forall x \in blockchain : x_t \Longleftrightarrow x_{t+1} \tag{9}$$

6.2 Formal Verification

Because of the immutability and availability of blockchain, the availability and traceability requirements are trivial. The rest of the security can be formally discussed as follows.

6.2.1 Integrity

For integrity in the proposed framework we should prove that $\forall m \in \mathcal{M} : m_t \Longrightarrow m_{t+1}$. In the proposed framework, upon arrival of data in the blockchain, its hash digest is computed and verified with the provided hash digest; and only if the two hash digests are matched, the data is stored off-chain in the distributed storage, and its hash digest and address are stored in the blockchain (m_0). In the blockchain, all the data are immutable; hence we can deduce that since the hash digests are stored in the blockchain, then the hash digests remain intact, i.e., $H(m)_t \Longrightarrow H(m)_{t+1}$. In order to alter a message in the distributed storage from m to $H(\hat{m})$, its hash digest should also be modified in the next timestep $H(\hat{m})_{t+1}$. Because of the immutability of blockchain and security features of hashing algorithm, the integrity of the message in the proposed framework can be formally proven as the followings:

$$\exists m, \hat{m} \in \mathcal{M} : m \neq \hat{m} \wedge H(m)_t \wedge H(\hat{m})_{t+1} \tag{10}$$

$$\begin{aligned}
\text{Equation (9)} : &\ \forall m \in \mathcal{M} : H(m)_t \Longrightarrow H(m)_{t+1} \\
\Longrightarrow &\quad H(\hat{m})_{t+1} = H(m)_{t+1} \\
\text{Equation (8)} : &\ \forall m_1, m_2 : H(m_1) = H(m_2) \Longleftrightarrow m_1 = m_2 \\
\Longrightarrow &\quad \hat{m} = m \\
\Longrightarrow &\quad m_t \Longrightarrow m_{t+1} \\
\Longrightarrow &\quad \forall m \in \mathcal{M} : m_t \Longrightarrow m_{t+1}
\end{aligned}$$

6.2.2 Confidentiality of Identification Information and Query

The identification information of users is never stored in the blockchain nor in the distributed storage. Based on the security of the homomorphic encryption scheme used for this encryption scheme, no user can gain any information about them. We use indirect proof for confidentiality: The contradictory assumption is that there is an unauthorized user $\hat{u}$ with access to the identification or query m (Eq. (11)).

Since we assumed the security of the owner, the unauthorized user have accessed the message from its cryptogram (Eq. (12)). However, due to the encryption scheme security characteristics formalized in Eq. (7), it narrows down the unauthorized user to the holder of the encryption key, which contradicts the assumptions of secure party handling the homomorphic encryption and proves the confidentiality.

$$\exists \hat{u}, m : Acc(\hat{u}, m) \wedge \overline{Own(\hat{u}, m)} \wedge u \neq q \tag{11}$$

$$\implies \exists \hat{u}, m : Acc(\hat{u}, c_m^{\pi}) \implies Acc(\hat{u}, m) \wedge \overline{Own(\hat{u}, m)} \wedge u \neq q \tag{12}$$

$$\text{Equation (7)} \quad : \quad \Big(Acc(\hat{u}, c_m^{\pi}) \implies Acc(\hat{u}, m)\Big) \implies Own(\hat{u}, \pi) \tag{13}$$

$$\implies \hat{u} = q \implies \perp \tag{14}$$

$$\implies \forall ID, \forall u : Acc(u, ID) \iff Own(u, ID) \vee u = q \tag{15}$$

6.2.3 Confidentiality of Metadata

The metadata is intended for only one recipient; hence, in the proposed framework, they are encrypted using normal asymmetric encryption schemes. The sender encrypts the metadata using the receiver's public key and stores it in the blockchain and distributed storage, and the receiver can only decrypt it using his/her private key. The confidentiality of metadata can be formally proved the same as the confidentiality of query information, presented in Sect. 6.2.2.

We use indirect proof for the confidentiality of data: The contradictory assumption is that there is an unauthorized user $\hat{u}$ with access to data m (Eq. (16)). Since we assumed the security of the client-sides and the fact the data is always encrypted in the proposed framework; hence, the unauthorized user has accessed the message from its cryptogram (Eq. (17)). However, due to the encryption scheme security characteristics formalized in Eq. (7), it narrows down the unauthorized user to the holder of the private keys. Since only the receiver has the private key of the encryption (Eq. (18)), it contradicts the security of the client-side. This contradiction proves the initial assumption is impossible and hence proves the confidentiality of data.

$$\exists \hat{u}, m : Acc(\hat{u}, m) \wedge \overline{Own(\hat{u}, m)} \wedge \overline{Send(\hat{u}, o, m)} \tag{16}$$

$$\implies \exists \hat{u}, d : Acc(\hat{u}, c_m^{P_u}) \implies Acc(\hat{u}, m) \wedge \overline{Own(\hat{u}, m)} \wedge \tag{17}$$

$$\text{Equation (7)} \quad : \quad \Big(Acc(\hat{u}, c_m^{P_u}) \implies Acc(\hat{u}, m)\Big) \implies Acc(\hat{u}, Pr_u)$$

$$\implies Send(o, \hat{u}, m) \tag{18}$$

$$\implies \perp$$

$$\implies \forall m, u : Acc(u, m) \iff Own(u, m) \vee Send(o, u, m)$$

6.2.4 Confidentiality of Data (Access Control)

The data is protected via the security of the broadcast encryption scheme. Hence, the confidentiality of data depends on the confidentiality of metadata, i.e., ACL, and the confidentiality of the encryption schemes, which are proved and assumed, respectively. We use indirect proof for the confidentiality of data: The contradictory assumption is that there is an unauthorized user $\hat{u}$ with access to data d (Eq. (19)). Since we assumed the security of the client-sides and the fact the data is always encrypted in the proposed framework; hence, the unauthorized user has accessed the message from its cryptogram (Eq. (20)). However, due to the encryption scheme security characteristics formalized in Eq. (7), it narrows down the unauthorized user to the holder of the encryption keys of broadcast or the receiver of ACL. Since in the proposed framework ACL is only shared with the data owner (Eq. (21)); it contradicts the confidentiality of metadata proved in Sect. 6.2.3. This contradiction proves the initial assumption is impossible and hence proves the confidentiality of data.

$$\exists \hat{u}, m : Acc(\hat{u}, d) \wedge \overline{Own(\hat{u}, d)} \wedge \overline{Allow(\hat{u}, o, d)} \tag{19}$$

$$\Longrightarrow \exists \hat{u}, d : Acc(\hat{u}, c_d^{acl}) \Longrightarrow Acc(\hat{u}, d) \wedge \overline{Own(\hat{u}, d)} \wedge \tag{20}$$

$$\text{Equation 7} \quad : \quad \left(Acc(\hat{u}, c_d^a cl) \Longrightarrow Acc(\hat{u}, d)\right) \Longrightarrow Acc(\hat{u}, acl)$$

$$\text{Section 6.2.3} \quad : \quad Acc(\hat{u}, ACL) \Longleftrightarrow Own(\hat{u}, d) \vee Send(o, \hat{u}, d)$$

$$\Longrightarrow Send(o, \hat{u}, m) \tag{21}$$

$$\Longrightarrow \hat{u} = o \Longrightarrow \bot$$

$$\Longrightarrow \forall d, u : Acc(u, m) \Longleftrightarrow User\ o : Own(o, m) \wedge Allow(o, u, m)$$

7 A Teleconsultation Use Case of the Proposed Secure Data-Sharing Framework

In order to depict the benefits of the proposed secure framework for data-sharing, we discuss a healthcare use case in this paper. Teleconsultation is complex and includes numerous members interacting and sharing data with each other; hence, a teleconsultation application, named HapiChain [16], is discussed as the use case of this proposed framework.

Since the proposed framework is dedicated to data-sharing, we explore the data used in teleconsultation and how they can be handled using the proposed framework.

- Doctors' information: the doctors might decide to publish their information, including their contact information and availabilities, publicly or only share it with their patients. In the latter case, the ACL for each doctor's information includes the patients of that doctor, resulting in fine-grained access control and confidentiality. Moreover, in both cases, because availability and integrity are in the proven requirements of the proposed framework, it is guaranteed that the doctors' information is available at all times and without the risk of unauthorized changes.
- Patients' medical information: the patients' medical information might be shared with their generalist, recurrent doctors, nurse, and visiting hospital and Hapicare. The data sharing might be in full or partial, and also permanent or temporary. For instance, the patient might prefer not to share his/her vital signs continuously with his/her doctor; but with Hapicare to avail him/her a telemonitoring service. Moreover, through the course of teleconsultation, he/she might want to allow the doctor to access their vital signs for remote measurement. This information can be securely shared with the above requirements using the proposed framework. The fine-grained access control enables any type of access based on the data type and time. Moreover, the proposed framework's integrity and availability characteristics guarantee the patients' data stay intact and available.
- Hapicare information: Besides the doctors' and patients' information, Hapicare information includes medical rules and monitoring reports. A doctor provides the former for a specific (group of) patient; hence, the medical rules are sent from doctors to Hapicare, with guaranteed security (confidentiality, integrity, and availability) in the proposed framework. Moreover, Hapicare generates monitoring reports and then shares them with the patient, his/her nurse, and doctor, with guaranteed security using the proposed framework.
- Transactions: The events occurring during teleconsultation are required for financial and legal aspects. The proposed framework enables an immutable and available logbook of transactions with guaranteed security.
- Call information: Although it is possible to use the proposed framework for sharing the data related to video calls, direct data-sharing of call information enables much higher performance. Moreover, a call's contents can grow huge to be stored in the proposed framework and usually are not useful. Hence, it is better to share the call events only, e.g., the patient started the call, the doctor joined the call, or the call is abandoned, using the proposed framework.

In this use case, we purposely put some information for the public to depict that the proposed framework can be used for public data sharing. However, based on the requirement, public data sharing, e.g., call information, can be transferred via secure data sharing. The summary of the above data and how they are handled is presented in Table 2.

Table 2 Summary of handling the teleconsultation data in the proposed secure blockchain data-sharing framework

Data	Data owner	Data consumer(s)	Security scheme	Most critical concern
Patients' info.	Patient	Doctors, nurses, hospitals, hapicare, etc.	BE[a] + DS[b]	Confidentiality
Hapicare rules	Doctor	Hapicare and patient	BE + DS	Confidentiality
Hapicare results	Hapicare	Patient, nurse, and doctor	BE + DS	Confidentiality
Transactions	Doctors, patients, hapicare, the framework itself	Everyone	None + DS	Integrity
Call info.	Patients and doctors	Doctors and patients	Externally	Availability
Call events	Patients and doctors	Doctors and patients	None + DS	Integrity

[a] Broadcast encryption
[b] Digital signature

7.1 Discussion

The added values of the proposed framework in the teleconsultation use case can be arguably discussed in two main categories: (1) the security and robustness and (2) the minimized overheads.

The use of encryption schemes protects the confidentiality of data in transit. Notably, patients' information which is often considered sensitive, is encrypted with secure encryption schemes.

Moreover, the use of digital signatures can ensure the integrity of data. Distinctly, the transactions are signed with a secure digital signature in this framework, ensuring this information's integrity. Consequently, all the actors can verify the correctness of the transactions.

On the other hand, using blockchain as the underlying platform provides unparalleled benefits in the robustness and availability of the system. The use of smart contracts ensures the correct execution of the processes in the framework.

Lastly, concerning the overheads: opting for broadcast encryption in this context allows single encryption for multiple receivers. For instance, if a patient's information is encrypted in traditional one-to-one encryption schemes, the number of encryption and, consequently, the storage complexity is linearly dependent on the number of receivers. However, with the use of broadcast encryption, these overheads in minimized. Although the receivers of each data for each patient might not be a considerable number, given the increasing number of data for each patient this save of overhead is highly beneficial. In other words, a ten-fold reduction of millions of encryption results in substantial savings in the process and storage consumption of the framework.

8 Conclusion

Because of the increasing use of IoT in modern technologies, the security of such applications has gained interest in recent years. To this end, in this paper, we have defined the security requirements of data sharing in IoT applications. We have proposed a secure data-sharing framework based on blockchain. For the robustness requirements of data sharing and storage, blockchain technology, distributed storage, and digital signature are used in the proposed framework. Moreover, the proposed framework uses a combination of cryptographic algorithms for providing secure data while avoiding undesired redundancy. It uses homomorphic encryption for privacy-preserving queries of information, asymmetric encryption for secure one-to-one communication, and broadcast encryption for secure one-to-many data sharing. The security requirements of data sharing and storage in IoT networks are defined, and their fulfilments are formally proved in the proposed framework. Additionally, the proposed framework is evaluated in a teleconsultation use case. For the future works, delegation of rights is an inspiring challenge; which can reduce the complexity and consequently improve usability, scalability, and manageability of access controls.

References

1. Benet, J.: IPFS - Content Addressed, Versioned, P2P File System p. 11
2. Bennett, B.: Using Telehealth as a Model for Blockchain HIT Adoption. Telehealth and Medicine Today **2**(4) (2018). DOI https://doi.org/10.30953/tmt.v2.25. URL http://telehealthandmedicinetoday.com/index.php/journal/article/view/25
3. Boneh, D., Silverberg, A.: Applications of Multilinear Forms to Cryptography. Tech. Rep. 080 (2002). URL http://eprint.iacr.org/2002/080
4. Cebe, M., Erdin, E., Akkaya, K., Aksu, H., Uluagac, S.: Block4Forensic: An Integrated Lightweight Blockchain Framework for Forensics Applications of Connected Vehicles. IEEE Communications Magazine **56**(10), 50–57 (2018). DOI https://doi.org/10.1109/MCOM.2018.1800137. URL https://ieeexplore.ieee.org/document/8493118/
5. Chowdhury, M.J.M., Colman, A., Kabir, M.A., Han, J., Sarda, P.: Blockchain Versus Database: A Critical Analysis p. 6
6. Dai, H.N., Zheng, Z., Zhang, Y.: Blockchain for Internet of Things: A Survey. IEEE Internet of Things Journal **6**(5), 8076–8094 (2019). DOI https://doi.org/10.1109/JIOT.2019.2920987. URL https://ieeexplore.ieee.org/document/8731639/
7. Dorri, A., Kanhere, S.S., Jurdak, R., Gauravaram, P.: LSB: A Lightweight Scalable BlockChain for IoT Security and Privacy. Journal of Parallel and Distributed Computing **134**, 180–197 (2019). DOI https://doi.org/10.1016/j.jpdc.2019.08.005. URL http://arxiv.org/abs/1712.02969. ArXiv: 1712.02969
8. Dorri, A., Steger, M., Kanhere, S.S., Jurdak, R.: BlockChain: A Distributed Solution to Automotive Security and Privacy. IEEE Communications Magazine **55**(12), 119–125 (2017). DOI https://doi.org/10.1109/MCOM.2017.1700879. Conference Name: IEEE Communications Magazine
9. Dwivedi, A., Srivastava, G., Dhar, S., Singh, R.: A Decentralized Privacy-Preserving Healthcare Blockchain for IoT. Sensors **19**(2), 326 (2019). DOI https://doi.org/10.3390/s19020326. URL http://www.mdpi.com/1424-8220/19/2/326

10. Fiat, A., Naor, M.: Broadcast Encryption. In: D.R. Stinson (ed.) Advances in Cryptology — CRYPTO' 93, pp. 480–491. Springer Berlin Heidelberg, Berlin, Heidelberg (1994)
11. Goyal, V., Pandey, O., Sahai, A., Waters, B.: Attribute-Based Encryption for Fine-Grained Access Control of Encrypted Data. In: Proceedings of the 13th ACM Conference on Computer and Communications Security, CCS '06, pp. 89–98. Association for Computing Machinery, New York, NY, USA (2006). URL https://doi.org/10.1145/1180405.1180418. Event-place: Alexandria, Virginia, USA
12. Hammi, M.T., Bellot, P., Serhrouchni, A.: BCTrust: A decentralized authentication blockchain-based mechanism. In: 2018 IEEE Wireless Communications and Networking Conference (WCNC), pp. 1–6. IEEE, Barcelona (2018). DOI https://doi.org/10.1109/WCNC.2018.8376948. URL https://ieeexplore.ieee.org/document/8376948/
13. Hasan, H.R., Salah, K., Jayaraman, R., Arshad, J., Yaqoob, I., Omar, M., Ellahham, S.: Blockchain-based Solution for COVID-19 Digital Medical Passports and Immunity Certificates **8**, 222093–222108 (2020). DOI https://doi.org/10.1109/ACCESS.2020.3043350
14. Huang, H., Lin, J., Zheng, B., Zheng, Z., Bian, J.: When Blockchain Meets Distributed File Systems: An Overview, Challenges, and Open Issues. IEEE Access **8**, 50574–50586 (2020). DOI https://doi.org/10.1109/ACCESS.2020.2979881. URL https://ieeexplore.ieee.org/document/9031420/
15. Jin, Z., Chen, Y.: Telemedicine in the Cloud Era: Prospects and Challenges. IEEE Pervasive Computing **14**(1), 54–61 (2015). DOI https://doi.org/10.1109/MPRV.2015.19. URL http://ieeexplore.ieee.org/document/7030248/
16. Kordestani, H., Barkaoui, K., Zahran, W.: HapiChain: A Blockchain-based Framework for Patient-Centric Telemedicine. In: 2020 IEEE 8th International Conference on Serious Games and Applications for Health (SeGAH), pp. 1–6. IEEE, Vancouver, BC, Canada (2020). DOI https://doi.org/10.1109/SeGAH49190.2020.9201726. URL https://ieeexplore.ieee.org/document/9201726/
17. Kordestani, H., Mojarad, R., Chibani, A., Osmani, A., Amirat, Y., Barkaoui, K., Zahran, W.: Hapicare: A Healthcare Monitoring System with Self-Adaptive Coaching using Probabilistic Reasoning. In: 2019 IEEE/ACS 16th International Conference on Computer Systems and Applications (AICCSA), pp. 1–8. IEEE, Abu Dhabi, United Arab Emirates (2019). DOI https://doi.org/10.1109/AICCSA47632.2019.9035291. URL https://ieeexplore.ieee.org/document/9035291/
18. Li, R., Song, T., Mei, B., Li, H., Cheng, X., Sun, L.: Blockchain for Large-Scale Internet of Things Data Storage and Protection. IEEE Transactions on Services Computing **12**(5), 762–771 (2019). DOI https://doi.org/10.1109/TSC.2018.2853167. Conference Name: IEEE Transactions on Services Computing
19. Mojarad, R., Attal, F., Chibani, A., Amirat, Y.: Automatic Classification Error Detection and Correction for Robust Human Activity Recognition. IEEE Robotics and Automation Letters **5**(2), 2208–2215 (2020). DOI https://doi.org/10.1109/LRA.2020.2970667
20. Mojarad, R., Attal, F., Chibani, A., Amirat, Y.: Context-aware Adaptive Recommendation System for Personal Well-being Services. In: Proceedings of 32nd International Conference on Tools with Artificial Intelligence (ICTAI) (2020)
21. Mojarad, R., Attal, F., Chibani, A., Amirat, Y.: A Context-aware Hybrid Framework for Human Behavior Analysis. In: Proceedings of 32nd International Conference on Tools with Artificial Intelligence (ICTAI) (2020)
22. Mojarad, R., Attal, F., Chibani, A., Amirat, Y.: A Context-based Approach to Detect Abnormal Human Behaviors in Ambient Intelligent Systems. In: Proceedings of the European Conference on Machine Learning and Principles and Practice of Knowledge Discovery in Databases (ECML-PKDD) (2020)
23. Mojarad, R., Attal, F., Chibani, A., Amirat, Y.: A Hybrid Context-aware Framework To Detect Abnormal Human Daily Living Behavior. In: Proceedings of IEEE World Congress on Computational Intelligence (WCCI) (2020)
24. Mojarad, R., Attal, F., Chibani, A., Fiorini, S.R., Amirat, Y.: Hybrid Approach for Human Activity Recognition by Ubiquitous Robots. In: IEEE/RSJ International Conference on

Intelligent Robots and Systems (IROS), pp. 5660–5665 (2018). DOI https://doi.org/10.1109/IROS.2018.8594173. ISSN: 2153-0866
25. Novo, O.: Blockchain Meets IoT: An Architecture for Scalable Access Management in IoT. IEEE Internet of Things Journal **5**(2), 1184–1195 (2018). DOI https://doi.org/10.1109/JIOT.2018.2812239. URL https://ieeexplore.ieee.org/document/8306880/
26. Paillier, P.: Public-Key Cryptosystems Based on Composite Degree Residuosity Classes. In: J. Stern (ed.) Advances in Cryptology — EUROCRYPT '99, pp. 223–238. Springer Berlin Heidelberg, Berlin, Heidelberg (1999)
27. Park, D., Zhang, Y., Rosu, G.: End-to-End Formal Verification of Ethereum 2.0 Deposit Smart Contract. In: S.K. Lahiri, C. Wang (eds.) Computer Aided Verification, pp. 151–164. Springer International Publishing, Cham (2020)
28. Rahulamathavan, Y., Phan, R.C.W., Rajarajan, M., Misra, S., Kondoz, A.: Privacy-preserving blockchain based IoT ecosystem using attribute-based encryption. In: 2017 IEEE International Conference on Advanced Networks and Telecommunications Systems (ANTS), pp. 1–6. IEEE, Bhubaneswar (2017). DOI https://doi.org/10.1109/ANTS.2017.8384164. URL https://ieeexplore.ieee.org/document/8384164/
29. Rivest, R.L., Adleman, L., Dertouzos, M.L.: On Data Banks and Privacy Homomorphisms. Foundations of Secure Computation, Academia Press pp. 169–179 (1978)
30. Rivest, R.L., Shamir, A., Adleman, L.: A Method for Obtaining Digital Signatures and Public-Key Cryptosystems. Commun. ACM **21**(2), 120–126 (1978). URL https://doi.org/10.1145/359340.359342. Place: New York, NY, USA Publisher: Association for Computing Machinery
31. Sharma, P., Jindal, R., Borah, M.D.: Healthify: A Blockchain-Based Distributed Application for Health care. In: S. Namasudra, G.C. Deka (eds.) Applications of Blockchain in Healthcare, vol. 83, pp. 171–198. Springer Singapore, Singapore (2021). DOI https://doi.org/10.1007/978-981-15-9547-9_7. URL http://link.springer.com/10.1007/978-981-15-9547-9_7. Series Title: Studies in Big Data
32. Shrestha, R., Kim, S.: Integration of IoT with blockchain and homomorphic encryption: Challenging issues and opportunities. In: Advances in Computers, vol. 115, pp. 293–331. Elsevier (2019). DOI https://doi.org/10.1016/bs.adcom.2019.06.002. URL https://linkinghub.elsevier.com/retrieve/pii/S0065245819300269
33. Weber, I., Gramoli, V., Ponomarev, A., Staples, M., Holz, R., Tran, A.B., Rimba, P.: On Availability for Blockchain-Based Systems. In: 2017 IEEE 36th Symposium on Reliable Distributed Systems (SRDS), pp. 64–73. IEEE, Hong Kong, Hong Kong (2017). DOI https://doi.org/10.1109/SRDS.2017.15. URL http://ieeexplore.ieee.org/document/8069069/

Reputation-Based Consensus on a Secure Blockchain Network

Manuel Sivianes, Teresa Arauz, Emilio Marín, and José M. Maestre

1 Introduction

Over the last decade there has been a great interest in cyber-security and, specifically in distributed problems where multiagent systems cooperate to solve a task in a reliable manner. In this regard, a core concept is that of consensus, which is defined as the agreement reached regarding a value of interest that depends on the state of the agents in a network [1]. In particular, the process that leads to this agreement through interaction rules is the consensus algorithm or protocol [2, 3], and its objective is to update the value of each node with the information received from its neighbours. A theoretical framework for consensus problems is presented in [3, 4], where different protocols for networks of dynamic cooperative agents are presented and assessed. For example, a commonly used method is the weight-based algorithm, where every agent is assigned a fixed weight that determines its impact on the consensus process [5–10]. Other works that analyze consensus problems in multiagent systems are [11–17]. In particular, in [11], a distributed control problem where a group of mobile autonomous agents must reach a determined formation is studied; in [12], some convergence results are established for the scenario of unbounded intercommunication periods; in [13], stationary and distributed consensus protocols are considered for multiagent networks under local information; in [14], nonsmooth gradient flows achieve the coordination goals in network consensus problems; in [15], a class of distributed iterative averaging algorithms that do not depend on global coordination or synchronization is presented; and, in [16], the consensus problem is considered for undirected networks of dynamic agents with communication delays. Nonetheless, consensus can be compromised under the presence of malicious or faulty agents. To deal with this issue several

M. Sivianes (✉) · T. Arauz · E. Marín · J. M. Maestre
Department of Systems and Automation Engineering, University of Seville, Seville, Spain
e-mail: mscastano@us.es; marauz@us.es; pepemaestre@us.es

Y. Maleh et al. (eds.), *Blockchain for Cybersecurity in Cyber-Physical Systems*,
Advances in Information Security 102, https://doi.org/10.1007/978-3-031-25506-9_9

strategies have been proposed: in works as [6, 18], it is used a variant of the so-called mean subsequence reduced algorithm to ignore extreme values from neighboring nodes; in [19], event-triggered updates rules are developed to mitigate the influence of malicious agents; in [20], the vulnerabilities of consensus-based distributed optimization protocols regarding nodes that deviate from the prescribed update rule are investigated; and, in [17], the so-called weighted mean subsequence reduced (W-MSR) algorithm is used to attain consensus under the presence of Byzantine adversaries and delays. In addition, different strategies are summarized in [21] that can be applied in distributed systems to guarantee cyber-security which could be adapted to the consensus problem, e.g., the software rejuvenation set-up [22], a cooperation based strategy [23, 24], a robust tree-based design [25, 26], or using learning approaches such as neural networks [27].

To deal with the presence of malicious entities in the consensus process and increase resilience against cyber-attacks, we propose to use blockchain. With the goal of removing intermediaries and making peer to peer (P2P) safe interactions possible, blockchain was created alongside with Bitcoin by Satoshi Nakamoto [28]. This technology drastically reduces the possibility of a breach and alteration of data by using cryptography and decentralization. A blockchain is conformed of data packages, called blocks, each of which contains multiple transactions. Every block is cryptographically linked to the previous one, except the first block, which is known as the *genesis block* [29]. With the addition of the smart contracts feature to the blockchain technology, distributed applications can be deployed. A smart contract is a group of source lines of code shared by every node within the blockchain. They contain immutable rules that trigger a certain logic when called by any node [30]. This feature takes over the requirement of a centralized authority executing and managing the application.

This work proposes a heuristic blockchain consensus algorithm based on reputation, which is defined as the trustworthiness of each agent given by the rest of the network [31] in a cyber physical context. The smart contract evaluates the behaviour of every agent by comparing the information shared with that expected from the consensus dynamics. The agent's reputation is bounded in a set and it is reduced if there is a mismatch greater than a specified threshold; otherwise, it increases. The goal of the algorithm is to decrease the impact of agents with a low reputation in the consensus result. In addition, agents communicate through the blockchain network to keep track of their status, hence avoiding an alteration of the information by a third party. The presented algorithm takes care of malicious or malfunctioning agents in cooperative cyberphysical systems, e.g., by overriding the impact of faulty sensors or attackers that participate in distributed state estimation processes such as environmental observation, position estimation, and temperature tracking. Other works that also combine blockchain and reputation are [32] and [33]. In contrast to the proposed work, these studies use reputation as a credibility indicator to decide which nodes are responsible of generating new valid blocks. In [32], there are four trusted states that correspond, depending on the reputation, to a higher or lower probability to be selected as a primary node; and, in [33], a two-stage security enhancement solution is given: in the first stage a reputation-based voting scheme is designed to ensure secure miner selection, whereas in the second stage the newly generated block is further verified and audited by standby miners.

The rest of the paper is organized as follows. Section 2 describes the mathematical formulation of the consensus problem. Section 3 presents the proposed reputation-based algorithm that operates using blockchain. Section 4 describes the blockchain implementation. Section 5 presents several simulations of different consensus algorithms and a comparison of their performance in different scenarios. Finally, conclusions are given in Sect. 6.

Notation The set of natural numbers is $\mathbb{N}$. The cardinality of a set $\mathcal{S}$ is denoted by $|\mathcal{S}|$. Given sets $\mathcal{S}_1, \mathcal{S}_2$, the reduction of $\mathcal{S}_1$ by $\mathcal{S}_2$ is denoted by $\mathcal{S}_1 \setminus \mathcal{S}_2 = \{x \in \mathcal{S}_1 : x \notin \mathcal{S}_2\}$.

2 Problem Formulation

Let us define a network of n >1 agents by the digraph $\mathcal{D} = (\mathcal{V}, \mathcal{E})$, where $\mathcal{V} = \{1, \ldots, n\}$ is the set of nodes or agents, and set $\mathcal{E} \subseteq \mathcal{V} \times \mathcal{V}$ contains directed connections (i, j) from node i to node j in $\mathcal{V}$. For each agent i, it is defined the group of *in-neighbours* by those agents with a directed edge towards i, i.e., $\mathcal{N}_i^+ := \{j \in \mathcal{V} \,|\, (k, i) \in \mathcal{E}\}$, and the group of *out-neighbours* as $\mathcal{N}_i^- := \{h \in \mathcal{V} \,|\, (i, h) \in \mathcal{E}\}$. A *path* is defined as a sequence of distinct vertices $i_0, i_1, \ldots, i_k$ such that $(i_l, i_{l+1}) \in \mathcal{E}, l = 0, 1, \ldots, k - 1$. The digraph $\mathcal{D}$ is said to be *strongly connected* if for every $i, j \in \mathcal{V}$, there exists a path starting at i and ending at j.

The following definitions apply for the network topologies considered in this work:

Definition 1 (r-Reachable, and (r,s)-Reachable Set [34]) Given a digraph $\mathcal{D}$ and a nonempty subset $\mathcal{S}$ of nodes of $\mathcal{D}$, $\mathcal{S}$ is an r-reachable set if it contains a node that has at least r in-neighbours outside of $\mathcal{S}$. $\mathcal{S}$ is an (r,s)-reachable set if there are at least s nodes in $\mathcal{S}$, each of which has at least r neighbors outside of $\mathcal{S}$. Hence, a (r,1)-reachable set is an r-reachable set.

Note that the reachability property is associated to set $\mathcal{S}$. The following definition extend the concept to the entire network.

Definition 2 (r-Robustness, and (r, s)-Robustness [34]) A nonempty, nontrivial digraph $\mathcal{D} = (\mathcal{V}, \mathcal{E})$ with n nodes ($n \geq 2$) is r-robust if for every pair of nonempty, disjoint subsets of $\mathcal{V}$, at least one of the sets is r-reachable. Defining $r, s \in \mathbb{N}$ such that $1 \leq s \leq n$, and the set $\mathcal{X}^r_{\mathcal{S}_m} = \{i \in \mathcal{S}_m : |\mathcal{N}_i^+ \setminus \mathcal{S}_m| \geq r\}$ for $m \in \{1, 2\}$, where $\mathcal{S}_1$ and $\mathcal{S}_2$ are nonempty, disjoint subsets of $\mathcal{V}$, the digraph $\mathcal{D}$ is (r, s)-robust if any of the following conditions holds: 1) $|\mathcal{X}^r_{\mathcal{S}_1}| = |\mathcal{S}_1|$; 2) $|\mathcal{X}^r_{\mathcal{S}_2}| = |\mathcal{S}_2|$; 3) $|\mathcal{X}^r_{\mathcal{S}_1}| + |\mathcal{X}^r_{\mathcal{S}_2}| \geq s$.

2.1 Weight-Based Consensus Algorithm

The weight-based consensus algorithm is defined next. Within this algorithm, each agent i updates its state as

$$x_i(k+1) = x_i(k) + \sum_{j \in N_i^+} \gamma_{ji}(x_j(k) - x_i(k)), \tag{1}$$

where $x_{\mathcal{D}}(k)$ represents the array of states for the digraph $\mathcal{D}$ at iteration k, i.e., $x(k) = [x_i(k)]_{i \in \mathcal{V}}$, and $\gamma_{ij} \in (0, 1)$ represents the updating weight, which satisfies $\sum_{j \in \mathcal{N}_i^+} \gamma_{ij} \leq 1$. In each iteration, the agent sends its state to the *out-neighbours* and receives the state from the *in-neighbours*.

The consensus dynamics of (1) can be modeled as

$$x_{\mathcal{D}}(k+1) = Ax_{\mathcal{D}}(k), \tag{2}$$

where A is the weight matrix and is defined as

$$A = \begin{bmatrix} 1 - \sum_{j \neq 1} \gamma_{j1} & \gamma_{21} & \dots & \gamma_{n1} \\ \gamma_{12} & 1 - \sum_{j \neq 2} \gamma_{j2} & \dots & \gamma_{n2} \\ \vdots & \vdots & & \vdots \\ \gamma_{1n} & \gamma_{2n} & \dots & 1 - \sum_{j \neq n} \gamma_{jn} \end{bmatrix}. \tag{3}$$

The objective is to reach a consensus by updating each $x_i(k)$ with the information available from its neighbours. Consensus is achieved once $x_i(k)$ converges for all $i \in \mathcal{V}$. However, if an agent j behaves maliciously, it may steer the whole consensus process away. For simplicity, let us consider additive attacks, so that the overall system dynamics of (2) become

$$x_{\mathcal{D}}(k+1) = Ax_{\mathcal{D}}(k) + Bu_{\mathcal{D}}(k), \tag{4}$$

where $u_{\mathcal{D}}(k)$ represents the array of perturbations for the digraph $\mathcal{D}$ at iteration k, i.e., $u_{\mathcal{D}}(k) = [u_i(k)]_{i \in \mathcal{V}}$, and $B \in \mathbb{R}^n$ is a column vector defined $\forall i \in \mathcal{V}$ as

$$B_i = \begin{cases} 0 & i \neq j, \\ 1 & i = j. \end{cases} \tag{5}$$

Example 1 Consider the four connected agents in Fig. 1. Setting $\gamma_{ji} = 0.2$ in (1) for all in-neighbours of $i \in \mathcal{V}$, we obtain

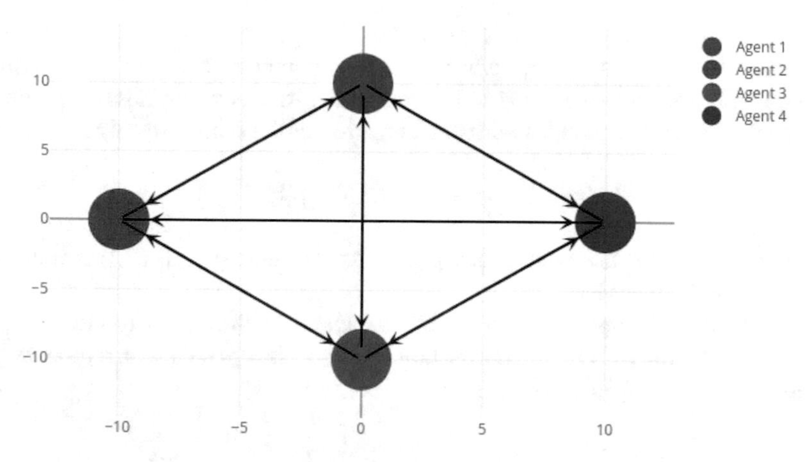

Fig. 1 Network with 4 nodes

$$A = \begin{bmatrix} 0.4 & 0.2 & 0.2 & 0.2 \\ 0.2 & 0.4 & 0.2 & 0.2 \\ 0.2 & 0.2 & 0.4 & 0.2 \\ 0.2 & 0.2 & 0.2 & 0.4 \end{bmatrix}, \quad \lambda^A = \begin{bmatrix} 1.0 \\ 0.2 \\ 0.2 \\ 0.2 \end{bmatrix}. \tag{6}$$

Convergence is attained because all eigenvalues (λ^A) are equal or lower than one. Assuming $x_{\mathcal{D}}(0) = \left[15,\ 5,\ 15,\ 5\right]^T$, agent 1 becomes an additive attacker with

$$u_1(k) = \begin{cases} 0 & k = 0, \\ 1 & k > 0, \end{cases} \tag{7}$$

it is observed that after 10 and 100 iterations we have

$$x_{\mathcal{D}}(10) = \begin{bmatrix} 13.18 \\ 11.93 \\ 11.93 \\ 11.93 \end{bmatrix}, \quad x_{\mathcal{D}}(100) = \begin{bmatrix} 35.68 \\ 34.43 \\ 34.43 \\ 34.43 \end{bmatrix}. \tag{8}$$

Notice that even after 100 iterations, there is no consensus. □

3 Proposed Reputation-Based Consensus Algorithm (RBCA)

The proposed reputation-based algorithm aims to mitigate the consequences derived from the presence of additive attackers. To this end, our proposal keeps track of the behaviour of every participant in the network verifying whether the consensus dynamics (2) is followed or has been altered, e.g., becoming (4). Therefore, a reputation value $r_i \in \{0, \overline{r}\}$, where $\overline{r}$ is a parameter that sets the maximum reputation, is assigned to every $j \in \mathcal{V}$ representing its trustworthiness. In particular, the agent's reputation changes with each iteration of the consensus algorithm as:

$$r_j(k+1) = \min(\overline{r}, \max(0, r_j(k) + \alpha)), \tag{9}$$

where $\alpha = \delta > 0$ if the agent j complies with the rules (i.e., Eq. (2) is followed), and $\alpha = \beta < 0$ otherwise.

The equation to update the state of the agents within each round using our algorithm is similar to (1), but the constant weight γ_j is replaced with a time-varying reputation-based weight γ_j^*, i.e.,

$$x_i(k+1) = x_i(k) + \sum_{j \in N_i^+} \gamma_j^*(k)(x_j(k) - x_i(k)), \tag{10}$$

$$\gamma_j^*(k) = \overline{\gamma_j} r_j(k)/\overline{r}, \tag{11}$$

where $\overline{\gamma_j}$ is the maximum weight that can be applied for each agent and is determined within each test. Consequently, the agent's best interest is to follow the rules to keep both reputation and weight as high as possible. For instance, in the network of Example 1, $\overline{\gamma} = 0.25$, and $\overline{r} = 100$.

The consensus dynamics of (10) can also be written following (2), but, in this case, the weight matrix A is time-dependent (for simplicity, we denote $A_k = A(k)$), i.e.,

$$x_{\mathcal{D}}(k+1) = A_k x_{\mathcal{D}}(k), \tag{12}$$

where

$$A_k = \begin{bmatrix} 1 - \sum_{j \neq 1} \gamma_j^*(k) & \gamma_2^*(k) & \dots & \gamma_n^*(k) \\ \gamma_1^*(k) & 1 - \sum_{j \neq 2} \gamma_j^*(k) & \dots & \gamma_n^*(k) \\ \vdots & \vdots & & \vdots \\ \gamma_1^*(k) & \gamma_2^*(k) & \dots & 1 - \sum_{j \neq n} \gamma_j^*(k) \end{bmatrix}.$$

Example 2 Consider Example 1, where $\gamma_j^*(0) = 0.2$ for $j \neq 1$, $r_j(0) = 100$ for $j \in \mathcal{V}$, $\overline{r} = 100$, and

$$A_k = \begin{bmatrix} 0.4 & 0.2 & 0.2 & 0.2 \\ \gamma_1^*(k) & 0.6 - \gamma_1^*(k) & 0.2 & 0.2 \\ \gamma_1^*(k) & 0.2 & 0.6 - \gamma_1^*(k) & 0.2 \\ \gamma_1^*(k) & 0.2 & 0.2 & 0.6 - \gamma_1^*(k) \end{bmatrix},$$

where agent 1 can be considered as a malicious agent, and its corresponding weight can change over time. Note that, while agent 1 follows the rules, $\gamma_1^*(0) = 0.2$ and matrix A_k remains constant and equal to matrix A of Example 1. Thus, convergence is obtained.

For example, considering additive attacks, the overall system dynamics of (12) becomes

$$x_{\mathcal{D}}(k+1) = A_k x_{\mathcal{D}}(k) + B u_{\mathcal{D}}(k). \tag{13}$$

Once agent 1 starts acting maliciously, its reputation $r_1^*(k)$ will be decreased each round (consider $\beta = -30$) until it becomes 0, and the other agents converge and attain consensus. The evolution of $\gamma_1^*(k)$ following (11), and the eigenvalues from reputation matrix A_k following (12) are shown below:

$$\gamma_1^*(0) = 0.2, \quad \lambda^{A_0} = [1.0,\ 0.2,\ 0.2,\ 0.2]^T$$
$$\gamma_1^*(1) = 0.14, \quad \lambda^{A_1} = [1.0,\ 0.26,\ 0.26,\ 0.26]^T$$
$$\gamma_1^*(2) = 0.08, \quad \lambda^{A_2} = [1.0,\ 0.32,\ 0.32,\ 0.32]^T$$
$$\gamma_1^*(3) = 0.02, \quad \lambda^{A_3} = [1.0,\ 0.4,\ 0.4,\ 0.4]^T$$
$$\gamma_1^*(4) = 0, \quad \lambda^{A_4} = [1.0,\ 0.4,\ 0.4,\ 0.4]^T.$$

Furthermore, the malicious agent modifies the consensus dynamics of the overall system, which becomes (13). The dynamics of each consensus round are analyzed below showing the influence of $u_{\mathcal{D}}(k)$ (for simplicity, $x_k = x_{\mathcal{D}}(k)$ and $u_k = u_{\mathcal{D}}(k)$):

$$\begin{aligned}
k = 0 &\rightarrow x_1 = A_0 x_0 + B u_0,\\
k = 1 &\rightarrow x_2 = A_1 A_0 x_0 + (A_1 + 1) B u_1,\\
k = 2 &\rightarrow x_3 = A_2 A_1 A_0 x_0 + (A_2 A_1 + A_2 + 1) B u_2,\\
k = 3 &\rightarrow x_4 = A_3 A_2 A_1 A_0 x_0 + (A_3 A_2 A_1\\
&\qquad + A_3 A_2 + A_3 + 1) B u_3,\\
k = 4 &\rightarrow x_5 = A_4 A_3 A_2 A_1 A_0 x_0 + (A_4 A_3 A_2 A_1\\
&\qquad + A_4 A_3 A_2 + A_4 A_3 + A_4 + 1) B u_4.
\end{aligned}$$

For $k > 4$, A_k stays unaltered since $\gamma_1^*(k) = 0$, i.e.:

$$\begin{aligned}
\forall k > 4 \rightarrow x_{k+1} =& A_4^{k-3} A_3 A_2 A_1 A_0 x_0 + (A_4^{k-3} A_3 A_2 A_1\\
&+ A_4^{k-3} A_3 A_2 + A_4^{k-3} A_3 + \sum_{i=1}^{k-3} A_4^i + 1) B u_k.
\end{aligned}$$

Finally, assuming $x_0 = \left[15, 5, 15, 5\right]^T$, two tests are done to analyze how sensitive the model is to u_k:

- If agent 1 complies with the consensus rules, the network state at round 8 of consensus is $x_{\mathcal{D}}(8) = \left[10.0, 10.0, 10.0, 10.0\right]^T$, meaning that agents have reached consensus as expected.
- If agent 1 misbehaves, the system dynamics are (13) and the following states are obtained:

$$x_{\mathcal{D}}(10) = x_{\mathcal{D}}(100) = \left[11.81\ 11.15\ 11.15\ 11.15\right]^T,$$

where agents 2 to 4 converge as expected, and agent 1's changes are not taken into account after its reputation drops to zero. The deviation of this solution compared to the one with no malicious agents is 1.15. Moreover, notice there is no need to reach 100 consensus rounds because results remain unaltered since agent 1 no longer affects the rest of the network. □

As shown in the previous example, even though the reputation of an agent decreases while it does not comply with the rules, its actions impact the consensus value. One possibility to eliminate the impact of the additive attacker is to run the algorithm for a second time because the malicious agent holds a null reputation and it will have a null weight during all the consensus process. Note that reputations across the network are saved over time and persist between consensus rounds.

4 Blockchain Implementation

This reputation-based consensus algorithm is executed within a blockchain network to provide full traceability and the ability to audit the process. This is achieved using Hyperledger Composer, a semi-private and permissioned blockchain where members have to be enrolled through a trusted membership service provider.

Hyperledger Composer grants a set of collaboration tools for building business network applications (BNAs) to create smart contracts.[1] The objective of this BNA is to provide a group of assets definitions, transactions, and queries that can be consumed from a client application to interact inside a blockchain network (see Fig. 2). Hence, an agent with the proper credentials can create an asset, interact with other agents in the network, and view the queries' responses. This blockchain is composed of three primary types of nodes:[2]

- Client node: used by the application to submit transactions to the network.
- Peer node: used for blockchain data synchronization across participants and nodes in the network. There are three sub-types of peer nodes:

[1] https://github.com/hyperledger-archives/composer.

[2] https://hyperledger-fabric.readthedocs.io/en/release-1.2/.

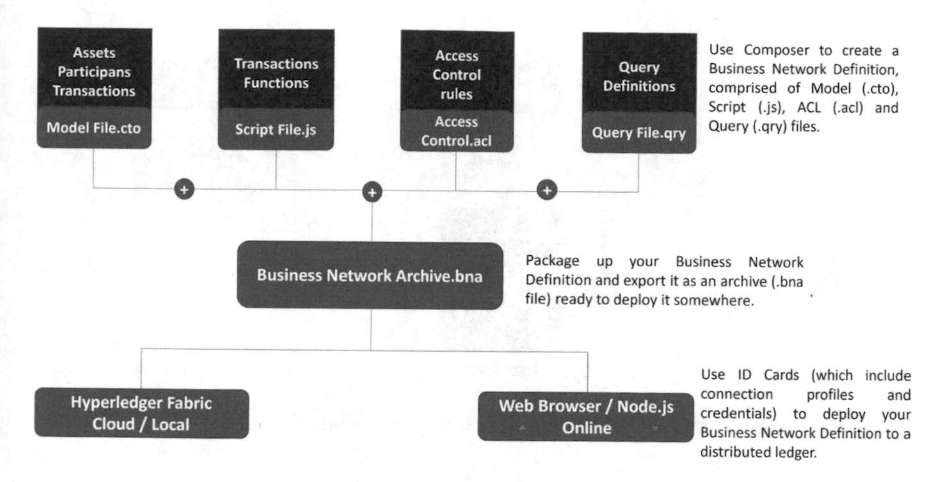

Fig. 2 Architecture of a business network application

  - General Peer: distributes information within the organization.
  - Anchor Peer: transactions submitted to the network are first sent to the anchor peers of each organization, which distribute them later to the general peers.
  - Endorser Peer: verifies transactions submitted by client nodes.

- Orderer node: maintains the state of the ledger across the organizations.

The deployed smart contract verifies whether each agent is following the consensus algorithm every time a measure is submitted. If there is an error greater than a specified threshold, the agent is either *lying* or malfunctioning, which leads to its reputation being reduced. On the other hand, if the measure is correct, the reputation is increased. Note that this step is run in a fully distributed fashion since the smart contract execution is carried out on the blockchain.

Figure 3 shows the process since a new transaction is submitted, e.g., a measure from a sensor is received, until the new block reaches every participant in the network:

1. The user invokes a transaction request through the web application connected to the network via the hyperledger software development kit or the representational state transfer server.
2. The application sends the transaction, e.g., the measure resulting from the reputation-based consensus algorithm, to an endorser peer.
3. This peer validates the certificates, simulates the chaincode, e.g., executes the smart contract code to update reputations, and retrieves the result.
4. If the transaction is approved, the application sends it to the orderer cluster for execution and block creation.
5. The orderer creates the block and sends it to the anchor peers on each organization, e.g., independent sensor clusters.

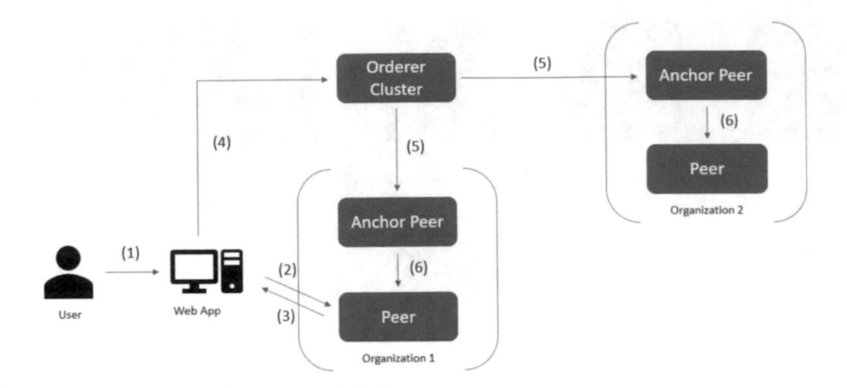

Fig. 3 Process for creation and distribution of a transaction within Hyperledger Fabric

6. Anchor peers broadcast the new block to all the peers inside the organization. All the peers are synced now, i.e., reputations are updated and available for each client.

5 Simulations

The benefits of using the blockchain reputation-based consensus are illustrated through numerical examples. Different scenarios are set up with differences in the network used, the malicious agents' behavior, and the type of consensus used.

Consider the network structures with seven agents of Figs. 4 and 5, which are tested from the initial state $x_0 = \left[15,\ 15,\ 5,\ 15,\ 5\ 15,\ 5\right]^T$, and $r_j(0) = 100$ for $j \in \mathcal{V}$. As can be seen, two different types of networks are studied:

- Network 1: strongly connected (Fig. 4).
- Network 2: (2,2)-robust (Fig. 5).

The participants must comply with the consensus rules to maintain their reputation high and reach an agreement, following (9) with $\beta = -30$ and $\delta = 5$. The maximum reputation $\overline{r}$ is set to 100.

The developed reputation-based algorithm is also compared via simulations with other two algorithms:

- The average consensus algorithm (ACA), where each agent calculates the average of its in-neighbours measures.
- The weight-based consensus algorithm (WBCA), where each agent computes a weighted average with $\gamma_{ji} = 0.20$ in (1) for all *in-neighbours* of $i \in \mathcal{V}$.

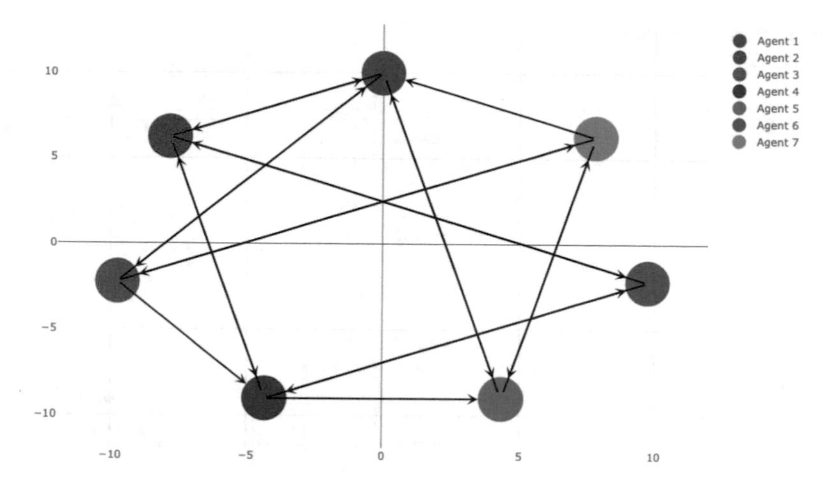

Fig. 4 Network 1: strongly connected graph with 7 agents

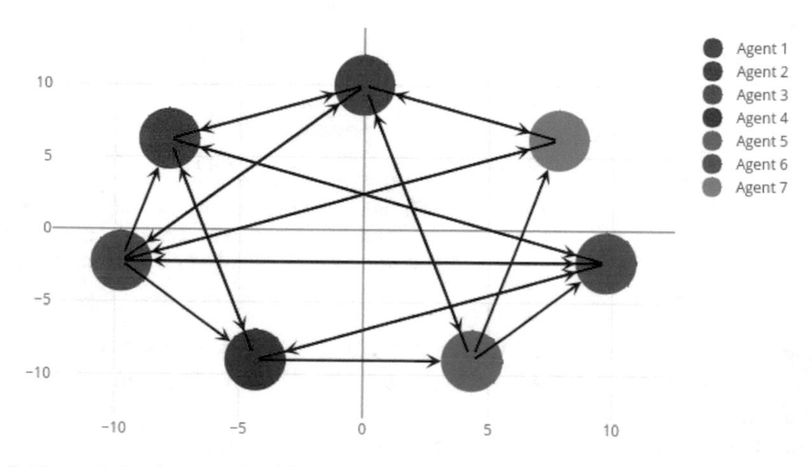

Fig. 5 Network 2: robust graph with 7 agents

Two different tests have been performed regarding agent 1's behaviour. For each test, six simulations are conducted to test the three consensus algorithms with each network. All results are summarized in Table 1 (Network 1) and Table 2 (Network 2).

5.1 Test 1: Agent 1 Shares Non-compliant Time-Varying States

In this test, agent 1 does not follow consensus rules switching its state every consensus round between 5 and 15. Simulations results with Network 1 are depicted in Figs. 6, 7, 8, and 9: Fig. 6 corresponds to the ACA, Fig. 7 to the WBCA, and

Table 1 Simulation results for Network 1

Test	Results			
	Consensus algorithm	Agreed value	No. of rounds	Error (%)
Test 1	ACA	–	–	–
	WBCA	–	–	–
	RBCA	8.755	71	4.976
Test 2	ACA	8.110	38	2.758
	WBCA	8.188	37	1.799
	RBCA	8.259	47	0.971

Table 2 Simulation results for Network 2

Test	Results			
	Consensus algorithm	Agreed Value	No. of Rounds	Error (%)
Test 1	ACA	–	–	–
	WBCA	–	–	–
	RBCA	9.212	18	9.955
		8.556[b]	20[b]	2.125[b]
Test 2	ACA	7.555	24	14.671
	WBCA	7.582	19	9.501
	RBCA	8.549	19	2.041
		8.409[b]	18[b]	0.370[b]

[b] The notation references the two times the algorithm was run. In the second attempt the rest of the network knows about the malicious agent

Figs. 8 and 9 to the proposed RBCA. In particular, Fig. 8 presents the state evolution of agents during the consensus process, and Fig. 9, the corresponding reputation evolution.

As can be seen, agents do not reach a consensus for both ACA and WBCA. For the former algorithm, the malicious agent rapidly affects the rest of the network due to its centrality, as is observed in Fig. 6. Similar behaviour is found using the WBCA, where agents that have the malicious agent as an in-neighbour present greater oscillations, as seen in Fig. 7.

In contrast with the previous consensus algorithms, the proposed RBCA is able to reach an agreement. Since agent 1 does not follow the rules, its reputation starts decrease, reaching the lower reputation limit by the fourth round of consensus (Fig. 9). This agent is able to manipulate the consensus for a short period; from round 5 until the end of consensus, the agents that have agent 1 as an in-neighbour ignore the measures received. The initial manipulation of the agent affected the final result, but it did not prevent the network from reaching consensus. Finally, note that once the reputation of agent 1 remains 0, the remaining agents attain consensus in 38 rounds.

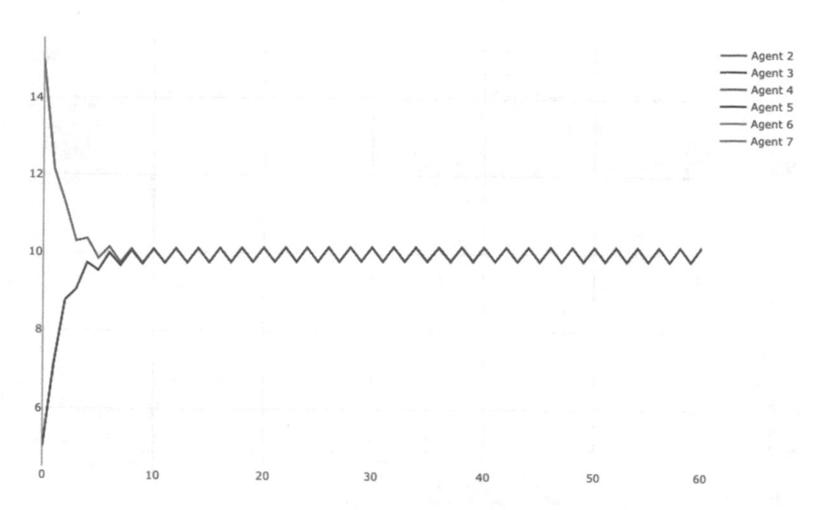

Fig. 6 Results of Test 1 using the ACA for Network 1

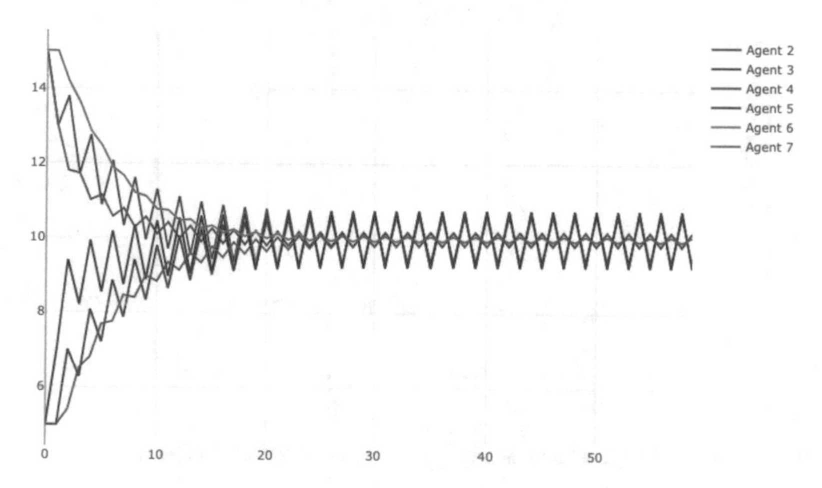

Fig. 7 Results of Test 1 using the WBCA for Network 1, γ_{ji}=0.20

5.2 Test 2: Agent 1 Shares Non-compliant Constant States

In this test, agent 1 malfunctions between the consensus round 2 to 15. During this period, its state remains constant, ignoring the inputs from its *in-neighbors*.

Simulation results with Network 1 are depicted in Figs. 10, 11, 12, and 13: Fig. 10 corresponds to the ACA, Fig. 11 to the WBCA, and Figs. 12 and 13 to the proposed RBCA. In particular, Fig. 12 presents the states evolution of agents during the consensus process, and Fig. 13, the corresponding reputation evolution.

Consensus is reached for every algorithm within this test with different performances. For the ACA, it is shown in Fig. 10 how the malfunction affects the overall

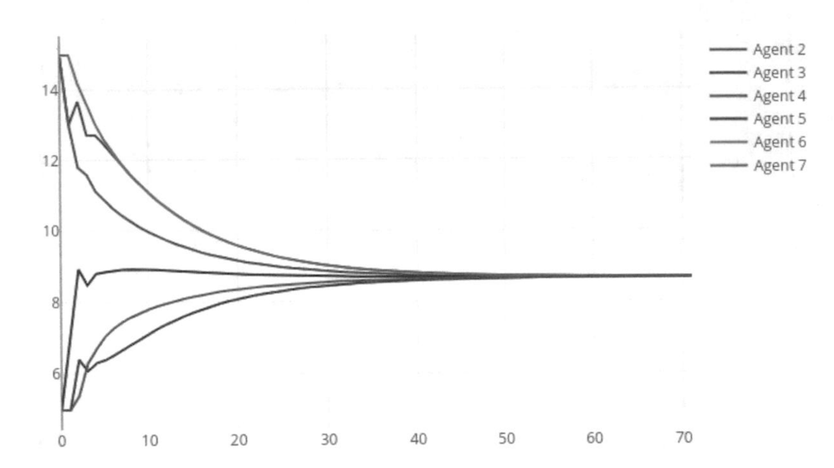

Fig. 8 Results of Test 1 using the RBCA for Network 1

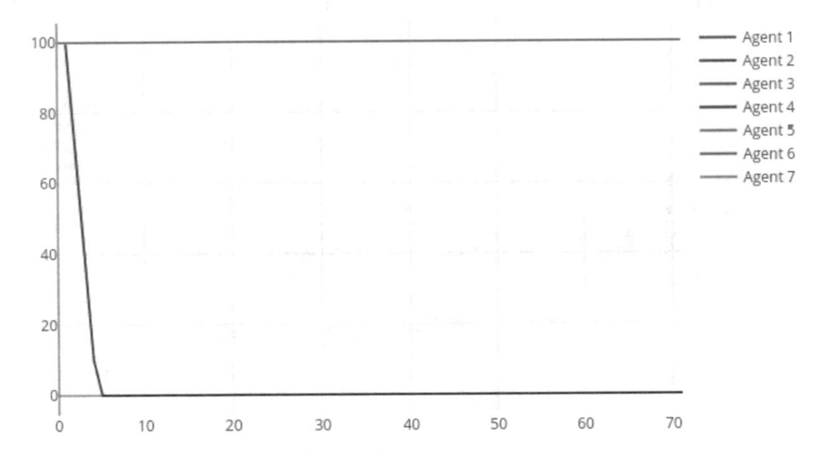

Fig. 9 Evolution of the agents' reputation of Test 1 using the RBCA for Network 1

consensus by bringing the agreed measure closer to the state in which the agent started malfunctioning. The same holds for the WBCA, where consensus is reached despite the malfunctioning period.

For the RBCA, the reputation of agent 1 decreases as long as it malfunctions, reaching the lower limit at round 7. Once it starts working as expected, its reputation increases until the higher limit is reached at round 36 (Fig. 13). Note that the impact that the malfunction generates in the agreement is much less severe than the observed in the other algorithms.

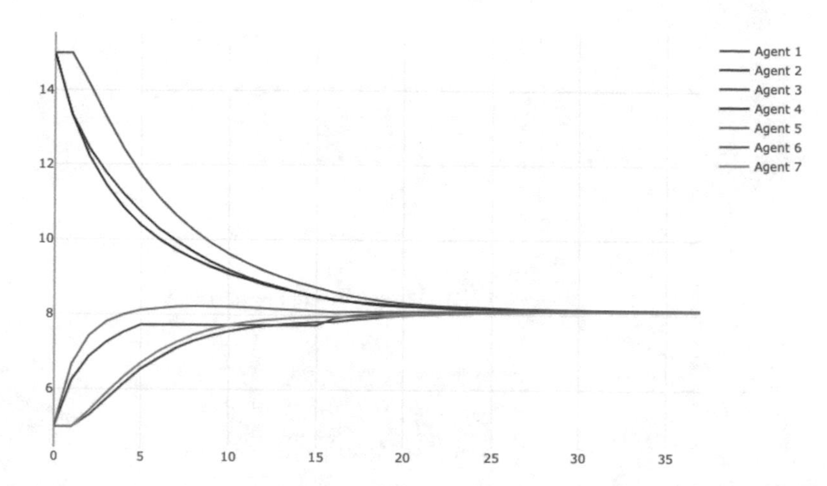

Fig. 10 Results of Test 2 where agent 1 malfunctions between rounds 2 to 15 using the ACA for Network 1

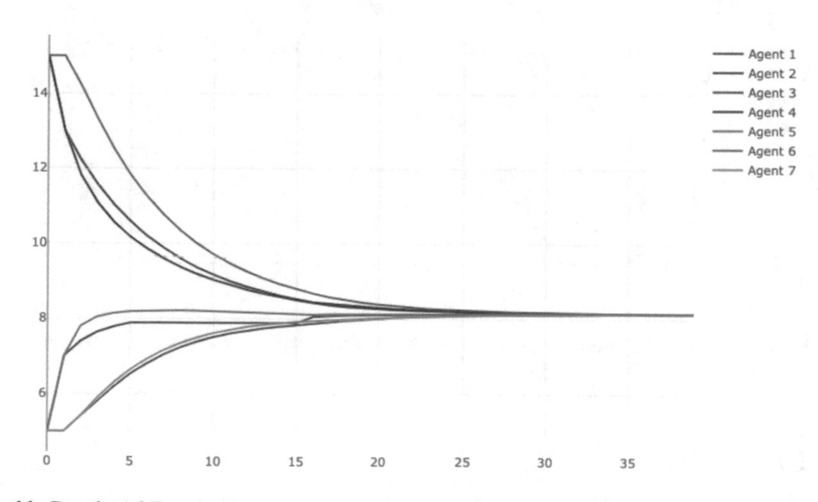

Fig. 11 Results of Test 2 where agent 1 malfunctions between rounds 2 to 15 using the WBCA for Network 1, $\gamma_{ji} = 0.20$

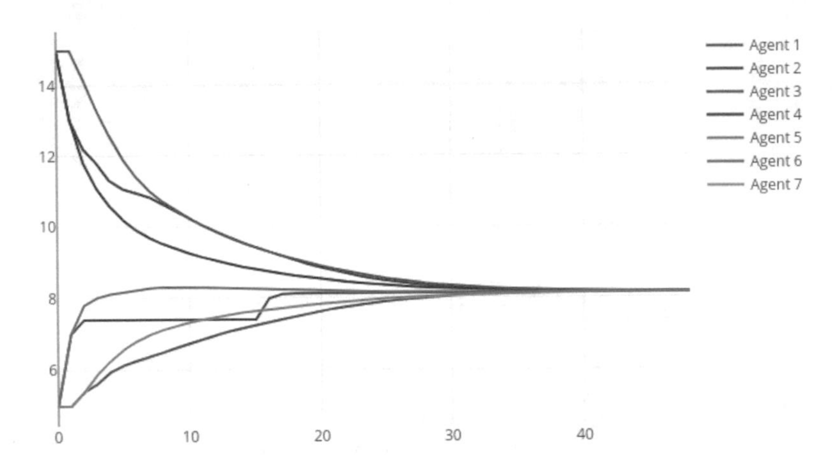

Fig. 12 Results of Test 2 where agent 1 malfunctions between rounds 2 to 15 using the RBCA for Network 1

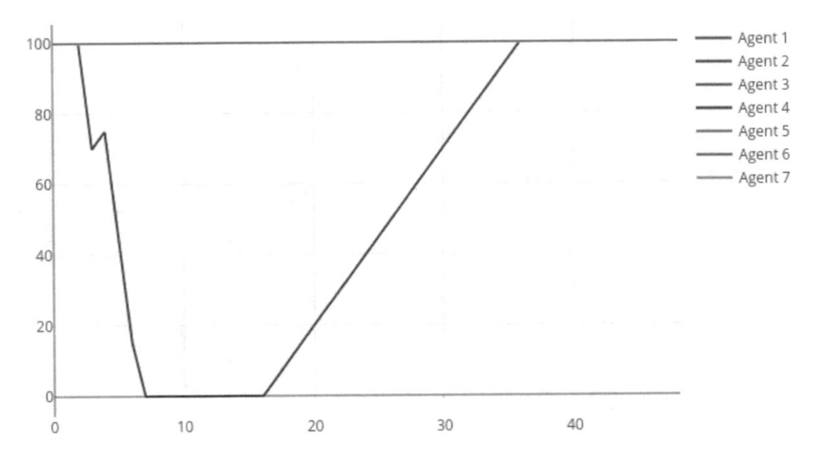

Fig. 13 Evolution of the agents' reputation of Test 2 where agent 1 malfunctions between rounds 2 to 15 using the RBCA for Network 1

6 Conclusions

This article proposes a reputation-based consensus algorithm that analyzes if the behaviour of the participants in the consensus is valid, i.e., if all agents follow the set of rules, and modifies the reputation linked to agents that violate them. These features are built in a private/hybrid blockchain network to take advantage of the security that it provides, among other benefits.

Thanks to the proposed reputation system, a consensus is always reached. In particular, when the malicious agent does not follow the rules, its reputation starts decreasing until zero after a few rounds. From that point, neighbours ignore the value received from that agent, continuing the consensus process without counting on it. In case of temporal malfunction, the impact on the final consensus value reached is less noticeable than when algorithms such as the weight-based consensus method are used. In addition, if the consensus were to be run twice, the effect would be even less perceptible.

Regarding future research, we plan to study the possibility of canceling an ongoing consensus process when a participant's reputation falls below a certain threshold to minimize its influence on the final consensus value. Also, it could be of interest using a public blockchain network, e.g, Ethereum, instead of a private/hybrid one to analyse its performance and complexity.

Acknowledgments This work has been funded by Ministerio de Ciencia e Innovación, Agencia Estatal de Investigación MCIN/AEI/ 10.13039/501100011033 under grant PID2019-104149RB-I00 (project SAFEMPC), the European Research Council Advanced Research Grant 769051-OCONTSOLAR, the Spanish MCE Project C3PO-R2D2 (PID2020-119476RB-I00), and the Spanish Training Program for Academic Staff (FPU19/00127).

References

1. R. Olfati-Saber, J. A. Fax, and R. M. Murray, "Consensus and cooperation in networked multi-agent systems," *Proceedings of the IEEE*, vol. 95, no. 1, pp. 215–233, 2007.
2. L. G. Mason, "Parallel and distributed computation—dimitri p. beertsekas and john," 1991.
3. R. O. Saber and R. M. Murray, "Consensus protocols for networks of dynamic agents," 2003.
4. R. Olfati-Saber and R. M. Murray, "Consensus problems in networks of agents with switching topology and time-delays," *IEEE Transactions on automatic control*, vol. 49, no. 9, pp. 1520–1533, 2004.
5. S. S. Kia, B. Van Scoy, J. Cortes, R. A. Freeman, K. M. Lynch, and S. Martinez, "Tutorial on dynamic average consensus: The problem, its applications, and the algorithms," *IEEE Control Systems Magazine*, vol. 39, no. 3, pp. 40–72, 2019.
6. S. M. Dibaji, H. Ishii, and R. Tempo, "Resilient randomized quantized consensus," *IEEE Transactions on Automatic Control*, vol. 63, no. 8, pp. 2508–2522, 2017.
7. K. Cai and H. Ishii, "Average consensus on general strongly connected digraphs," *Automatica*, vol. 48, no. 11, pp. 2750–2761, 2012.
8. W. Ren and R. W. Beard, *Distributed consensus in multi-vehicle cooperative control*, vol. 27. Springer, 2008.
9. Y. Cao, W. Yu, W. Ren, and G. Chen, "An overview of recent progress in the study of distributed multi-agent coordination," *IEEE Transactions on Industrial informatics*, vol. 9, no. 1, pp. 427–438, 2012.
10. W. Ren, R. W. Beard, and E. M. Atkins, "Information consensus in multivehicle cooperative control," *IEEE Control systems magazine*, vol. 27, no. 2, pp. 71–82, 2007.
11. Z. Lin, M. Broucke, and B. Francis, "Local control strategies for groups of mobile autonomous agents," *IEEE Transactions on automatic control*, vol. 49, no. 4, pp. 622–629, 2004.
12. V. D. Blondel, J. M. Hendrickx, A. Olshevsky, and J. N. Tsitsiklis, "Convergence in multiagent coordination, consensus, and flocking," in *Proceedings of the 44th IEEE Conference on Decision and Control*, pp. 2996–3000, IEEE, 2005.
13. D. Bauso, L. Giarre, and R. Pesenti, "Distributed consensus in networks of dynamic agents," in *Proceedings of the 44th IEEE Conference on Decision and Control*, pp. 7054–7059, IEEE, 2005.

14. J. Cortés, "Achieving coordination tasks in finite time via nonsmooth gradient flows," in *Proceedings of the 44th IEEE Conference on Decision and Control*, pp. 6376–6381, IEEE, 2005.
15. M. Mehyar, D. Spanos, J. Pongsajapan, S. H. Low, and R. M. Murray, "Distributed averaging on asynchronous communication networks," in *Proceedings of the 44th IEEE Conference on Decision and Control*, pp. 7446–7451, IEEE, 2005.
16. P.-A. Bliman and G. Ferrari-Trecate, "Average consensus problems in networks of agents with delayed communications," *Automatica*, vol. 44, no. 8, pp. 1985–1995, 2008.
17. L. Yuan and H. Ishii, "Asynchronous approximate byzantine consensus via multi-hop communication," in *2022 American Control Conference (ACC)*, pp. 755–760, IEEE, 2022.
18. S. M. Dibaji and H. Ishii, "Resilient consensus of second-order agent networks: Asynchronous update rules with delays," *Automatica*, vol. 81, pp. 123–132, 2017.
19. Y. Wang and H. Ishii, "Resilient consensus through event-based communication," *IEEE Transactions on Control of Network Systems*, vol. 7, no. 1, pp. 471–482, 2019.
20. S. Sundaram and B. Gharesifard, "Consensus-based distributed optimization with malicious nodes," in *2015 53rd Annual Allerton Conference on Communication, Control, and Computing (Allerton)*, pp. 244–249, IEEE, 2015.
21. T. Arauz, P. Chanfreut, and J. Maestre, "Cyber-security in networked and distributed model predictive control," *Annual Reviews in Control*, 2021.
22. T. Arauz, J. M. Maestre, R. Romagnoli, B. Sinopoli, and E. F. Camacho, "A linear programming approach to computing safe sets for software rejuvenation," *IEEE Control Systems Letters*, vol. 6, pp. 1214–1219, 2021.
23. P. Velarde, J. M. Maestre, H. Ishii, and R. R. Negenborn, "Scenario-based defense mechanism for distributed model predictive control," in *2017 IEEE 56th Annual Conference on Decision and Control (CDC)*, pp. 6171–6176, IEEE, 2017.
24. P. Chanfreut, J. M. Maestre, and H. Ishii, "Vulnerabilities in distributed model predictive control based on jacobi-gauss decomposition," in *2018 European Control Conference (ECC)*, pp. 2587–2592, IEEE, 2018.
25. T. Pierron, T. Árauz, J. M. Maestre, A. Cetinkaya, and C. S. Maniu, "Tree-based model predictive control for jamming attacks," in *2020 European Control Conference (ECC)*, pp. 948–953, IEEE, 2020.
26. T. Arauz, J. M. Maestre, A. Cetinkaya, and E. F. Camacho, "Model-based pi design for irrigation canals with faulty communication networks," in *2021 European Control Conference (ECC)*, pp. 1236–1242, IEEE, 2021.
27. P. Chanfreut, A. Sánchez-Amores, J. M. Maestre, and E. F. Camacho, "Distributed model predictive control based on dual decomposition with neural-network-based warm start," in *2021 European Control Conference (ECC)*, pp. 1969–1974, IEEE, 2021.
28. S. Nakamoto and A. Bitcoin, "A peer-to-peer electronic cash system," *Bitcoin.–URL:* https://bitcoin.org/bitcoin.pdf, vol. 4, 2008.
29. M. Nofer, P. Gomber, O. Hinz, and D. Schiereck, "Blockchain," *Business & Information Systems Engineering*, vol. 59, no. 3, pp. 183–187, 2017.
30. V. Buterin *et al.*, "Ethereum: A next-generation smart contract and decentralized application platform," 2014.
31. F. Gai, B. Wang, W. Deng, and W. Peng, "Proof of reputation: A reputation-based consensus protocol for peer-to-peer network," in *International Conference on Database Systems for Advanced Applications*, pp. 666–681, Springer, 2018.
32. K. Lei, Q. Zhang, L. Xu, and Z. Qi, "Reputation-based byzantine fault-tolerance for consortium blockchain," in *2018 IEEE 24th International Conference on Parallel and Distributed Systems (ICPADS)*, pp. 604–611, IEEE, 2018.
33. J. Kang, Z. Xiong, D. Niyato, D. Ye, D. I. Kim, and J. Zhao, "Toward secure blockchain-enabled internet of vehicles: Optimizing consensus management using reputation and contract theory," *IEEE Transactions on Vehicular Technology*, vol. 68, no. 3, pp. 2906–2920, 2019.
34. H. J. LeBlanc, H. Zhang, X. Koutsoukos, and S. Sundaram, "Resilient asymptotic consensus in robust networks," *IEEE Journal on Selected Areas in Communications*, vol. 31, no. 4, pp. 766–781, 2013.

AI and Blockchain for Cybersecurity in Cyber-Physical Systems: Challenges and Future Research Agenda

Kamini Girdhar, Chamkaur Singh, and Yogesh Kumar

1 Introduction

A Cyber-physical system (CPSs) means incorporating physical techniques into the real world and control software into the cyber-physical world. These two words are interchangeably used to share information [1]. The paths through which an attacker can invade the CPS increase as the CPS's connectivity grows and becomes more complex [2]. External attackers are particularly vulnerable to the networks that connect the physical systems and the control software that aims to penetrate the CPS and cause physical system malfunctions [3]. When an attacker gains access to a network, control-critical software can be disrupted. Artificial intelligence (AI) is increasingly being used in computer security. Maintaining physical security in the cyber-physical system is one of the most common issues in cyber-physical space [4]. Cyber-physical security protects devices, software, and networks from cyber-physical attacks [5–7]. The adversaries who carry out these attacks are primarily interested in modifying/accessing confidential information, laundering money from users, and disrupting normal business operations [8–10]. Various AI technologies detect network interference in vulnerability management analysis [11]. Deep learning is a well-known technique for detecting network intrusions. To detect network anomalies, several researchers have used machine learning methods such as convolutional neural networks [12] and help vector machines [13]. Hence, to ensure sensor security, some physical authentication methodologies are needed. The authors in [5] analyzed the side effects of threats on the actuators. The primary two

K. Girdhar · C. Singh
Chandigarh Group of Colleges, Mohali, Punjab, India
e-mail: kamini.3651@cgc.edu.in; chamkaur.4032@cgc.edu.in

Y. Kumar (✉)
Department of CSE, School of Technology, Pandit Deendayal Energy University, Gandhinagar, Gujarat, India

Y. Maleh et al. (eds.), *Blockchain for Cybersecurity in Cyber-Physical Systems*, Advances in Information Security 102, https://doi.org/10.1007/978-3-031-25506-9_10

attacks are a) Finite Energy Attack, related to the loss and alteration of packets b) Finite-Time Attack, also known as Bounded-Attack, which causes the suppression of control signals. The actuator's security control is related to the passive or active mode of operation where no action is taken without proper procedures. Authors in [14] have discussed distributed attacks on various computing resources, including Trojans, Viruses, Worms, and DoS attacks. The latest security advancement is essential to study the in-depth protection of CPS.

In contrast to traditional cyber-physical security, Cyber-physical security is an extension that considers the physical components [15]. Even though ongoing study on blockchain cyber security in the literature indicates that IoT security could be reinvented if it is supported by blockchain technology, security in IoT networks has been claimed as a pressing need of the industry and has received the highest priority for improvement and enforcement [16]. It is vital to understand how opponents react to them for designers of security classification. This is vital to the testing community in determining better ways to predict deployment effectiveness [17].

Working with the physical components leads to security issues contributing to the study of the CPS system [18–21].

- The widespread cyber-attacks and vulnerabilities to IoT devices.
- Modeling the security threats
- Designing the fault-tolerant system for the prevention of cyber and physical attacks.

However, prior research on AI-based cyber-security systems methods has made limited attempts to encapsulate actual knowledge by utilizing literature reviews holistically [22, 23]. For example, the authors in the study [24] have discussed the approach based on machine learning to detect the application-layer CPS attack. The authors have also disclosed the pattern-based model of graph-based segmentation and dynamic programming [25]. Research done by authors in [26] has also explored the machine learning approach for CPS attack surfaces. That analysis also provides a natural way for reasoning attack & threat models. Another research done by authors in [27–29] has discussed machine learning methods for biological data that are further integrated for detecting Cyber-Physical attacks in cyber manufacturing systems (CMS). The experts in the study [6] have analyzed strategies for three significant cyber-physical security issues: interruption detection, malware investigation, and spam detection. The authors have investigated a few issues that impact the application of ML to cybersecurity. We relate to this need by conducting a detailed survey on implementing AI methods in a cyber-physical system [30]. This conceptual study may help summarize existing information in a field of research and enable the detection of emerging information gaps and, as a result, future research directions [19].

Therefore, technological development must be initiated to achieve security, privacy, integrity, and confidentiality in the CPS [31–34]. The aim of this chapter is an existing review of artificial intelligence-based cyber-security systems and solutions, which various authors have proposed for the different types of cyber-attacks. This research covers security-related matters due to the cyber-physical system and discusses inherent security challenges in cyber-physical systems that attackers can potentially exploit [35–39]. In this chapter, we shortlisted documents published by Springer, IEEE Access, Elsevier, and ACM between 2009 and 2021 to classify the detection and prevention of cyber-physical attacks.

The rest of this chapter is organized into different sections. Section 2 describes the methodologies, highlighting the selection criteria for choosing the chapters to conduct the extensive survey. Section 3 includesthe framework of AI in CPS security, along with the network and threat model. Moreover, Sect. 4 is devoted to the methodologies of AI in the CPS system. Section 5 presents the comparative analysis from different research chapters on Cyber-Physical systems. In Sect. 6, a future perspective is mentioned, and finally, the conclusion of the CPS security is given in Sect. 7.

2 Research Methodology

This section presents the methodology considered to conduct the literature survey. In this study, we have applied various methods of a systematic literature review followed by existing studies on AI-based processes for the CPS systems. In Fig. 1, we describe our survey procedure in the following three steps: preparation, execution, and reporting. The research database sources are used to address questions RQ1, RQ2, RQ3, and RQ4. Some of the RQs and their objectives for the literature review are listed in Table 1. EBSCO, PubMed, Scopus, and other publications draw on all four databases.

In this chapter, we used Web of Science to develop this survey for a broad spectrum of articles. This article was based on research using specific inclusion and exclusion criteria [40]. We have emphasized to make this review paper attractive and meaningful, the research papers of 2009 to 2021 are included. In this paper, we have downloaded two hundred eighty-nine research chapters from databases like Scopus,

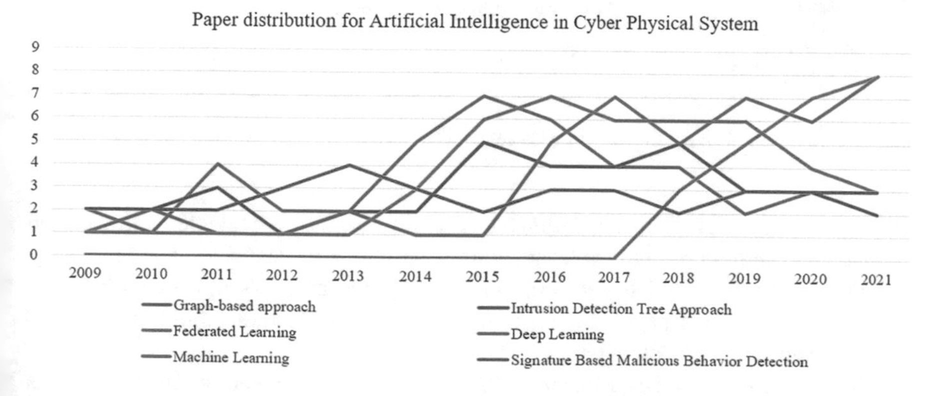

Fig. 1 Distribution of papers reviewed for AI in the cyber-physical system

Table 1 Research questions and their objectives

Q.No Identified research questions	Objective
RQ.1 Which AI-based approaches have been used extensively in cyber-physical systems?	It aims to design the security framework of AI to identify vulnerabilities
RQ.2 What are the different types of cyber security attacks, and what is the existing dataset used to mitigate from attack?	It targets to explore the security attacks and identify the existing datasets
RQ.3 What are the different methods used to measure the performance of the cyber-physical system?	It targets to provide information on AI methods for cyber security systems
RQ.4 Discuss the comparative analysis of various AI-based Models and future prospective?	It aims to compare federated, deep, and machine learning models in the secure Cyber-Physical system

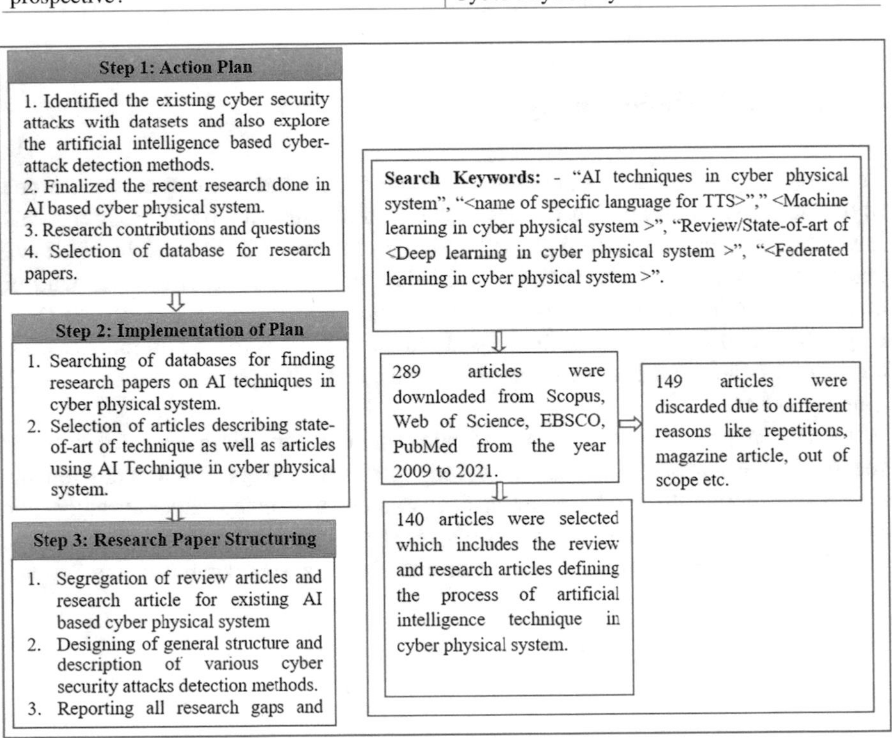

Fig. 2 The mechanism for the literature survey

web of science, EBSCO, and PubMed. The filtering of the selected information is based on parameters like irrelevant, out-of-scope, and magazine papers. Finally, we identified articles related to artificial intelligence methods in a cyber-security system [41–43]. This paper describes the research on Artificial Intelligence (AI) based methodologies with cyber-physical techniques.

The database search keywords with their finding are presented in Fig. 2. Whereas Table 1 shows that four suitable keyword options are available — "Artificial

Table 2 Database search keywords

Database	Keywords	Total hits appeared	Abstract reads	Full text down-loaded
EBSCO	"Artificial Intelligence Technique in Cyber-Physical System"	41	41	27
	"Blockchain in Cyber-Physical System"	31	37	25
	"Deep Learning in Cyber-Physical System"	41	41	19
	"Federated Learning in Cyber-Physical System"	47	47	11
PubMed	"Artificial Intelligence Technique in Cyber-Physical System"	51	51	24
	"Machine Learning in Cyber-Physical System"	41	41	23
	"Deep Learning in Cyber-Physical System"	46	46	21
	"Federated Learning in Cyber-Physical System"	41	41	11
Scopus	"Artificial Intelligence Technique in Cyber-Physical System"	73	73	22
	"Machine Learning in Cyber-Physical System"	37	37	15
	"Blockchain-based security for Cyberphysical systems "	51	51	18
	"Federated Learning in Cyber-Physical System"	45	45	10
Web of Science	"Blockchain Technique in Cyber-Physical System"	49	49	20
	"Machine Learning in Cyber-Physical System"	38	38	17
	"Deep Learning in Cyber-Physical System"	45	45	16
	"Federated Learning in Cyber-Physical System"	43	43	10

Intelligence Technique in Cyber-Physical System ","Machine Learning in Cyber-Physical System ","Deep Learning in Cyber-Physical System ","Federated Learning in Cyber-Physical System ", (see Table 2). The steps involved in the survey are shown in the diagram in Fig. 3.

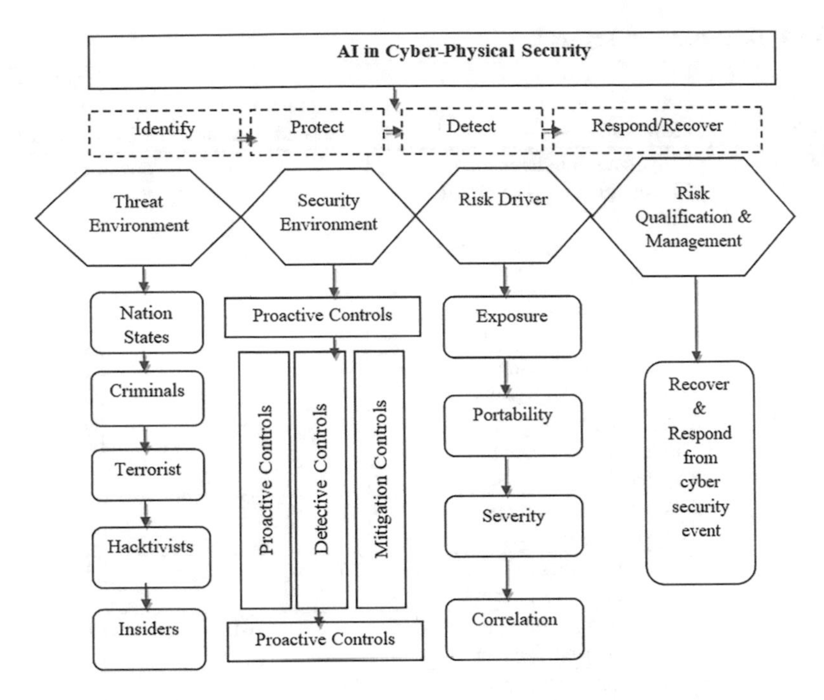

Fig. 3 AI framework of cybersecurity

3 Background Details

This section explored different techniques and methods for modeling threats, Human-**machine** teaming protection, domain vulnerabilities, and security resources. We have also presented the structured framework for artificial intelligence in cybersecurity: identification, protection, detection, response, and recovery. We have also mentioned the types of attacks along with Blockchain for AI data security-based applications. We highlight types of datasets, their format and tags, and their year of origin. We will also discuss the reasons for writing this chapter and the aims and priorities.

3.1 AI-Framework for Cyber-Physical Security

Due to increased knowledge of how susceptible AI segments are to malicious activity, worries about the CPS integrity of the entire information-handling pipeline in which AI segments are deployed have been addressed. The various application of the CPS system is described below.

- Industrial and political campaigns
- Smart Grid and its services

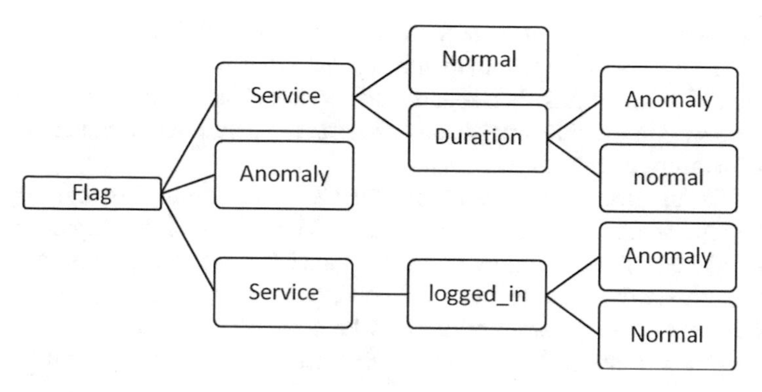

Fig. 4 Intrusion detection tree approach

- Transportation system
- Healthcare and ambient assisted living

Multiple applications can be affected because of secret dependencies in the pipeline. When using AI as a part of a system, research is required to develop a strategy, engineering principles, and industry standards [44]. Threat modeling, security techniques, domain weaknesses, and acquiring human-machine learning should be included. These models must allow for iterative concepts of attacks and improvements, be built with the help of an AI professional, and take into account data availability and integrity, access controls, network orchestration and activity, conflict resolution, privacy, and a complex policy setting [45–47]. Engineering concepts should be focused on science, group experience, and AI component functionality study that involve redundancy and other mechanisms to make AI-enabled systems more trustworthy. Understanding the environments, risks, territories, and constraints is an essential but secondary objective. Traditional CPS structure and robust system design may reduce the impact [48, 49], allow for more redundancy and diversity to be built in. and explore domain-specific countermeasures, bounds, and protection defaults [50].

The primary purpose of identification in AI is to develop an organizational way to control cybersecurity risks associated with CPS systems, people, assets, data, and capabilities [51–54]. In business, various resources are used to support risk-related functions and maintain consistent operations to support multiple business needs and risk management strategies associated with cybersecurity risks management to focus and define its efforts [44]. As shown in Fig. 4, the first stage in the AI framework is identifying vulnerabilities applied in various fields, including CBT systems [46], intelligent production, and grid systems. The other stage, known as protection, describes the necessary vital points to ensure the mode of delivery for infrastructure-related services. To support the protection function, it must retain & limit the ability for potential cybersecurity events. To keep the detection & discovery of various cybersecurity time-based events and check how it is responsive to action-related events for the cybersecurity attack detections. Different action plans are required in the response function, identifying activities for recovering cybersecurity incidents and storing unused & abandoned services. The recovery function is also paving the way to provide timely and recoverable operations to reduce the impact of the cybersecurity incident.

The CPS architecture mainly covers the Cyber layer and Physical Layer. The variables represent the data and control signals. The set point is the expected value of certain variables. The controllers find the distance between the process and control unit. After this, offset is calculated for solving complex operations. Furthermore, controllers and operators are used in the CPS system to send the measurements to control central systems and to be aware of the current status of the controlled objects.

3.2 Network, Threat, and Design Goal of CPS

A. **Network Model**

The vital advantage of the network model is that it provides a trusted data network for storing the information generated on the internet. This information includes the identities of CPS devices. Cyber-physical users are actors who directly interact with the devices. These systems help retrieve the data from the trusted networks to perform the cryptographic identities connected with the trusted model. CPS devices have the capabilities of computations and cryptographic identities using blockchain, so communication becomes easy with sending and receiving data signals.

B. **Threat Model**

A cyber-physical attack is a violent and transparent attempt by an organization or entity to breach the information system of another person or organization. Eavesdropping involves monitoring non-secure CPS network traffic to gain sensitive data. Denial of Service attacks (DDOS) are distributed system attacks targeting cyber-physical resources. They are frequently carried out by Botnets, which consist of a large number of infected devices that are hijacked by DDOS attacks. Another cyber threat effect is confidentiality required for the security of private information. The next threat model is integrity, where data can not alter without proper permission. The availability means the resource available in case of a system and another hardware failure. The non-repudiation action would have to occur at the time interval. Accountability is where an entity is responsible for its work.

Ransomware is a kind of pernicious programming (malware) that takes steps to distribute or impede admittance to information or a PC framework, typically by encoding it, until the casualty pays the expense to the assailant. Such security dangers will have the most wrecking impacts as autonomous vehicles, robots, and PC frameworks can be hacked by artificial consciousness. Artificial intelligence can likewise close down power supplies, public safety frameworks, and clinics' safeguards for trust systems.

C. **Design Goals**

One of the most prominent approaches is decentralization to achieve effectiveness and rapid decision-making in real-world applications. The decentralization of networks is necessary to maintain the security-related issues of CPS. A CPS network should no longer rely on a centralized entity because it may cause performance issues. In CPS distributed network environment, the entities must be unable to estimate each other's transactions. Therefore, uncertainty is highly required in distributed CPS, where privacy is a matter of concern.

The authentication process is a critical goal in real-world cyber-physical systems where unauthorized access can quickly enter bogus data into a CPS device. For this, the blockchain-based system must assure the authenticity of the data from the trusted network. Identity management can help track and trace device information and status in a legitimate CPS network. Consequently, for the Ethereum-based cyber-physical things, secure authentication methods are a must (CPTS). This process guarantees that neither a CPS, blockchain nor an actor can deny any activity they have taken. CPS devices typically transmit data daily, so to avoid this, a method must be in place to verify that data received from a CPS device is accurate.

4 Reported Work

This section describes the various artificial intelligence approaches for Cyber-Physical systems. We have also represented the comparative study of different AI-based techniques such as graph-based approach, intrusion detection tree approach, other procedures and methods of deep learning, Blockchain, machine learning, and federated learning. Finally, the results obtained from different techniques have been demonstrated through equations and algorithms.

4.1 Graph-Based Approach

The graph-based approach may be a multi-graph representation of the information with middles compared to CPS objects or concepts and edges interfacing concepts that share similitude. The chart regularly contains both named information and unlabelled information. HTTP requests are sent to a web server by a client. Costa et al. [7] presented a graph-based system that uses a set of standard expressions that represent typical web requests sent by clients to a web application. In this case, the map G = (V, E) is an undirected chart with vertices vi V and edges (vi, vj) E connecting the adjacent vertices. The vertices in the form correspond to the HTTP message (HTTP Ask sort, URL, parameters). An example of what happens after an HTTP GET request (given by Eq. 1) is as follows:

$$GEThttp : //URL.addressparam1 = value1 \wedge param2 = value \quad (1)$$

A non-negative degree of difference between vertices $\boldsymbol{vi}$ and $\boldsymbol{vj}$ is required for each edge (vi, vj) E. The difference is known as the edge weight and is written as $\boldsymbol{w(vi, vj)}$. The problem of constructing a set of standard expressions for displaying an ordinary HTTP request is formalized as chart division. Vertices comparable to each other are permitted to be allocated to the same component is given in Eq. 2.

$$Ci \in S = (C1, \ldots., Ck) \quad (2)$$

The number k denotes the total number of components. Each Ci component is given a standard expression at the end of the division strategy in Eq. 2 [27]. The chart division algorithm employs a technique comparable to the calculation shown as Eq. 3.

$$G = (v, e) \quad (3)$$

A map is used in the calculation using $\boldsymbol{n}$ vertices and $\boldsymbol{m}$ edges as input and outputs. A segmentation component is defined in Eq. 4.

$$S = (C1, \ldots, Cr) \quad (4)$$

Algorithm 1: Implementation Steps for the Graph-Based Approach

1. for each (vi, vj) a ? E calculate edge weights w (dissimilarity between vertices vi and vj).
2. Arrange edges ascending according to their weights w values.
3. Begin with segmentation $S0$, where each vertex v is assigned to its component.
4. Iterate over the sorted set of edges for $q = 1, \ldots., m$ and perform the following steps:
   - Let $Cq - 1i$ be the component of $Sq - 1$ containing and $Cq - 1j$ be the component of $Sq - 1$ containing j.
   - Assign $Sq = Sq - 1$
   - Merge $Cq - 1i$ and $Sq - 1$
   - Segments whenever dissimilarity between them falls within the predefined threshold and update Sq accordingly.
5. Return Sm as a segmentation result S.

The above algorithm presented a graph-based system that uses a set of standard expressions that represent typical internet requests sent by clients to a web application. In this case, the map $\mathbf{G} = (\mathbf{V}, \mathbf{E})$ is an undirected chart with vertices vi V and edges (vi, vj) E connecting the adjacent vertices. The vertices in the form correspond to the HTTP message (HTTP Ask sort, URL, parameters). Dixitl et al. [10] proposed a graph-based method for defining a common behavior in connected

CPS devies, IoT devices. The source IP and the destination IP are also called nodes. The operation stream between the source and destination IP addresses is an edge between the hubs given in Eq. 5.

$$\boldsymbol{Gs} = \{\boldsymbol{Gi}, \boldsymbol{Gi} + \mathbf{1}, \ldots., \boldsymbol{Gt} > +\mathbf{1}, \ldots.\} \tag{5}$$

Let be a chart stream where each Gi indicates a chart at each one minuscule short-term. The inventor has considered a graph as given in Eq. 6.

$$\boldsymbol{Gi} = (\boldsymbol{v}, \boldsymbol{e}, \boldsymbol{f}) \tag{6}$$

as a generic undirected heterogeneous graph is given in Eq. 7.

$$\exists \boldsymbol{v} \in \boldsymbol{A}\,(\boldsymbol{x}, \boldsymbol{y}) \in \boldsymbol{A} \wedge \boldsymbol{x} = \boldsymbol{y}\ \boldsymbol{and} \forall \boldsymbol{e} \in (\boldsymbol{x}, \boldsymbol{y}) \tag{7}$$

An edge goes from vertex x to vertex y and (x, y) is an unordered match. At whatever point a modern graph Gi arrives within the stream, a biased-random walk of a settled length l is performed from each hub in Gi, extricating a walkway p is defined as Eq. 8.

$$\boldsymbol{Gi} = \{\boldsymbol{v1}, \boldsymbol{v2}, \ldots., \boldsymbol{vl}\} \tag{8}$$

Here, the n-shingles are then constructed from a walking path. Nguyen et al. [30] also have proposed a lightweight strategy for recognizing IoT botnet attacks for the cyber guards, which is based on extricating high-level highlights from function–call charts, called PSI-Graph, for each executable record. The researchers recognized the PSIs as a characteristic of a bot's life cycle and generated a PSI-Graph chart. PSI-Graph attempts to illustrate the life cycle of an IoT botnet through the interconnections between such PSIs, which represent the activities at any level of a botnet's life cycle [30]. This highlights the adequacy when managing the multi-architecture issue while maintaining a strategic distance from the complexity of the control stream chart examination utilized by most existing strategies.

4.2 *Intrusion Detection Tree Approach*

Intrusion Detection Tree Approach is the foremost practical framework that can handle the interruptions of the computer environments by activating alarms to create the investigators take activities to close down this interruption on the CPS network. Sarker et al. [41] suggested an Intrusion Finding Tree approach for Cyber-Physical attack detection. When the security highlights are ready, the developer has created a tree-like display to enable users to develop a data-driven, intelligent decision interruption discovery environment. Rather than using all of the security highlights available within the provided dataset, the authors consider the security highlights, calculated by their significance score and ranking. Experts began with a root hub to plan a tree-like demonstration. It incrementally generates a related tree department by breaking down a given preparing dataset into smaller subsets. Hao et al. [48] have also defined the "Gini Index" as used to discern the property for the root hub in each step. Quality with a lower Gini list is chosen [37]. As a result, the authors have increased the size of the tree by adding the necessary number of branches, which are made up of inside and leaf hubs, as well as their interfacing edges or curves [37]. In this study, either tree leaf is named with a target path, irregularity, or type, and each inside hub is detailed with the security highlights chosen previously [37–39]. Each of the possible values of the related function is called on the curves of the tree that emerge from a hub. The result may be a multi-tier tree with several leaf hubs that imparts inconsistency or standard [41]. The intrusion detection shows IntruDTree considers only two properties:

Reducing the highlight measurements by deciding the include significance and positioning, reducing highlight measurements by agreeing on the importance and placement of the highlights, and building a multi-level tree with the chosen imperative highlights in mind. In a given interruption dataset, an example of an IntruDtree considers a few highlights such as f slack, gain, duration, logged in, and their test values. Calculation lays out the general procedure for constructing an anIntruDTree as defined in Eq. 9.

$$\boldsymbol{DS = X1, X2, \ldots., Xm} \tag{9}$$

Where m is the information estimate, given a preparing interruption dataset. N-dimensional highlights speak to each occasion. Srivastava al [48]. have provided different cyber-attack groups CA = normal. An anomaly has a position in preparing information as well. For example, an illustration running the display with a single highlight could be "if the counter value is RSTR, then the result is an abnormality. "The result is an IntruDTree, which may be a DS-related rule-based classification tree [48]. For example, an illustration running the display with a single highlight could be "if the flag value is RSTR, then the result is an anomaly." Also, "if hail esteem is SF, the gain is FTP, and term = 4, then the result is an anomaly" could be another run the show with different highlights [42, 43]. As a result, by navigating the generated IntruDTree (in Fig. 5) several security rules can be extracted, which can be used to decide if the test case is typical or unique.

4.3 *Federated Learning for Cyberattack*

This section gives a survey of the commitments made by distinctive analysts within the field of federated learning in cyber-Physical security in conjunction with a comparative investigation based on assaults, the strategy utilized, and the related challenges in the secure planning stage. Malomo et al. [27] have discussed an open presentation to the common thought of unified learning, conjointly proposed a few conceivable applications in 5G systems, and depict key specialized challenges and

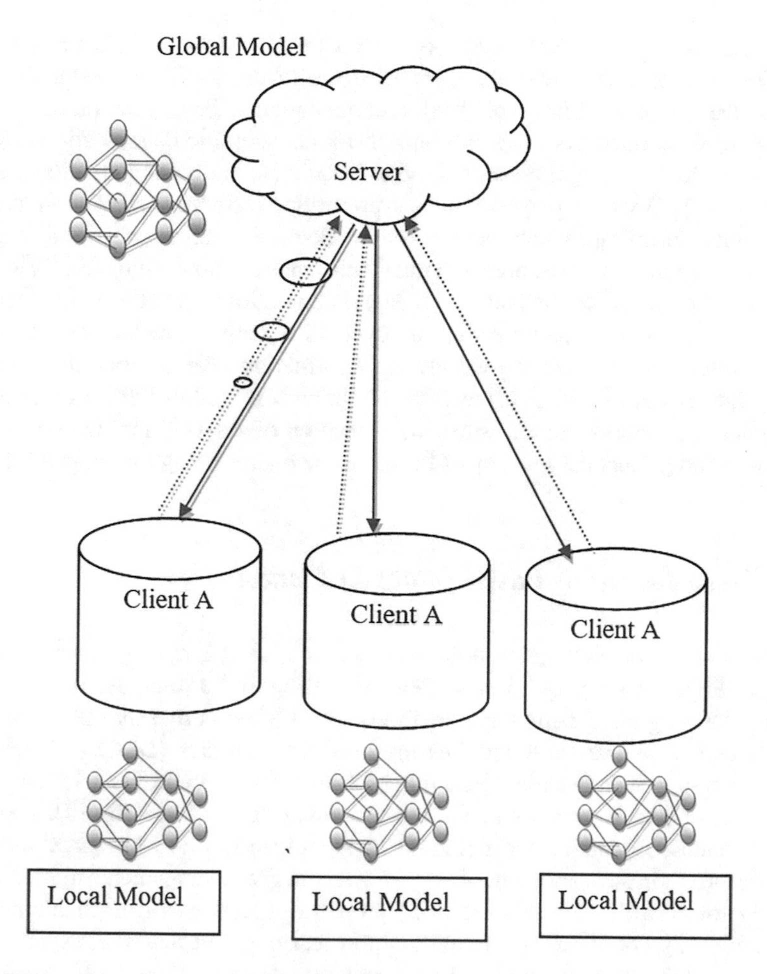

Fig. 5 Federated learning model for data security

open issues for future inquiries about unified learning within the setting of wireless communications [13]. Afterward, Liuet al [25]. proposed to expand combined learning with nearby representation learning on each gadget to memorize valuable and compact representations from crude information.

Panigrahi et al. [31] found it successful against nearby models, which adapt to managing heterogeneous information and can be adjusted to memorize reasonable representations that muddle traits such as race, age, and sexual orientation. Chhetri et al. [6] have also explored and actualized the combined learning system as a profound learning system for privacy-preserving and parallel preparation with physical devices. Singh et al. [54] presented a novel system to execute disseminated federated learning (FL) calculations inside an unmanned airborne vehicle (UAV) swarm that comprises a driving unmanned ethereal vehicle (UAV) and a few taking after unmanned airborne vehicle (UAVs) [55].

The fetched of the show may be the misfortune work esteem or recreation mistake, such that the lower the esteem of the taken a toll, the better the show occurrence obtained. Liu et al. [24] recommended a brief presentation on the concept of Combined Learning and one-of-a-kind scientific classification covering risk models and two major assaults, such as harming assaults and induction assaults. Geetha et al. [12] also displayed a privacy-preserving restorative NER strategy based on federated learning, which utilized the labeled information in several stages to boost the preparation of restorative named substance acknowledgment (NER).

Figure 5 outlines the fundamental design and relationship between the local and global model of a federated learning life cycle [30]. Bolts show that, as it were, the totaled weights are sent to the worldwide information lake, as contradicted to the nearby information itself, as is the case in routine ML models [29]. As a result, FL makes it conceivable to realize superior utilization of assets, minimize information exchange and protect the security of those whose data is being exchanged [45].

4.4 Deep Learning-Based Attack Detection

In this section, work is done by various authors in the fields of deep learning models in cyber-Physical security. Li et al. [23] have proposed a non-specific system for veering DL cognitive computing methods into Cyber Forensics (CF) from now on, alluded to as the Profound Learning Cyber Forensics (DLCF) System. DL employments a few machine learning procedures to fathom issues through the utilization of neural systems that recreate human decision-making [7]. Based on these grounds, DL holds the potential to significantly alter the space of Cyber Forensics (CF) in an assortment of ways as well as give arrangements to measurable investigators [40]. Sebastian et al. [43] proposed a strategy of mechanizing post-exploitation by combining profound support learning and the PowerShell Realm, which is popular as a post-exploitation framework. Rieke et al. [37] also displayed a novel assault strategy leveraging the assault vector, which makes profound learning expectations not diverse from irregular speculating by debasing the precision of the forecasts. Li et al. [25] have proposed a deep setting displaying engineering (DCM) for multi-turn reaction determination by utilizing BERT as the setting encoder. The experts in [41] also portrayed that a profound setting demonstrating engineering (DCM) is defined as a four-module design, specifically relevant encoder, utterance-to-response interaction, highlights conglomeration, and reaction selection. Rieke et al. [37] have classified the reinforcement Learning (RL) papers within the writing into seven categories based on their zone of application and services to the physical security and CPS user. The pundit $\boldsymbol{Qw(s,a)}$ is parameterized by w and learned utilizing the Bellman condition, by minimizing the observational Bellman leftover. Thus, Eq. 10 is defined as follows.

$$\boldsymbol{L\,(w) = Es\mathfrak{t}pb, a\mathfrak{t}b\,[(Qw\,(st, at) - yt)\,2]\,, where\ yt = rt + y\,Qw} \tag{10}$$

a and b may be behavior approaches utilized to gather tests. These so-called value-based agents store the value function and base their decisions on it. The authors have investigated a few issues that impact the application of ML to cybersecurity—too, talked about the familiar information sources in SOC, their workstream, and how to leverage and handle these information sets to construct a compelling machine learning framework. The authors have defined two measures of show goodness in Conditions in Eqs. 11 and 12.

$$ModelDetectionrate = \frac{Numberofriskyhostincertainpredictions}{totalnumberofriskyhosts} x100\% \tag{11}$$

$$ModelLift = \frac{Proportionofriskyhostincertainpredictions}{total Proportuinrofriskyhosts} \tag{12}$$

Riyaz et al. [38] have moreover secured phishing locations, organized interruption locations, analyzing protocol security attributes, as demonstrated by pattern matching, cryptography, human interface proofs, spam recognition in social networks, digital energy utilization profiling, and challenges with the safety of machine learning procedures themselves.

In Fig. 6, a machine learning-based cybersecurity detection engine received all relevant input, network, and program binaries by matching with signature values from the extracted features, including IP addresses and network data. This static

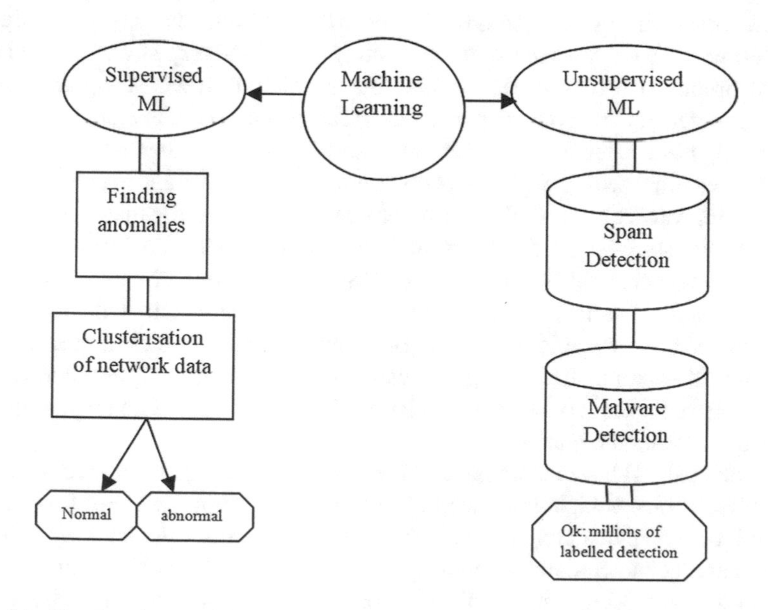

Fig. 6 Machine learning model for cyber security

input-based matching process is used to identify the attacks after a long system process. The threat detection-based process becomes dynamic to recognize new threats based on prior learning & experiences by incorporating methodologies based on artificial language and machine learning. In addition, various known artificial intelligence is used for pattern recognition and anomaly detection.

4.5 Signature-Based Malicious Behavior Detection

Malware has threatened computers, systems, and foundations since the eighties. There are two significant advances to protect against this, but most organizations depend nearly only on one fair approach, the decade's ancient signature-based technique. The more progressed strategy of identifying malware utilizing behavior examination picks up fast footing but is still generally new. When an anti-malware arrangement supplier distinguishes a question as pernicious, its signature is included in a database of known malware. These storehouses may contain hundreds of millions of marks that distinguish pernicious objects. This strategy of determining malicious objects has been the essential method utilized by malware items and remains the base approach used by the most recent firewalls, mail, and organized gateways. Sen et al. [44] have recommended an approach based on input design and coordinating relationship signatures with the sifted logs from the overstep. Signatures are carefully created for tall discovery exactness.

Moreover, implied consent ask commands were recorded as marks for identifying necessary consents at runtime. Ferdowsi et al. [11] have proposed signature-based behavior-based discovery methods that utilize API calls that are promising for discovering malware variations. A signature-based discovery strategy matches an already produced set of marks against the suspicious tests. A signature could be an arrangement of bytes at particular areas inside the executable, a standard expression, hash esteem of double information, or any other groups made by malware examiner which ought to recognize malware occasions precisely and their relationship. Maeda et al. [26] have recommended a show for malware discovery based on signature-based discovery and created CBSigIDS, a bland system for collaborative blockchain signature-based IDSs, which receives blockchains to assist in incrementally sharing and constructing a trusted signature database. Within the assessment, our exploratory comes about in both mimicked and real IoT situations, illustrating that CBSigIDS can upgrade the strength and adequacy of signature-based discovery beneath ill-disposed scenarios (e.g., flooding assaults) by irrefutably sharing the marks.

Buczak et al. [4] have recommended the approach of the programmed location of cyber assaults on a vehicle framework. To decrease the significance of irregularities that tend to happen in a non-attack environment, the creator calculates the weight of each signature characteristic irregularity test $\boldsymbol{w(cij}$), within the taking after way: for some n scenarios S,1 $\leq\leq$ n, we take the complement of the cruel of each signature characteristic irregularity test ***cij***(), where i speaks to an information source and j speaks to a signature characteristic is given in Eq. 13.

$$\boldsymbol{w\,(cij) = 1 - ci,\, j = 1 - n = 1cij\,(\,)\,n} \tag{13}$$

Which produces the weight of an anomaly sample for the signature characteristic? This permits the framework to determine a more exact score considering the peculiarities that tend to be less characteristic of an assault as the results hold on in non-attack conditions.

5 Comparative Analysis

The cybersecurity framework discussed in this survey represents the various attacks using deep learning, machine learning, and federated learning algorithm. Cybersecurity problems can be found in various places, including mail, operating systems, cars, entertainment, banks, businesses, financial institutions, and online data storage, to name a few. We chose a deep, machine learning, and federated learning-based cybersecurity attack detection concept for this research study. Approximately 58 papers relevant to the survey subject have been selected. Various criteria are used to estimate and relate the efficiency and performance in the reported output. The comparative study for types of attack, technique, and challenges is discussed in this Table 3.

6 Future Prospective

Artificial intelligence is required to solve complex problems, and the cyber-physical security area falls under that category. AI is best suited to solving some of the world's most challenging issues, and cyber-physical security is one of them. Machine learning and artificial intelligence are becoming increasingly important with today's ever-increasing cyber-attacks and the proliferation of gadgets. Artificial intelligence (AI) can keep track of digitized risk discovery and respond more quickly than conventional software-based approaches.

- **Detection and Response Time**

Identifying simple matters can quickly speed up through cross-referencing various attentions, and different ways of securing data are possible through AI today. Up to this time, Expertise assists in improving CPS security by providing different mechanisms with the needs of occurrences of events. On the other hand, AI frameworks also provide a way to enhance actions based on plans-based recommendations.

Table 3 Comparison of various Artificial Intelligence model

Authors	Types of attack	Techniques used	Challenges	Reported output
Dixit et al. [10]	Model-poisoning	Poisoning Resilience Defence	Bandwidth Communication, noise interference &Robustness was not discussed in this reported study	Error performance = 0.2
Ferdowsi et al. [11]	Privacy, training issues, real-world tasks	FEDAVG, Local only, MTL, LG-FEDAVG	Methods for learning fair representations were not discussed in the reported work	Cross-entropy loss 100-piece batch 0.01% learning rate 0.9 Momentum 0.0005 learning rate decay The number of global epochs equals 100
Geetha et al. [12]	Model-inversion attacks, man-in-the-middle attacks, Insider attacks	Deferentially private stochastic gradient descent algorithm, classification, and segmentation	Lack of speed up the process and reduced the amount of data exchanges	Speedup = up to 9* Reduced the amount of data exchanged = up to 34% Performance = 4:5%
Gupta et al. [13]	UAV antenna angle, fading, and transmission delay modifications	Joint power allocation and scheduling design	There are delay constraints at uplink and downlink	Bandwidth = 35%
Li et al. [23]	Poisoning attacks and inference attacks	Game-theoretic research	The robustness of the federated learning system is still a significant concern	Optimal accuracy has been achieved
Hannah et al. [14]	Encryption of model submissions, secure authentication of all parties, traceability of actions, differential privacy, verification systems, model confidentiality, execution honesty as well as precautions	Medical NER Model Bi-LSTM and CRF, CNN	Formal security verification and evaluation of the proposed methods were not discussed	Accuracy = 84%

Hassan et al. [15]	Data poison attack, user Data reconstruction attack	Quantified indicator, memorization management	Eliminating unexpected memorization is still not addressed by the authors	Accuracy = 87.49%
Dangi et al. [8]	Active or passive attacks, Model poisoning	Gradient Subset, Robust aggregation Homomorphic Encryption, SMC, Dropout	Effectiveness and Defensive methods and strategies against attack were not discussed	Accuracy = 76.8%
Muniyandi et al. [29]	Jamming attack	Client group Prioritization	Architecture, framework, reliability, and techniques for decentralized the global model was not discussed	Accuracy = 82:01% Outperforms = 49:11%
Ibrahim et al. [17]	Privacy leakages semi-honest adversaries	TernGrad. Homomorphic encryption	There is the need to focus on resisting the more powerful adversary	Better in communication and computation and more Accurate design
Hitaj et al. [16]	Malicious adversary, passive attack, model inversion	Generative Adversarial Network (GAN)	Lack in the evaluation of and study of the high-dimensional dataset and complex neural The network was missing	Functionality and accuracy 90.8%, 91.5%
Karie et al. [20]	Data poisoning attack, adversarial attacks	Reputation-based scheme	More accurate and efficient validation schemes were not discussed	Accuracy = 76.12 percent thresholds = 1.6
Ashok et al. [2]	Alter global model parameters	Blockchain-based FL, game-theoretic incentive mechanisms can	Performance evaluation of memory, accurate data for the proposed algorithms were not performed	response time =optimal CPU-frequency = maximize utility function = maximize the
Zoph et al. [51]	Adversarial attacks	Lagrange Coded Computing (LCC), secrete sharing, key agreement, and public key infrastructure	A fair comparative analysis of the existing algorithm was missing	FedAvg = 93.19
Srinivasu et al. [47]	differential privacy	Differentially private (FNAS). private gradient sharing or Gaussian mechanism	gradient compression, periodic updates, and diverse example selection methods were missing	Variance of Noise = = 0.5 Validation error (%) = 14.0 ± 0.32

(continued)

Table 3 (continued)

Authors	Types of attack	Techniques used	Challenges	Reported output
Ijaz et al. [18]	Malicious threats and data contamination attacks	Uniform probability, Federated Averaging algorithm	The model's size, bandwidth, and reliability of client connections were not addressed.	Weight decay = $4 \times 10{-}4$ Accuracy = 40.3%
Karimipour et al. [21]	Poisoning attacks, backdoor attacks	Fine-pruning, Byzantine-tolerant distributed learning	The design of the Robust system in federated learning was not discussed	Accuracy = 99%
Bhagoji et al. [3]	Poisoning attack	Layer-wise Relevance Propagation (LRP) techniques	More sophisticated detection strategies at the server-side were not discussed	Accuracy = 91.7%
Singh et al. [54]	Data leakage and misuse, end-users privacy	Communication-efficient federated learning techniques	Optimization methods for low-frequent high-volume communication were not discussed	Sparsity rate = 0.001 accuracy = 55%
Karie et al. [20]	Cyber-based attacks	Clustering techniques, Deep learning cognitive computing techniques	The authors did not report digital forensic investigating methods in detail	Accuracy=optimum
Muniyandi et al. [29]	Vulnerabilities, patching issues	Deep reinforcement learning and the PowerShell Empire	The training environment, Frameworks, and methodology used by the authors are not appropriate	Probability = 60% Efficiency=maximum
Rao et al. [35]	Mind Control attack	TensorFlow, CNTK, and Caffe running on CUDA	GPU function vulnerabilities were not reported in this study	Accuracy = 0.496
Sahu et al. [39]	Abundant yet noisy contextual information, back-channeling	Retrieval-based methods, generation-based methods	The performance enhancement technique was not addressed	Pushing recall = 86.8% E-Commerce Dialogue corpus = 68.5% MAP and MRR = 61.6% and 64.9%

Perera et al. [33]	Dispatch problem	The model with such a white box. Value-Aware Model Learning (VAML), and Policy-Aware Model are manifestations of data-driven models (PAML)	An adequate test case to validate performance was not reported by the authors	Improvement = 10–20% considerable effort = 13%
Dixit et al. [10]	Phishing, spear-phishing, password attack, and denial of service attacks are all examples of phishing	CNN-Convolutional Neural Network, AE- Auto Encoder, DBN- DeepBelief Network, RNN, GAN, and DIL-Deep Reinforcement Learning	A practical algorithm and robust design for cyber security were not discussed	Accuracy = 99.85%.
Saltzer et al. [40]	Adversarial attacks	Deep Reinforcement Learning (DRL)	Lack in the Dataset and defensive measures	Qmax = 80 Reward = 10 Policy = Epsilon Greedy Epsilon = 0.1
Jiang et al. [19]	Cyber threats, ransomware, and other forms of cybercrime are all on the rise	Firewalls, cryptographic encryption and decryption methods, anomaly detection of intrusion	Architectures and Algorithm for cyber security use cases were not reported	Accuracy = 0.968 Precision = 0 .814 Recall = 0.984 F1-score = 0.891
Dash et al. [9]	Cyber-attack	N-fold cross-validation	Lac in the performance on different datasets	Accuracy = 99(%) DR = 99.27 (%) FAR = 0.85 (%)
Wu et al. [50]	Cyber-attack	Unsupervised deep learning, K-means	More accurate and efficient validation schemes not discussed	Accuracy = 100(%)
Singh et al. [52]	Adversarial attacks	RF (Shallow Learning) and another based on FNN (Deep Learning)	Have not provided solutions to mitigate detecting specific threats	F1-score = 0.90 Precision = 0.91 Recall = 0.73

(continued)

Table 3 (continued)

Authors	Types of attack	Techniques used	Challenges	Reported output
Cetinkaya et al. [5]	Malicious flow detection	Tree-Shaped Deep Neural Network (TSDNN), oversampling method, and the under-sampling Method	Effectiveness and Defensive methods and strategies against attack were not discussed	Accuracy = 99.63% Precision = 85.4%
Moustafa et al. [28]	Adversarial attacks	Complicated data processing and AI-based techniques	Various use case application involving IoT system was not reported in this study	f1-score = 99.33% FPR = 0.23%
Dangi et al. [8]	Malicious Socket address	Support Vector Machines (SVM)	The study was limited to malicious URL (IP) and port address	Accuracy = 95.6%
Katzir et al. [22]	Anomaly-based IDS detection	IDTL (Incremental Decision Tree Learning), Incremental Linear Discriminant Analysis (ILDA)and Mahalanobis distance	Incremental learning system methodology neither has included by the authors.	IDTL accuracyNSL-KDD-75.71 SAME -96.04 Phishing-91.05
Li et al. [23]	DGA Detection and Network Intrusion Detection	RF classifiers, FNN, DNN	Not considering the concepts of adversarial learning	Intrusion detection F1-score- 0.7985 DGA detection classifiers F1-score- 0.8999
Ferdowsi et al. [11]	Safety logs, alert data information, and analysis of insights to identify risky user	Multi-layer Neural Network (MNN), Random Forest (RF), SVM	Other learning algorithms should be considered to improve the detection accuracy further	Average lift = 20%
Buczak et al. [4]	Phishing detection, Network intrusion detection	Hierarchical Clustering (HC), K-Medoids (KM)	Spam detection, virus detection, and surveillance camera robbery were not discussed in the study	Success rate For segmentation = 66.2 %

Sen et al. [44]	Unauthorized access, destruction, theft, or damage	BPNN architecture, Neural Network	Execution time can be further minimized	Accuracy = 97%
Chowdhury et al. [31]	Data security, Cyberthreat, Malware	Binary associative memory (BAM), Multilayer perceptron (MLP)	High False Positive Rate	Accuracy = 98.6%
Reddy et al. [36]	Risk associated with storing data	Interest Groups (IGs), Proof of Common Interest (PoCI)	Faced Generative Adversarial Network problem	Optimum Accuracy Achieved
Panigrah et al. [32]	Cyber attacks, TCP SYN flooding	Federated Network Traffic Analysis Engine (FNTAE), K Nearest Neighbor (KNN) Classifier	Real-time intrusion detection experiments using Live net traffic monitoring and analysis The authors did not report it	Accuracy = 98.198%
Sarkar et al. [41]	Network Attack	K-NN (K Nearest Neighbor)	Lack in the framework, architecture, and design used for the effectiveness of the system	Accuracy = 85.69%
Sathyanaryan et al. [42]	Unknown attack	KYOTO 2006+ data set, Machine Learning technique	The designing of a Robust system in machine learning was not discussed	Accuracy = 97.23% Instances = 97.23% High true positive rate (99%)
Sen et al. [44]	Military and commercial sectors cyber attack	KDD data set, Chi square, Information Gain and Relief	The authors did not address major attacks in the KDD dataset	Accuracy = 95.0207
Saltzer et al. [40]	Cyber-Physical attack	k-Nearest Neighbors, Computer Numerical Control (CNC)	Various detection methods and data process used in the system was not appropriate	Accuracy = 93.8%
Sicari et al. [41]	Denial of service (DOS), User to Root(USR), Remote to Local attack (R2L)	KDD Cup 1999 Dataset, ISOT (Information Security and Object Technology) Dataset	New datasets for solving various national and international cyber-attacks were not addressed	Command Execution = 23% SQL Injection = 18% Path Traversal = 18%

- **Network Security plans and recognition**

The two significant aspects of system security are the development and methods for CPS assembly of security plans and recognition of the system's geographic link. These practice units are pretty tedious, so utilizing the AI procedures would speed up. It will observe and learn traffic styles as suggested security measures make preparations. That does not save time in either situation, nor does it hold a significant amount of work and money that are prepared to or can relate to areas of a mechanical flip of cases and development.

- **Controlling Phishing Detection and Prevention**

Phishing is a commonly used computerized attack technique in which software engineers communicate their payload using a phishing trap. Phishing messages are prevalent; one out of every ninety-nine messages may be a phishing attempt. Fortunately, AI-ML can agree to persuasion work in phishing prevention and avoidance for CPS. More than 10,000 active phishing sources will be identified and followed by a computer-based insights metric capacity unit, responding and correcting loads faster than humans. To boot, AI-enabled machine learning works at sifting phishing threats from all over the planet. There is no impediment in its comprehension of phishing endeavors in a chosen soil science region. Computer-based insights have made it conceivable to partition between a fake web location and a real chop-chop.

- **Secure & Stable Authentication**

Passwords have effectively been associated with highly acute administration in terms of confidence. Physical identifiable confirmation is the most common method of secure CPS confirmation, in which AI uses different components to say apart from a private. To encourage search inside, a phone will use unmistakable finger impression scanners and facial affirmations. The process includes the software analyzing precious information from all over the world and using fingers to acquire it if the login is correct. Aside from that, AI will look at various factors to determine whether or not the buyer is authorized to check into some particular device. The system looks at how fast you type and how many mistakes you make when ordering.

- **Behavioral Analytics**

Another critical application of AI in cyber-physical security is its ability to investigate the actions of CPS. The particulars would incorporate everything from regular login times and data preparation to writing and reviewing examples of the AI calculations to pick up on unusual exercises or other behaviors that are not typical. It would be flagged as having been bundled up by a suspicious buyer or probably sq. the buyer. It works out that the stamp of the AI calculations is frequently anything from large online transactions sent to addresses other than a sudden spike in report exchange from archived envelopes or a sudden change in composing pace.

- **Blockchain for AI data security**

In the modern computing ecosystem, data is captured from various sources and transmitted among devices (e.g., IoT) through networks. Artificial Intelligence (AI) and its derivatives have been used as powerful tools to analyze and process the captured data to achieve effective reasoning in addressing security issues. The deceptive analysis would be produced when damaged, or dishonest data is purposefully or unintentionally integrated by a harmful third party based on adversarial inputs, even though AI is vital and may be used with distributed computing. Blockchain, a well-known ledger technology, has the potential to be used in a variety of online contexts. Due to its qualities, such as decentralization, verifiability, and immutability for ensuring the validity, trustworthiness, and integrity of data, blockchain aims to reduce transaction risks and financial fraud. AI can generate more secure and reliable results when the veracity and dependability of the data are guaranteed. The investigation of blockchain technology for the security of AI data in B2B and M2M environments may be a future research focus.

7 Conclusion

CPSs will likely play the main role in the plan and development of the forthcoming engineering system. The primary focus of CPS is on design assurance and cyber-physical security for the complex CPS system. This paper provides a definition and background of CPS. The fast and significant development of the CPS system affects people's way of life and enables a broader range of facilities. The security framework and network threat model were also discussed in detail. This chapter presents our research outcome in AI-based methodologies in cyber-physical systems. We have started with the recognition that can address the network's security, integrity, and privacy. However, using AI methodologies in CPS has its challenges. Here, Controllers also direct the received quantities to the core control servers and perform the selected commands. In CPS, system operatives should be alert to the current position of the controlled objects. Thus, we have started with the framework of AI-based methods in CPS, as AI is the latest technology that can potentially improve CPS's security.

To highlight the various AI-based methods relevant to the current security of CPS systems and discuss them. The comparison table provides the main contribution of authors in the areas of types of attack detection, techniques used, and challenges faced by current security architecture, discussed in the study and future prospective of each chapter.

We have also considered that cyber resilience depends on adequate security controls that protect authenticity, confidentiality, reliability, resilience, and integrity. Finally, we emphasize providing a future research inclination and unique features in this field. We have also identified the betterment of CPS security by providing different mechanisms, CPS assembly of security plans and recognition, for ensuring the prevention and avoidance for CPS, to determine the common method of secure CPS confirmation and the development of the security protocol. This research work would help researchers and academicians in CPS security.

References

1. Abeshu, A., & Chilamkurti, N. (2018). Deep learning: The frontier for distributed attack detection in fog-to-things computing. *IEEE Communications Magazine*, *56*(2), 169–175. https://doi.org/10.1109/mcom.2018.1700332
2. Ashok, A., Hahn, A., & Govindarasu, M. (2014). Cyber-physical security of Wide-Area Monitoring, Protection and Control in a smart grid environment. *Journal of Advanced Research*, *5*(4), 481–489. https://doi.org/10.1016/j.jare.2013.12.005
3. Bhagoji, A. N., Chakraborty, S., Mittal, P., & Calo, S. (2018). Analyzing federated learning through an adversarial lens. In *arXiv [cs.LG]*. http://arxiv.org/abs/1811.12470.
4. Buczak, A. L., & Guven, E. (2016). A survey of data mining and machine learning methods for cyber security intrusion detection. *IEEE Communications Surveys & Tutorials*, *18*(2), 1153–1176. https://doi.org/10.1109/comst.2015.2494502
5. Cetinkaya, A., Ishii, H., & Hayakawa, T. (2017). A probabilistic characterization of random and malicious communication failures in multi-hop networked control. In *arXiv [cs.SY]*. http://arxiv.org/abs/1711.06855
6. Chhetri, S. R., Lopez, A. B., Wan, J., & Al Faruque, M. A. (2019). GAN-sec: Generative adversarial network modeling for the security analysis of cyber-physical production systems. 2019 Design, Automation & Test in Europe Conference & Exhibition (DATE).
7. Costa, G. (2016). A Methodological Approach for Assessing Amplified Reflection Distributed Denial of Service on the Internet of Things. 4, 11–18.
8. Dangi, C. S. (n.d.). Cyber Security Approach in Web Application Using SVM. International Journal of Computer Ap-Plications, 2012(2).
9. Dash, S., Verma, S., Kavita, Khan, M. S., Wozniak, M., Shafi, J., & Ijaz, M. F. (2021). A hybrid method to enhance thick and thin vessels for blood vessel segmentation. *Diagnostics (Basel, Switzerland)*, *11*(11), 2017. https://doi.org/10.3390/diagnostics11112017
10. Dixit, P., & Silakari, S. (2021). Deep learning algorithms for cybersecurity applications: A technological and status review. *Computer Science Review*, *39*(100317), 100317. https://doi.org/10.1016/j.cosrev.2020.100317
11. Ferdowsi, A., & Saad, W. (2019). Generative adversarial networks for distributed intrusion detection in the Internet of Things. In *arXiv [cs.CR]*. http://arxiv.org/abs/1906.00567
12. Geetha, R., & Thilagam, T. (2021). A review on the effectiveness of machine learning and deep learning algorithms for cyber security. *Archives of Computational Methods in Engineering. State of the Art Reviews*, *28*(4), 2861–2879. https://doi.org/10.1007/s11831-020-09478-2
13. Gupta, D., Rani, S., Ahmed, S. H., Verma, S., Ijaz, M. F., & Shafi, J. (2021). Edge caching based on collaborative filtering for heterogeneous ICN-IoT applications. *Sensors (Basel, Switzerland)*, *21*(16), 5491. https://doi.org/10.3390/s21165491
14. Hannah, J., Mills, R., Dill, R., & Hodson, D. (2021). Traffic collision avoidance system: false injection viability. *The Journal of Supercomputing*, *77*(11), 12666–12689. https://doi.org/10.1007/s11227-021-03766-9
15. Hassan, M. U., Rehmani, M. H., & Chen, J. (2018). Differential privacy techniques for cyber physical systems: A survey. In *arXiv [cs.CR]*. http://arxiv.org/abs/1812.02282
16. Hitaj, B., Ateniese, G., & Perez-Cruz, F. (2017). *Deep Models under the GAN: Information Leakage from Collaborative Deep Learning. arXiv*. *28*, 78–82.

17. Ibrahim, A., Valli, C., McAteer, I., & Chaudhry, J. (2019). Correction to: A security review of local government using NIST CSF: a case study. *The Journal of Supercomputing*, *75*(9), 6158–6158. https://doi.org/10.1007/s11227-019-02972-w
18. Ijaz, M. F., Attique, M., & Son, Y. (2020). Data-driven cervical cancer prediction model with outlier detection and over-sampling methods. *Sensors (Basel, Switzerland)*, *20*(10), 2809. https://doi.org/10.3390/s20102809
19. Jiang, F., Fu, Y., Gupta, B. B., Liang, Y., Rho, S., Lou, F., Meng, F., & Tian, Z. (2020). Deep learning based multi-channel intelligent attack detection for data security. *IEEE Transactions on Sustainable Computing*, *5*(2), 204–212. https://doi.org/10.1109/tsusc.2018.2793284
20. Karie, N. M., Kebande, V. R., & Venter, H. S. (2019). Diverging deep learning cognitive computing techniques into cyber forensics. *Forensic Science International. Synergy*, *1*, 61–67. https://doi.org/10.1016/j.fsisyn.2019.03.006
21. Karimipour, H., Dehghantanha, A., Parizi, R. M., Choo, K.-K. R., & Leung, H. (2019). A deep and scalable unsupervised machine learning system for cyber-attack detection in large-scale smart grids. *IEEE Access: Practical Innovations, Open Solutions*, *7*, 80778–80788. https://doi.org/10.1109/access.2019.2920326
22. Katzir, Z., & Elovici, Y. (2018). Quantifying the resilience of machine learning classifiers used for cyber security. *Expert Systems with Applications*, *92*, 419–429. https://doi.org/10.1016/j.eswa.2017.09.053
23. Li, L., Li, C., & Ji, D. (2021). Deep context modeling for multi-turn response selection in dialogue systems. *Information Processing & Management*, *58*(1), 102415. https://doi.org/10.1016/j.ipm.2020.102415
24. Liu, H., & Lang, B. (2019). Machine learning and deep learning methods for intrusion detection systems: A survey. *Applied Sciences (Basel, Switzerland)*, *9*(20), 4396. https://doi.org/10.3390/app9204396
25. Liu, K., Dolan-Gavitt, B., & Garg, S. (2018). *Defending against Backdooring Attacks on Deep Neural Networks. arXiv [cs.CR].*
26. Maeda, R., & Mimura, M. (2021). Automating post-exploitation with deep reinforcement learning. *Computers & Security*, *100*(102108), 102108. https://doi.org/10.1016/j.cose.2020.102108
27. Malomo, O. O., Rawat, D. B., & Garuba, M. (2018). Next-generation cybersecurity through a blockchain-enabled federated cloud framework. *The Journal of Supercomputing*, *74*(10), 5099–5126. https://doi.org/10.1007/s11227-018-2385-7
28. Moustafa, N., Slay, J., & Creech, G. (2019). Novel geometric area analysis technique for anomaly detection using trapezoidal area estimation on large-scale networks. *IEEE Transactions on Big Data*, *5*(4), 481–494. https://doi.org/10.1109/tbdata.2017.2715166
29. Muniyandi, A. P., Rajeswari, R., & Rajaram, R. (2012). Network anomaly detection by cascading K-means clustering and C4.5 decision tree algorithm. *Procedia Engineering*, *30*, 174–182. https://doi.org/10.1016/j.proeng.2012.01.849
30. Nguyen, H.-T., Ngo, Q.-D., & Le, V.-H. (2020). A novel graph-based approach for IoT botnet detection. *International Journal of Information Security*, *19*(5), 567–577. https://doi.org/10.1007/s10207-019-00475-6
31. Panigrahi, R., Borah, S., Bhoi, A. K., Ijaz, M. F., Pramanik, M., Jhaveri, R. H., & Chowdhary, C. L. (2021). Performance assessment of supervised classifiers for designing intrusion detection systems: A comprehensive review and recommendations for future research. *Mathematics*, *9*(6), 690. https://doi.org/10.3390/math9060690
32. Panigrahi, R., Borah, S., Bhoi, A. K., Ijaz, M. F., Pramanik, M., Kumar, Y., & Jhaveri, R. H. (2021). A consolidated decision tree-based intrusion detection system for binary and multiclass imbalanced datasets. *Mathematics*, *9*(7), 751. https://doi.org/10.3390/math9070751
33. Perera, A. T. D., & Kamalaruban, P. (2021). Applications of reinforcement learning in energy systems. *Renewable and Sustainable Energy Reviews*, *137*(110618), 110618. https://doi.org/10.1016/j.rser.2020.110618

34. Radanliev, P., De Roure, D., Van Kleek, M., Santos, O., & Ani, U. (2021). Artificial intelligence in cyber physical systems. *AI & Society*, *36*(3), 783–796. https://doi.org/10.1007/s00146-020-01049-0
35. Rao, N. T. (n.d.). of Computer Science and EngineeringVignan's Institute of Information Technology (A). *Int. J. Adv. Res. Comput. Inf. Secur*, *2019*(1), 1–8.
36. Reddy, S., & Shyam, G. K. (2022). A machine learning based attack detection and mitigation using a secure SaaS framework. *Journal of King Saud University - Computer and Information Sciences*, *34*(7), 4047–4061. https://doi.org/10.1016/j.jksuci.2020.10.005
37. Rieke, N., Hancox, J., Li, W., Milletari, F., Roth, H., Albarqouni, S., Bakas, S., Galtier, M. N., Landman, B., Maier-Hein, K., Ourselin, S., Sheller, M., Summers, R. M., Trask, A., Xu, D., Baust, M., & Cardoso, M. J. (n.d.). *The Future of Digital Health with Federated Learning. arXiv. 2020*, 1–8.
38. Riyaz, S., Sankhe, K., Ioannidis, S., & Chowdhury, K. (2018). Deep learning convolutional neural networks for radio identification. *IEEE Communications Magazine*, *56*(9), 146–152. https://doi.org/10.1109/mcom.2018.1800153.
39. Sahu, S., & Mehtre, B. M. (2015). Network intrusion detection system using J48 Decision Tree. 2015 International Conference on Advances in Computing, Communications and Informatics (ICACCI).
40. Saltzer, J. H., & Schroeder, M. D. (n.d.). The Protection of Information in Computer Systems, Proc. Proc. IEEE 2010, 63, 1278–1308.
41. Sarker, I. H., Abushark, Y. B., Alsolami, F., & Khan, A. I. (n.d.). Machine Learning Based Cyber Security Intrusion Detection Model. Symmetry (Basel) 2020. 12.
42. Sathyanarayan, V. S., Kohli, P., & Bruhadeshwar, B. (2008). Signature generation and detection of malware families. In Information Security and Privacy (pp. 336–349). Springer Berlin Heidelberg.
43. Sebastian, O., & Ackere Ann, R, L. E. (2017). Interdependencies in Security of Electricity Supply. Interdependencies in Security of Electricity Supply. Energy, 598.
44. Sen, R., Chattopadhyay, M., & Sen, N. (2015). An Efficient Approach to Develop an Intrusion Detection System Based on Multi Layer Backpropagation Neural Network Algorithm: IDS Using BPNN Algorithm. In Proceedings of the 2015 ACM SIGMIS Conference on Computers and People Research. ACM.
45. Sheth, A., Anantharam, P., & Henson, C. (2013). Physical-Cyber-Social Computing: An Early 21 St Century Approach. IEEE Intell. Syst, 28(1), 78–82.
46. Sicari, S., Rizzardi, A., Grieco, L. A., & Coen-Porisini, A. (2015). Security, privacy and trust in Internet of Things: The road ahead. *Computer Networks*, *76*, 146–164. https://doi.org/10.1016/j.comnet.2014.11.008
47. Srinivasu, P. N., SivaSai, J. G., Ijaz, M. F., Bhoi, A. K., Kim, W., & Kang, J. J. (2021). Classification of skin disease using deep learning neural networks with MobileNet V2 and LSTM. *Sensors (Basel, Switzerland)*, *21*(8), 2852. https://doi.org/10.3390/s21082852
48. Srivastava, A., Morris, T., Ernster, T., Vellaithurai, C., Pan, S., & Adhikari, U. (2013). Modeling cyber-physical vulnerability of the smart grid with incomplete information. *IEEE Transactions on Smart Grid*, *4*(1), 235–244. https://doi.org/10.1109/tsg.2012.2232318
49. Walker-Roberts, S., Hammoudeh, M., Aldabbas, O., Aydin, M., & Dehghantanha, A. (2020). Threats on the horizon: understanding security threats in the era of cyber-physical systems. *The Journal of Supercomputing*, *76*(4), 2643–2664. https://doi.org/10.1007/s11227-019-03028-9
50. Wu, M., Song, Z., & Moon, Y. B. (2019). Detecting cyber-physical attacks in CyberManufacturing systems with machine learning methods. *Journal of Intelligent Manufacturing*, *30*(3), 1111–1123. https://doi.org/10.1007/s10845-017-1315-5
51. Zoph, Barret, Vasudevan, V., Shlens, J., & Le, Q. V. (2017). Learning Transferable Architectures for Scalable Image Recognition. In *arXiv [cs.CV]*. http://arxiv.org/abs/1707.07012
52. Singh, H., Arora, D., & Kumar, V. (2022). Variational approach for intensity domain multi-exposure image fusion. Retrieved from http://arxiv.org/abs/2207.04204

53. Singh, H., Cristobal, G., Bueno, G., Blanco, S., Singh, S., Hrisheekesha, P. N., & Mittal, N. (2022). Multi-exposure microscopic image fusion-based detail enhancement algorithm. Ultramicroscopy, 236(113499), 113499. https://doi.org/10.1016/j.ultramic.2022.113499
54. Singh, S., Singh, H., Mittal, N., Singh, H., Hussien, A. G., & Sroubek, F. (2022). A feature level image fusion for Night-Vision context enhancement using Arithmetic optimization algorithm based image segmentation. Expert Systems with Applications, 209(118272), 118272. https://doi.org/10.1016/j.eswa.2022.118272
55. S. Dhir and Y. Kumar, "Study of Machine and Deep Learning Classifications in Cyber Physical System," 2020 Third International Conference on Smart Systems and Inventive Technology (ICSSIT), 2020, pp. 333–338, https://doi.org/10.1109/ICSSIT48917.2020.9214237.

Assessing the Predictability of Bitcoin Using AI and Statistical Models

Keshanth Jude Jegathees, Aminu Bello Usman, and Michael ODea

1 Introduction

There has been a huge interest in Cryptocurrency within the investment, academic and media communities in recent times. A cryptocurrency is a form of online currency that circulates without the need for a central monetary authority, such as a bank. It is a decentralized financial network, it is popular with investors, but it can also be used to exchange goods and for services.

The most well known cryptocurrency is Bitcoin (BTC), which was created in 2009, by a pseudonymous hacker called Satoshi Nakamoto. It was the world's first completely virtual and decentralized currency [9]. It has become increasingly important as an electronic payment tool and a speculative financial asset, which makes bitcoin (BTC) a digital gold [20]. Bouoiyour and Selmi [9], suggest that Bitcoin is detached from macroeconomic fundamentals and behaves as a 'speculative bubble'. Bitcoin lacks the underlying value of real currency, meaning its importance cannot be derived from either consumption or production (such as, gold).

Bitcoin has exponentially grown over the past decade, making it an attractive asset. The first block in the blockchain was mined on Jan. 3, 2009. During this age, bitcoin had no real monetary value. The first bitcoin transaction occurred in May 2010, when 10,000 bitcoins were used to buy $25 worth of pizza. In 2011, Bitcoin passed its first threshold, earning $1. The next couple of years sailed smoothly, and finally, in 2017 bitcoin was worth over $3000. The following years had high fluctuations, during June 2019 the price surpassed $10,000 and immediately surged to $6,635.84 by mid-December. Finally, after the pandemic caused the economic shutdown, bitcoin resurfaced into activity. It started at $6,965.72. and by December

K. J. Jegathees · A. B. Usman (✉) · M. ODea
Department of Computer Science, York St John University, York, UK
e-mail: jega@yorksj.ac.uk; a.usman@yorksj.ac.uk; m.odea@yorksj.ac.uk

Y. Maleh et al. (eds.), *Blockchain for Cybersecurity in Cyber-Physical Systems*,
Advances in Information Security 102, https://doi.org/10.1007/978-3-031-25506-9_11

it increased by over 400% in value. It reached its all-time high in mid-April 2021 over $63,000, which followed another surge that caused the prices to drop by 50%. Since then, the prices have experienced extreme fluctuations [39]. This paper also discusses the possible factors that cause this effect in the chapters ahead.

Generally, the forecasting approaches can be categorized into two classes: Artificial Intelligent (AI), and Statistical models. Common examples of AI models include MLP (Multilayer Perceptron), ANN (Artificial Neural Network), and RNN (Recurrent Neural Network). Some of the statistical models that kept repeating in all literature were, Exponential Smoothing, ARIMA (Autoregressive Integrated Moving Average), and GARCH (Generalized Autoregressive Conditional Heteroskedasticity). [26].

1.1 A Blockchain-Enabled Cyber-Physical System

Cyber-Physical Systems (CPS) connect the physical world of machines and manufacturing facilities to cyberspace. Implementing CPS in the real world requires connectivity and a computational platform.

Bitcoins' underlying technology, blockchain, is a distributed ledger that can authenticate financial transactions and thwart fraud. Blockchain mitigates the risks of a centralized architecture by decentralizing the information validation process across network peers. To put it another way, the most secure form of validation is the most effective approach to speed up the provision of financial services, giving customers greater autonomy and control. Bitcoin has increased exponentially over the last decade, making it an appealing asset. On January 3, 2009, the first block in the blockchain was mined. During this time, bitcoin had no monetary value; the first bitcoin transaction occurred in May 2010, when 10,000 bitcoins were used to purchase a $25 pizza. Bitcoin crossed its first boundary in 2011, reaching a value of $1; the next several years were smooth sailing, and by 2017, bitcoin was valued at more than $3000. The following years saw significant changes; in June 2019, the price topped $10,000 and quickly dropped to $6,635.84 by mid-December. Finally, after the epidemic caused an economic shutdown, bitcoin emerged into activity, starting at $6,965.72 and increasing in value by nearly 400% by December. It reached an all-time high of $63,000 in mid-April 2021, following another surge that led prices to tumble by 50%. Prices have fluctuated dramatically since then [39]. This study will also address the probable variables that generate this impact in the following chapters.

This chapter focuses on investigating the predictability of Bitcoin price using AI and Statistical models and identifying the approach/model which provides the best results in predicting the closing price of bitcoin. Additionally, the impact of other external factors such as blockchain, mining, social attraction, and macroeconomics is also used in conjunction with the bitcoin historical data to improve the accuracy of the predictions, creating multivariate time series data. However, making accurate forecasts of this type is challenging due to the inherently noisy and non-stationary nature of bitcoin price data.

2 Literature Review

Several studies revolve around achieving a similar goal to this literature. This section of the literature addresses them by the data used, the types of approaches, and their results.

Many works of literature fall into one category. Some papers implement hybrid models. However, there are very few studies that compare both two forecasting approaches. For instance, a study by authors Lee et al. (2007), compares the prediction performance of a NN and SARIMA model in forecasting the Korean Stock Exchange (financial time series data). Their results showed both models had robust forecasts. While the SARIMA model provided accurate forecasts for the Korea Composite Stock Price Index (KOSPI), the neural network model proved to predict its returns better. Xie et al. [28] implement a hybrid model with LSTM and MLP to forecast loads in the operation management of a power system. Khashei et al. [32] modeled a hybrid of ARIMA and PNN (probabilistic neural network) to obtain higher accuracy in time series prediction. Pai and Lin [33] combine the ARIMA and SVM (Support vector machine) models to exploit the unique strengths of each model in forecasting stock price

Greaves and Au [18] use a blockchain network feature to predict bitcoin prices. The researchers conducted supervised machine learning to achieve this. However, they reported an average accuracy of only 55% using an ANN model. They concluded blockchain data alone is inadequate as means of predicting bitcoin prices. Research by McNally et al. [13] suggests the most commonly used stock model, and other time series prediction, Multilayer Perceptron (MLP) is not as effective as a Recurrent Neural Network (RNN) model. Agarwal and Sastry, (2015) prove this claim by successfully implementing predicting stock returns with a combined RNN and genetic optimization algorithm. Although, a study by Giles, [23] supporting this claim conducted exchange rate prediction and obtained predictable results with a forty percent error rate, which is a significant gap compared to the literature by Valencia et al. [24]. They compare three neural networks to predict four important cryptocurrency price movements, out of which MLP outperformed with an accuracy rate of over seventy-two percent. Similar to the comparison made above, research by Raudys and Mockus [42] suggests the statistical model ARMA (which is comparatively less efficient than ARIMA) method is much faster than the MLP.

Some works in literature follow a statistical model approach; Cermak, [27] and apply the GARCH (1,1) model to analyze Bitcoin's volatility concerning macroeconomic variables. The results proved that Bitcoin appeared to be an attractive asset and behaves similarly to fiat currency in certain countries, especially China. Ariyo et al. [12] conducted a study in which the authors used an ARIMA model to predict stock price. Their research revealed the ARIMA model has a strong potential for short-term prediction. Their paper also argues that ARIMA models perform more efficiently forecasting financial time series data, especially in near future predictions, than most popular ANN models.

Similarly, Nochai and Nochai, [31] apply the ARIMA model in forecasting palm oil prices in Thailand. Wang (2011), compares ARIMA and the Fuzzy time series method in forecasting Taiwan exports, in which ARIMA showed minor prediction errors and closer predictions for prolonged sample periods. In contrast, the fuzzy method performed well with smaller sample periods.

According to [17], the price of a bitcoin is determined by supply and demand. When demand for bitcoins increases, the price increases, and when the market falls, the price drops. Kristoufek's [6] research identifies the main drivers of bitcoin price. The author conveniently categorizes each factor into areas such as, the Bitcoin Price Index (BPI), Macroeconomics & Policies, Technical Drivers, and Interest/Popularity. The author uses a first-order AR model to identify correlation with coherent wavelet analysis. His literature mentions two significant relationships between the Bitcoin price, firstly, is a negative relationship between the Bitcoin price and estimated output volume, i.e., an increase in output volume brings a drop in Bitcoin price, notably in the long run. Secondly, bitcoin mining inherently leading its price. Studies also indicate mining difficulty positively correlates with the price [6]. Similarly, research done by, Ciaian et al. (2016), tests the significance of the Market forces of bitcoin. It showed that the number of bitcoins in circulation and transaction volume significantly impacted the price. Unlike the fiat currency issued and supervised by the nation's central bank, the value of a Bitcoin is determined by how much the investors are willing to pay for it. [1]. The relationship between investors and their interest stimulates the price of bitcoin. During episodes of volatile prices, this interest drives prices further up, and during rapid declines, it pushes them further down [6].

A study conducted by Aggarwal et al. [40] discusses the social factors that govern the cryptocurrency market and analyzes the impact of these factors. The authors conclude there is no strong correlation between cryptocurrency and social aspects. On the contrary, a study by Wołk [14] focused on predicting the prices of cryptocurrencies like Bitcoin, Ethereum, and more in relation to social media, specifically Twitter and google search trend data. The study made use of six advanced predictive and descriptive models and sentiment analysis in conjunction to identify the correlation between social influence and its effect on crypto prices. Similarly, many other studies have supported social media data and its significance. Tandon et al. [15] use the popular ARIMA model to predict the price of Bitcoin and Dogecoin in correlation to Tweets by popular tech idol Elon Musk. The authors concluded by indicating no one person can control the utter volatile world of cryptocurrencies. Similarly, authors (Matta et al. [30]) researched the cross-correlation between the bitcoin price and the volumes of tweets and web search data using sentiment analysis. Overall, the results indicate a significant positive relationship between the social factors and the price change.

Google's Chief Economist Hal Varian suggested the potential of search data in speculating interests of various economic activities in real-time. Choi and Varian,

[7] support this claim by providing evidence that search data can predict home sales, automotive sales, and tourism, which are all-time series data. Although Search Volume Index (SVI) does allow the opportunity to find patterns in the attraction of a cryptocurrency, it does not link to investor interests. It is difficult to identify the motives of internet users searching for information about Bitcoin [6]. The search term used in this and many other works of literature is "bitcoin". Therefore it is difficult to differentiate the effect, whether it is positive or negative [19]. SVI patterns analyzed by Da et al. [8] claim that it primarily affected the decisions of individual or retail investors, meaning that more sophisticated retail investors do not show signs of correlation by these trends. Ciaian et al. (2016) also support this claim. In their research, they conclude that search data (Wikipedia/Google) had a stronger impact in the beginning years because when Bitcoin was little known, queries were made online about it (views on Wikipedia).

There is only limited literature that considers macroeconomics when forecasting bitcoin prices. Most of the works of literature analyze the significance of the effect it causes. Studies show that bitcoin appears to react to macroeconomic changes. For instance, in 2017, China suspended users from withdrawing bitcoins amid concerns about cryptocurrency replacing its fiat currency. In reaction to this, bitcoin drastically dropped by over 30% within a week. Since its recovery and peak, denial in the application for the Bitcoin Trust ETF by the SEC caused a 24% price crash [27]. Bouoiyour and Selmi [9] research describes the relationships between macroeconomic variables and bitcoin price. In his paper he finds no significant relationship between gold prices and Bitcoin prices. In support of this, Zhu et al. [34] analyzed the macro factors that affect the Bitcoin price index (BPI) and concluded gold price has no impact on the BPI, in the long run. The paper also identifies that the effect of the US Dollar Index on the BPI is highly significant in the long run.

The reviewed literature shows evidence that both approaches have been successfully implemented and which exogenous variables have the most significance in the predictions. However, predicting cryptocurrency is still a widely researched topic. Some papers approve and disapprove claims of another. Therefore, this paper will implement the strategies that were proven successful previously. In the aspect of data, a number of variables discussed in the literature will be used in conjunction because cryptocurrency, bitcoin especially, had grown in different directions from when most papers were written.

The primary goal is to assess the predictability of the closing price of bitcoin and, by doing so, identify which approach provides a prediction with higher accuracy and few complex variables. Finally, the paper's testable hypothesis is that Bitcoin price can be predicted, and AI can be used to predict the price. In this case, the Null Hypothesis, or H0, would be that AI cannot be used to predict cryptocurrency.

3 Methodology and Experiment

3.1 Models Used in the Study: LSTM and ARIMA

This chapter attempts to answer the research question whether using an AI and a statistical approach, namely LSTM (Long Short-Term Memory) and ARIMA is more accurate at prediction of Bitcoin values. Previous studies used machine learning techniques such as ANN and MLP to solve similar problems. Similarly, ARIMA demonstrated enormous potential in numerous papers and has been widely used in price prediction problems [3].

The LSTM (Long Short-Term Memory) was first introduced by Hochreiter & Schmidhuber in 1997. LSTM is a type of recurrent deep neural network architecture, meaning that the output at one step is fed back into the system, creating a loop [21]. LSTM is a subset of RNN; however, LSTM can effectively solve the problem of long-term dependency, also known as the vanishing gradient problem in the RNN model [25].

A typical LSTM network comprises four major components, the cell state and a series of 'gates' that control how the information in a sequence of data comes into, is stored in, and leaves the network, as shown in the image below. Namely, the forget gate, input gate, and output gate, respectively. The first step is deciding whether the information stored in the cell state needs to be disregarded or stored. This is achieved by implementing a sigmoid layer [22]. The forget gate. Secondly, the input gate layer decides whether to store the given information. After that, the cell state is stored, and the information that needs to be forgotten is replaced with the current state and a filtered version of the cell state [35]. Figure 1 shows an example of the LSTM Network.

Box and Jenkins [43] introduced the ARIMA model. The acronym ARIMA stands for Auto-Regressive Integrated Moving Average. The term "Autoregressive" implies that past values are used as predictors for the current values. This is also referred to as lagged values. In contrast, the "Moving average" component uses the

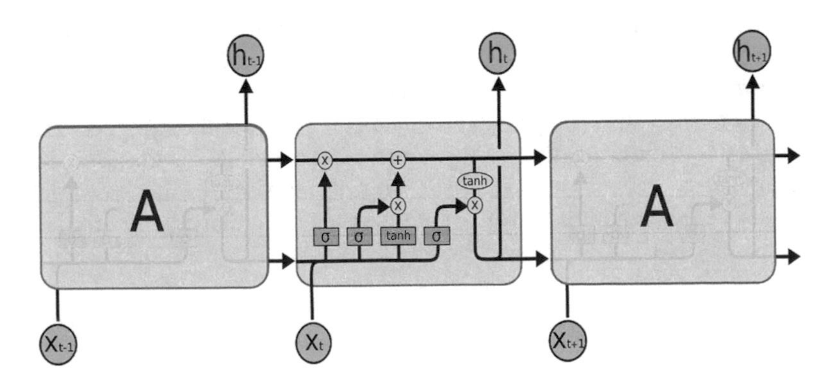

Fig. 1 LSTM Network

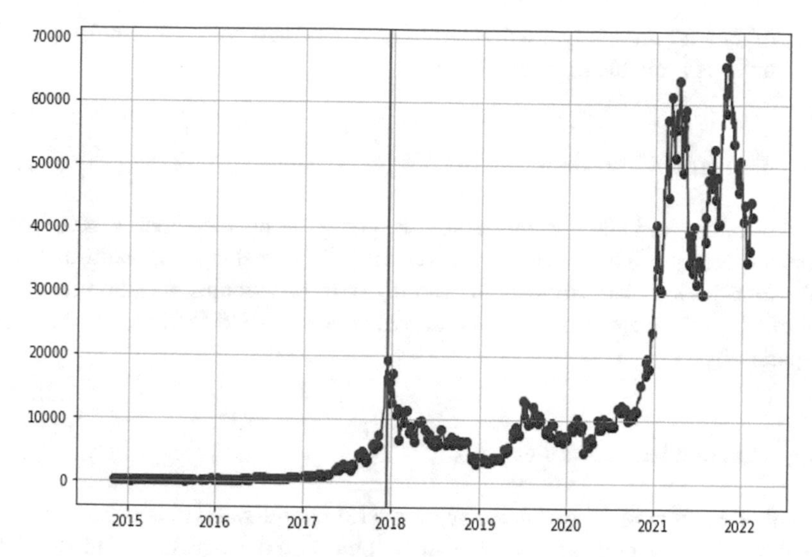

Fig. 2 Data selection based on BTC peaks

lagged values prediction errors as its predictors. Finally, "Integration" simply means differencing the stationary time series.

A general form of an ARIMA model is ARIMA(p,d,q)(P,D,Q), p refers to the order of the autoregressive process, d indicates the order of integration/differencing, and q is the order of moving average/error process. In each case the uppercase variables define the seasonal level of the model, while the lower case indicates non-seasonal levels [36].

A general equation that defines the ARIMA model was presented in Eq. (1). Where Yt is the actual value, Et is the random error at tϕi and θj are the coefficients, p and q are integers often referred to as Autoregressive (AR) and Moving Average (MA), respectively. [12].

$$y_t = \varphi_0 + \varphi_1 \mathrm{Y_t} - 1 + \varphi_2 \mathrm{Y_{t-2}} + \cdots + \varphi_\mathrm{p} \mathrm{Y_t} - \mathrm{p} + \varepsilon_\mathrm{t} - \varphi_1 \varepsilon_{\mathrm{t}-2} \cdots - \varphi_1 \varepsilon_{\mathrm{t}-\mathrm{q}} \quad (1)$$

3.1.1 Data and Variables

Choosing your sample points that define your time series data is crucial. The graph below shows bitcoin began its growth in 2017–2018. Therefore, data following that period should provide a good direction in learning Bitcoin's price movement. The sample period spans five years, from January 2017 to January 2022. Figure 2 shows a data selection based on BTC peaks.

Data used to train the models will be extracted from online resources previously used by similar studies and have proved the authenticity of the sources. The primary source of data for this study is extracted from blockchain.com. More information on data sources is available in Appendix A.

3.1.2 Dependent Variable

The dependent variable for this study is going to be the closing price of the cryptocurrency, as the goal is to predict the market value of Bitcoin at the end of the time period. The stock of the money base of Bitcoins is denominated in a traditional government-controlled fiat currency. Such as US Dollars, for the sake of comparability.

3.1.3 Independent Variables

The volatile nature of cryptocurrencies render it complicated to accurately forecast its value daily. Several factors need to be considered when trying to predict its market value. These variables were previously used in similar studies; by Lamothe-Fernández et al. [4], Balcilar et al. [5], Da et al. [8], Bouoiyour and Selmi [9], McNally et al. [13], Greaves and Au [18], Kristoufek [6], Guo et al. [25].

1. Demand and Supply – Although Bitcoin is usually considered a purely speculative asset, standard fundamental factors—usage in trade, money supply, and price level—play a role in Bitcoin price over the long term [6]. This also includes the number of mined bitcoins circulating on the network, the number of transactions per day, cost per transaction and miner's revenue, Exchange Trade volume, mining difficulty, and hash rates, following Ciaian et al. [44] and Lamothe-Fernández et al. [4], Bouoiyour and Selmi, [9].
2. Trends and Attractions – Search trends aggregated from the past to the present accumulatively depict the level of attraction a cryptocurrency has brought forth. A study done by Da et al. [8] states the representation of internet search, most internet users use search engines to make these queries.
3. Macroeconomics – External factors such as gold price and the exchange rate of USD to CNY are considered to have a significant effect on the bitcoin price. Because Gold and Bitcoin have many similarities. As for the exchange rates, historically, the cryptocurrency looks to have achieved a positive performance during bouts of weakness in the Chinese currency. Table 1 lists all the selected independent variables.

3.1.4 Feature Selection

Feature selection is a crucial aspect of any data analytics project. Irrelevant or partially relevant features can negatively impact model performance [16]. In order to select only the most significant IV and to achieve an optimal model complexity,

Table 1 Independent variables

Variable	Description
(a) Demand and supply	
Number of bitcoins	Number of mined bitcoins currently circulating on the network
Transaction value	Value of daily transactions
Transaction volume	Number of transactions per day
Exchange trade volume	The total USD value of trading volume on major bitcoin exchanges.
Mining difficulty	Level of difficulty in mining a bitcoin block
Hash rates	Times a hash function can be calculated per second
Cost per transaction	Miners' revenue divided by the number of transactions.
Output value	The total value of all transaction outputs per day
(b) Trends and attractions	
Google & Wikipedia Trends	Number of searches/views accumulated by the term "bitcoin".
(c) Macroeconomics	
Gold prices	Gold price in US dollars per troy ounce
Exchange rate	The exchange rate between USD to CNY

we use measures of predictive accuracy. These measures are carried out in addition to calculating the p-values. J Hyndman, and Athanasopoulos, argue p-values could be misleading when two or more IV are correlated to one another. Statistical significance does not always indicate predictive value.

Various measures calculate the predictive accuracy between the dependent and independent variables, such as R2, AIC (Akaike's Information Criterion), BICe (Schwarz's Bayesian Information Criterion). Cross-Validation and there are also other techniques such as the Chi-Squared Test, and RFE (Recursive Feature Elimination) [38].

3.1.5 Unit Root Testing

Many time series include trends, cycles, and seasonality. When choosing a forecasting method, we will first need to identify the time series patterns in the data and then choose a method that is able to capture the patterns properly (J [10]).

Firstly, the data must be tested to check whether it is stationary. Not checking the stationarity degrees of the series, may result in incorrect findings [2]. When using models like ARIMA, it requires its dataset to be stationary. Unit root tests must be conducted on time series data to identify non-stationarity or trends. If it is present, then it must be removed before analysis, as it will have no predictable patterns in the long term (J [10]). Therefore, stationarity will be tested using ADF (Augmented Dickey Fuller) test.

4 Results, Findings, Analysis, and Discussions

4.1 Time Series Analysis

Preparing a multivariate time-series dataset is an intricate procedure. Merging multiple time series datasets without losing or overwriting data.

A shortcoming that occurred while collecting data was that the frequency of data available on the google search hits (SVI) was weekly, while the rest of the time series data were collected daily. To cover the lack of daily data for the SVI variable, the dataset was modified with the feature's missing values interpolating with its mean values. Below is a table showing the summary statistics for the dataset. This sample is fragmented into portions which will be used for training (70%), validation (10%), and testing (20%). Figure 3 presents a summary statistics for daily data.

4.2 Feature Engineering

This paper conducted and compared the results from an AIC, Chi-squared, and RFE test. The model with the lowest value for AIC and the top score of the other two tests is considered the best model for forecasting.

Summary Statistics

	N	Minimum	Maximum	Mean	Std. Deviation
close	1826	778.8397800	67553.94893	15470.65612	16954.38725
open	1826	778.8397830	67554.84000	15446.30799	16943.04089
high	1826	822.8602390	68990.90000	15900.04676	17426.13078
low	1826	748.6983680	66316.00000	14932.58260	16377.80915
estimatedtransactionvolumeusd	1826	112142904.0	1.46426E+10	2004008560	2112128161
ntransactions	1826	124640.0	490644.0	282361.699	55605.4601
hashrate	1826	2147763.141	198514005.7	74789091.54	54229901.87
difficulty	1826	3.17688E+11	2.50465E+13	1.03452E+13	7.52477E+12
costpertransaction	1826	5.612592289	300.3105491	72.06656832	58.82795010
Goldprice	1826	867.4000000	1888.700000	1377.316553	217.8559908
outputvolume	1826	421940.2071	24528670.35	1858919.703	1334891.637
tradevolume	1826	.000000000	4956849516	382889165.1	439863463.9
USDCNYPrice	1826	6.269000000	7.178600000	6.729909853	.2037990750
SVI	1826	4.000000000	100.0000000	18.57471264	5.753845318
Wikiviews	1826	1538	131165	8577.15	9960.120
Valid N (listwise)	1826				

Fig. 3 Summary statistics for daily data

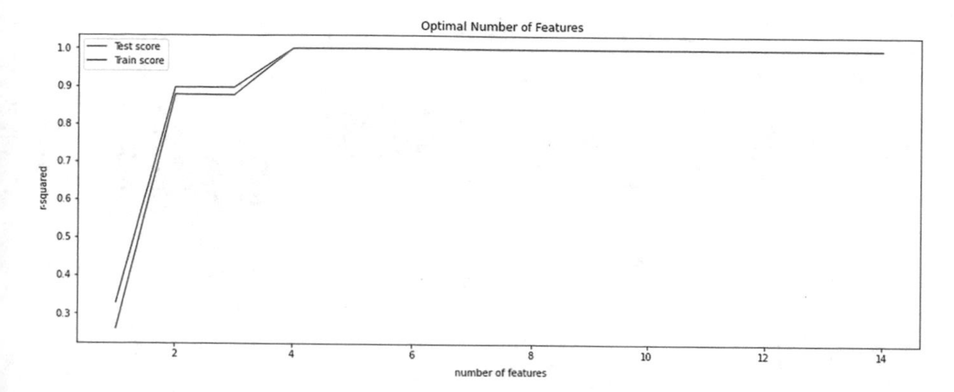

Fig. 4 Feature selection: optimal feature graph

A cross-validation grid search algorithm provided by the Sci-Kit-Learn package was used, with the RFE model as its estimator and r2 as a score metric. The mean scores showed improvement after six features and remained constant after nine features (mean score = 0.999462). Similarly, a cross-validation score with n number of parameters revealed the mean scores peaked at nine features (mean score = 0.999435) and remained constant. This means a minimum of six features should be sufficient for accurate results (Fig. 4).

Following this, a regression model was fit with the nine predictors that ranked first in the two tests (chi-squared and RFE). AIC score and the p-values for each predictor were calculated (AIC = 2.703e+04, Adj.R2 = 0.99, Total features = 14). The tests selected six features (high, open, low, n-transactions, cost-per-transaction, and estimated-transaction-volume-USD), while other features scored above the significance level.

However, as discussed above, disregarding all variables whose p-values are greater than 0.05 is not always the right choice. A completely useless variable can provide a significant performance improvement when taken with others [11]. This is why visual, and statistical tests must be conducted before fitting the model.

4.3 Feature Visualization

A study by Buchholz et al. [37] suggested that visualizing the relationship between two variables should be a good estimate of significance. Below are a few graphs displaying the relationship between the BPI (Bitcoin price) and some of the predictor variables that proved significant. The selected predictors have also been tested by previous literature, and the results appeared to be the same.

Figure 5 shows subplots of three features that show correlation to the dependent variable when visualizing. These three features were chosen regarding its high significance in previous literature. The first plot shows the inverse correlation

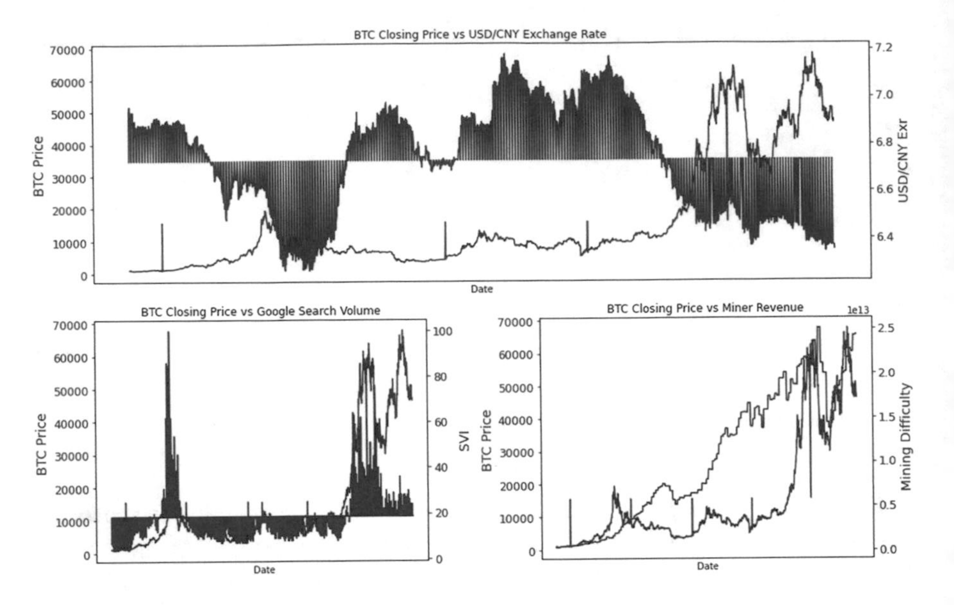

Fig. 5 Correlation plots for features

between the bitcoin price and the USD/CNY exchange rate. Notably, these two features are negatively correlated. In the second row, the first subplot depicts the Google Search hits (SVI) against the Bitcoin Price (BPI). Although the SVI does not indicate whether a search emotion is positive or negative, it is evident that both SVI and BPI displayed a positive correlation. The plot to its right shows the effect mining difficulty has on the bitcoin price. This was also discovered by Kristoufek [6]. As mentioned in the author's literature, Mining difficulty is positively correlated to the bitcoin price

It is also important to understand the difference between correlation and causation. A variable may be useful to forecast another variable. However, that does not necessarily mean x is causing y. The above variables may not necessarily cause the bitcoin price to go up/down, but their change seems to be relative to the Bitcoin price

4.4 Unit Root Result (Fig. 6)

All columns were firstly visualized with ACF and PACF plots to understand the autocorrelation and to determine the number of differences required to make the series stationary. Both plots should exhibit a decaying pattern, indicating the data is stationary. Unfortunately, all the columns displayed a growing/reducing or stable pattern, indicating the data is non-stationary.

ADF has a null hypothesis of a unit root ($d = 1$) against the alternative of no unit root ($d < 1$) [19]. All columns in the dataset were tested for stationarity. A

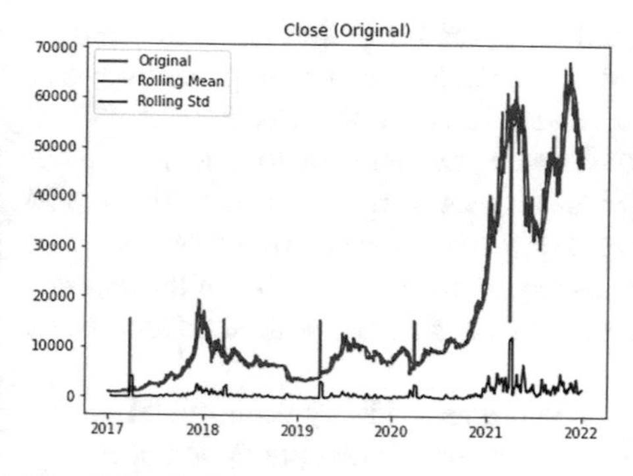

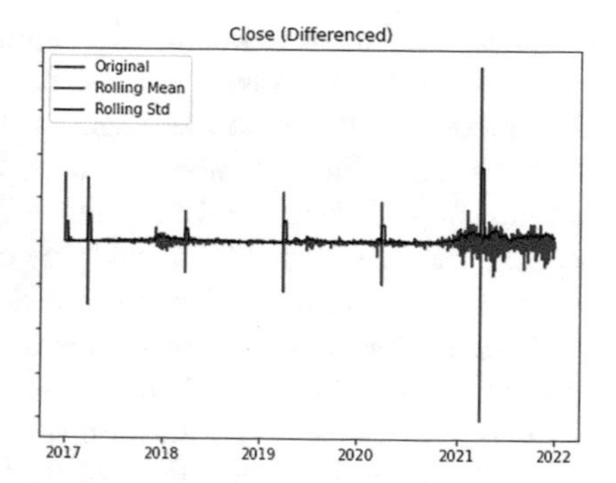

Fig. 6 Original vs. differenced close price

table in Appendix C shows the p-values before and after differencing the data. Most features had p-values greater than the significance level before differencing. After a first order differencing, the p-values were significant (<0.05). Therefore, a first-order differencing should make the dataset stationary.

Further data pre-processing done for the models is explained in Appendix B.

4.5 Time Series Prediction

The following metrics are used in evaluating the accuracy of the model; Mean Absolute Error (MAE), Root Mean Squared Error (RMSE), and Mean Absolute Percentage Error (MAPE). Smaller values of MAPE and RMSE generally indicate more accurate predictions and good performance of a model. In the context of BTC price.

- MAE of 5 means that the predicted price is ± USD 5 from the actual price.
- MAPE quantifies the error in terms of percentage. For example, a MAPE of 3% can mean either USD 3 or 30 depending on whether the actual price is USD 100 or 1000, respectively
- RMSE indicates the spread of the forecast errors. A model that predicts erratic values will have a higher RMSE value, although it may still have had lower MAE or MAPE. Thus, the models should be evaluated concerning all three metrics. [41]

4.6 LSTM Results

Four models were built and tested in search of the model with the best scores and lowest parameters. All four models were compiled with the Adam optimizer and mean squared error as the loss function. Selecting the prediction timestep (number of days to forecast) was an iterative process. The models seemed to perform poorly in windows that were too far ahead or too short. Its most effective timestep found was 30 days. For a time-series task, two layers are enough to find non-linear relationships among the data [13]. LSTM models converged between 20 and 100 epochs, but it was found that anything over 20 epochs showed no improvements or was prone to overfitting. This may be due to the relatively small size of the dataset.

Figure 7 below shows the results obtained by each model with validation (test) data in terms of MAE and RMSE.

Figure 8 shows a visualization of the predictions made by each model on the validation data. It can be observed that mode3 has the closest prediction accuracy. Model 3 outperformed the other models and had the lowest number of trainable parameters (20,289).

The predictions made for the next 30 days showed drastic changes in accuracy metrics compared to validation data (MAPE = 5.756%, RMSE = 2696.141). The graph below shows the predictions made ahead of time compared to the actual prices during that time period. The predictions had an accuracy of 84.32%. The present data were collected using the Yahoo Finance API in python. The vertical dotted line shows the period where the training data ends, and the predicted and present prices begin (Fig. 9).

All models, except one (Model 2 with 4 hidden layers) had only 3 layers, Input, Hidden, and the Output layer. Model 2 evidently took the longest time to train, only to be outperformed by models with fewer layers and complexity. Model 4 and Model 3 were built identically. However, Model 4 was only trained with the features which were found as significant

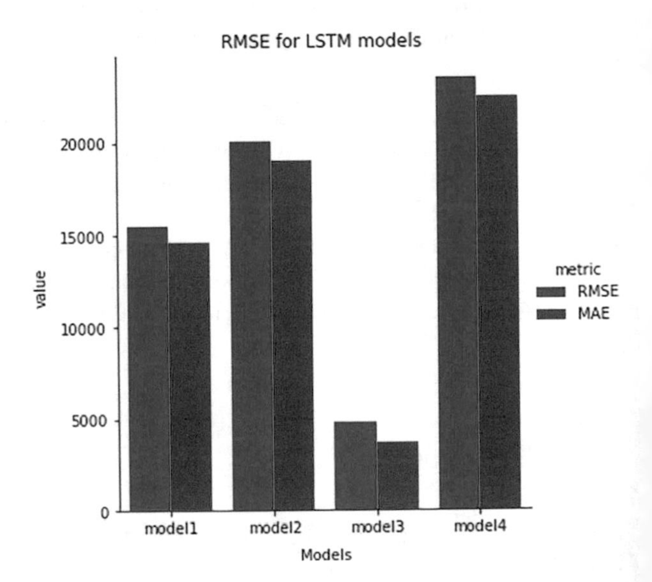

Fig. 7 Accuracy metrics: LSTM

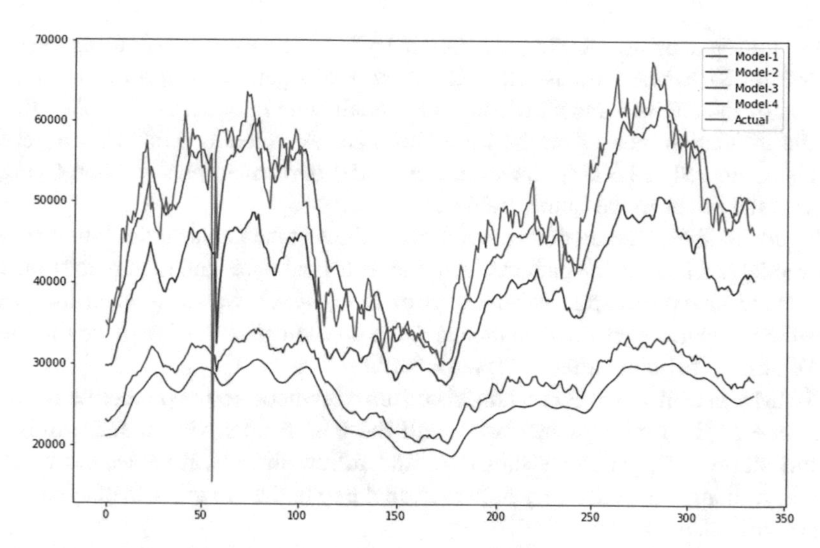

Fig. 8 LSTM model predictions comparisons

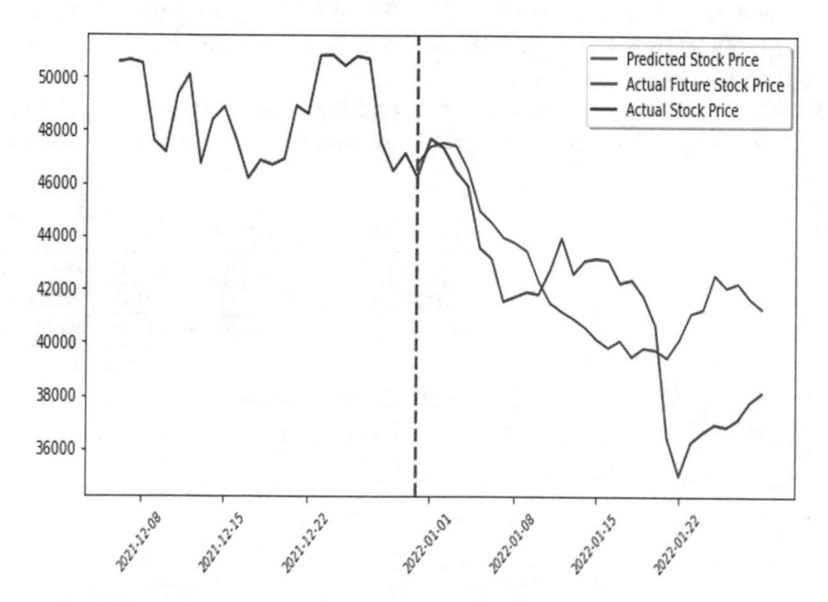

Fig. 9 LSTM future prediction graph

4.7 *ARIMA Results*

The models were configured using the best parameters discovered from a Grid search procedure. This procedure fits a model with different combinations of orders and the model with a desirable AIC score is selected.

In addition to using ARIMA, two SARIMAX models were also fit using the same procedure. SARIMAX (Seasonal ARIMA with eXogenous factors) is an updated version of the ARIMA model where it also deals with exogenous variables. Below are the evaluation metrics on the validation data. All models produced impressive results compared to LSTM. However, the model performed poorly when trying to forecast the prices in the future (Table 2).

Figure 10 shows the model's prediction of the test data against the actual values. The model has learned the patterns from the historical data with high precision. The confidence intervals started to widen from June, which was also when the prices began to fluctuate heavily, which indicates the problem discussed in much literature, the volatility of Bitcoin affects its predictability

ARIMA is well suited for making short-term predictions, the predictive accuracy worsened as the timesteps increased, and the best results were found within 10-30 timesteps. All predictions start high and follow the actual values with decent accuracy, though there is a very high marginal gap in the prices, as displayed by the large confidence interval.

The graph below shows comparisons for n days of forecast against the actual values. Forecast 30 had a lower error (MAPE = 18.141%) compared to forecast 60 (MAPE = 48.994%). Although forecast 5 had the overall lowest MAPE, it is too short of a time frame (Fig. 11).

30 days is an adequate time frame and performs comparatively well. However, Fig. 12 shows a graph of the present prices compared with the predicted values with

Table 2 ARIMA results

Model (p,d,q)	MAPE	RMSE	AIC
ARIMA (1, 1, 2)	2.12%	1255.57	18872.110
SARIMAX (2, 1, 3)	1.79%	1019.90	18883.238
SARIMAX (1, 1, 2)	3.662%	1991.73	19270.158

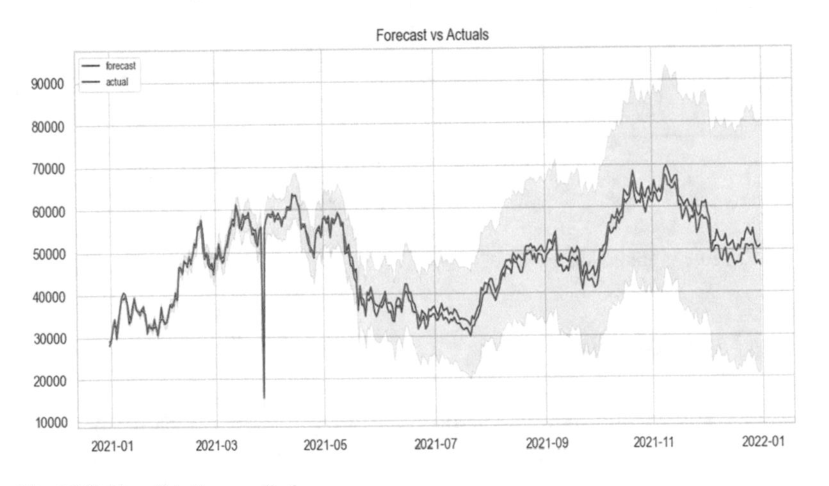

Fig. 10 ARIMA validation prediction

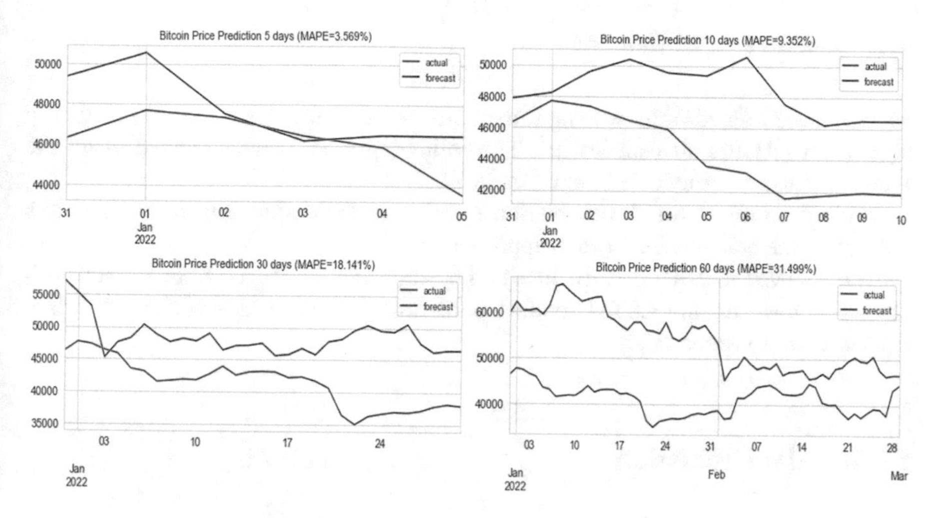

Fig. 11 Prediction 5/10/30/60 days ARIMA

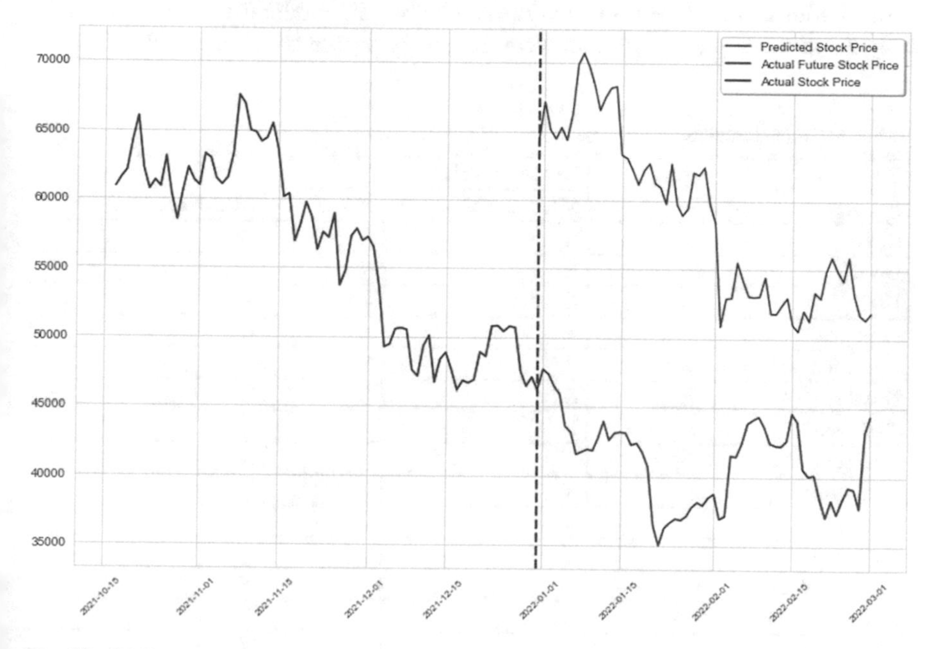

Fig. 12 ARIMA future predictions graph

the historical data. The predictions appear to have identified the direction of change in the prices but were predicted marginally higher than the actual values (Approx. MAE ± USD 17794.66).

Comparing predictive accuracy is not an easy task. Its difficulty is why there are very few agreements as to which model produces the best bitcoin price predictions.

4.8 Feature Significance

Models were trained/fit in two ways, using the features that were found statistically significant (sf) and all features (tf). The following models were trained using the features marked in the table below (Table 3).

Table 4 above shows the evaluation metrics on validation data by the features used from the best models in each approach.

The ARIMA model's performance had no substantial changes when trained with sf or tf. However, the LSTM models performance improved significantly when trained with tf instead of sf.

5 Results Discussion

The graph below shows the Prediction-1, and Prediction-2 from the LSTM models trained with *sf* and *tf,* respectively. It can be seen the model that was trained with only the sf made predictions that were extremely higher than the actual values but

Table 3 Trained models

	LSTM Models		
Feature	M1, M2, M3	M4	ARIMA Models
Close	X	X	X
Open	X	X	X
High	X	X	X
Low	X	X	X
Estimated trans. Vol. USD	X	X	
N-trans.	X	X	X
Hash rate	X		
Cost per trans.	X		X
Gold Price	X		X
Output Vol.	X		X
Trade Vol.	X		
USD-CNY Exr.	X		X
SVI	X	X	X
Wiki views	X		X

Table 4 Evaluation metrics

	Sf		tf	
Model	MAPE	RMSE	MAPE	RMSE
LSTM: Model 3	34.833%	14601.12	5.756%	2696.14
ARIMA: ARIMA (1,1,2)	3.662%	1991.739	4.242%	2282.727

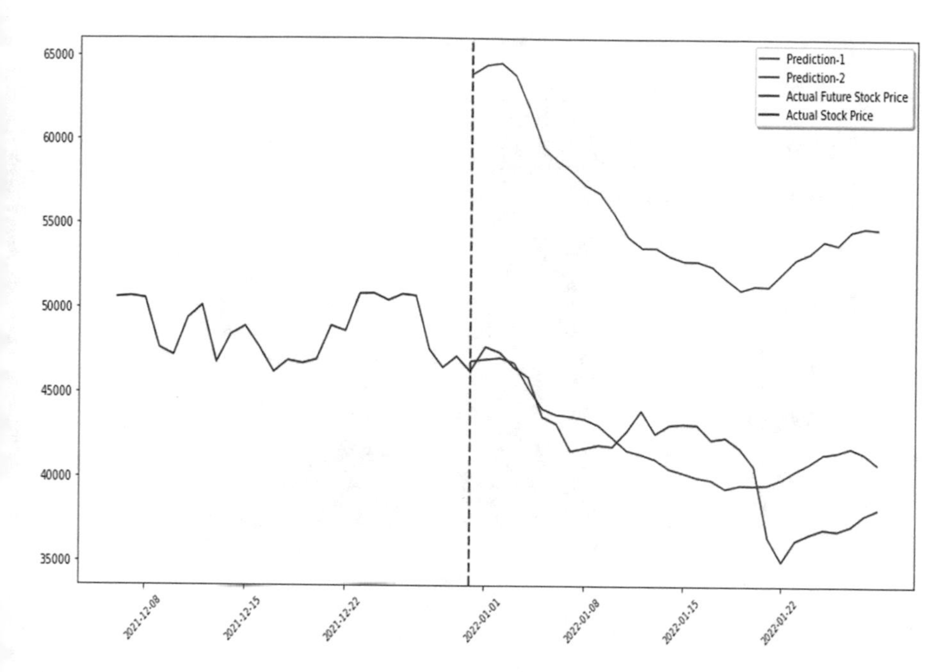

Fig. 13 LSTM prediction with sf and tf

appeared to have identified the direction of change, while the other model produced significantly better predictions that fit closer to the actual price (Fig. 13).

This proves that the selected features help the model learn well. Unfortunately, it has a high error rate (Fig. 14).

Figure 15 displays the evaluation metrics for all models, out of which the LSTM model produced the results with the highest accuracy. On the other hand, ARIMA's evaluation had an overall lower error than LSTM. Additional findings are discussed in Appendix D.

The results based on the training data show that both models could successfully learn and predict the price directions ahead of a certain temporal frame. ARIMA performed significantly better on validation data compared to the LSTM models. The LSTM achieved a MAPE of 8.08% on validation data, while the ARIMA achieved 3.6%. Contrastingly, LSTM outperformed ARIMA when forecasting prices in the future.

Although ARIMA had obtained the overall lowest metrics, it does not indicate the model has good performance but has done a decent job in identifying the price's direction. A study by McNally et al. [13] also obtained excellent numbers in terms of error. However, its predictive accuracy was imbalanced. Upon further analysis of the results of ARIMA, it was evident the model's predictions after the 20 days began to replicate the patterns and noise from the historical data with marginal errors. Figure 16 shows the predictions made for 90 days ahead using the past 90 days in the

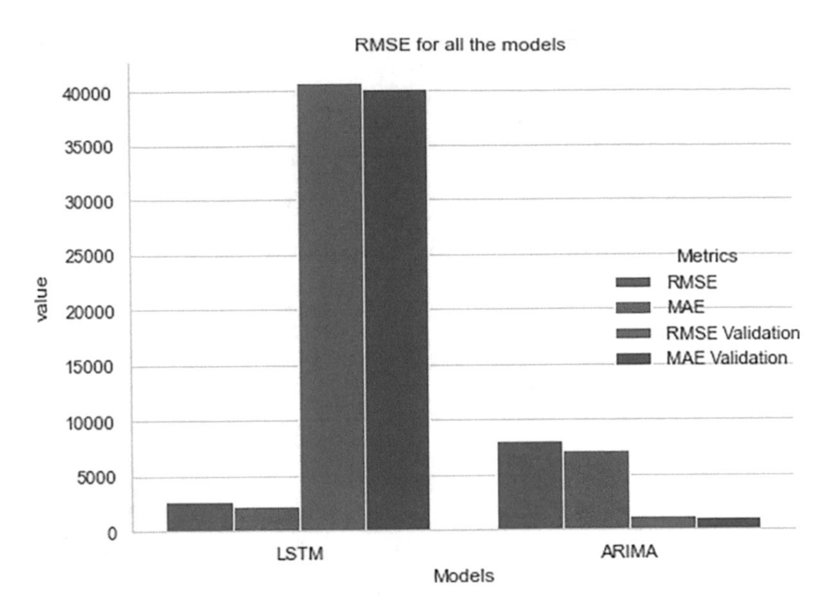

Fig. 14 Evaluation metrics for all models

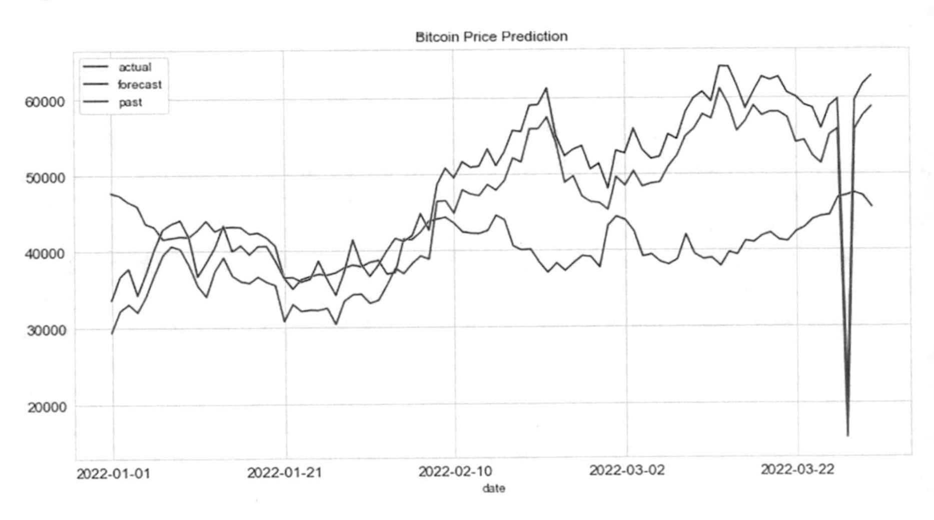

Fig. 15 ARIMA Comparison chart for predicted/actual/past prices

dataset. The green line shows the past prices relative to the given future dates, the forecast replicates the patterns and noises too closely.

Finally, the graph below shows both model's predictions and the present price of Bitcoin. Both models were given the past 30 days as input to forecast the next 30 days. ARIMA was unsuccessful in predicting the direction of the price ahead. Instead, it has made a forecast that looks much like the data it was trained with.

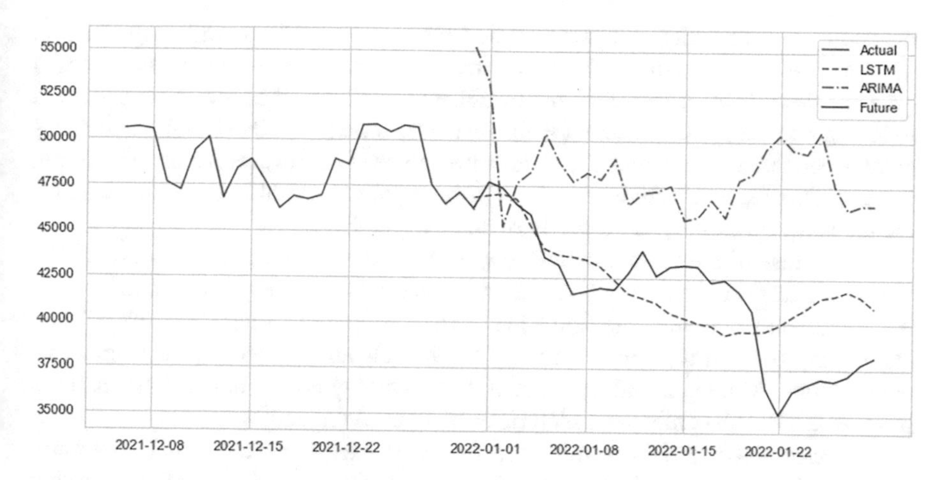

Fig. 16 Model prediction comparison graph

LSTM, on the other hand, has done an excellent job in finding the direction of the price. Though it is not precise, the line follows the actual price's direction

6 Conclusion and Future Work

The results show it is possible to predict the Bitcoin price with low error rates, therefore rejecting the null hypothesis. However, prediction of time series problems is inherently difficult, especially the bitcoin price index, as it is affected by various external factors. The primary concern when trying to predict the said type of data, its limitations must be taken into consideration, which is: the models (1) do not account for exogenous variable uncertainty, (2) do not account for the fact that forecast-error variances vary across time [29].

In this experiment, both models complement each other extremely well. LSTM is best suited to make long-term forecasts. Contrastingly, ARIMA is most useful when making short-term forecasts. LSTM produced the best results when forecasting the future horizon (94.2% predictive accuracy, 82.7% validation accuracy) compared to ARIMA (82.7% predictive accuracy, 98.13% validation accuracy), ARIMA was not very good at making forecasts with past data,. However, it learned the historical data with high accuracy, LSTM, on the other hand, was able to identify the direction of change in the price successfully. However, ARIMA performed much better in shorter time frames (Day 10 Accuracy = 90.65%).

BTC prices are stochastic, and no given set of features can provide a complete forecast [41]. With that being said, many other researchers have previously used internal and external factors to classify the increase/decrease of BTC price. It was discovered the arrival of new information impacts the Bitcoin Price positively. For instance, the time series analysis revealed the queries made online about Bitcoin (SVI and Wiki views) had a stronger impact in the early stages and decreased gradually after Bitcoin became more established. Various other social factors could help improve the model's accuracy, for instance, Twitter sentiment data, which has already been used in previous literature with debate about its usefulness. In addition to search trends, user activity in trading/forum sites could also be collected. This feature was also used by Lamothe-Fernández et al. [4].

In conclusion, the results of this paper show it is possible to forecast BTC's direction with decent error rates. At the same time, it is extremely difficult to forecast its rise/fall in price with precision [41]. The results are satisfactory overall and can be improved further by fine-tuning the models and increasing the number of observations. Although, making forecasts for a highly erratic and volatile asset is prone to errors and comes with high risks when used for trading.

Future research on this study could focus on integrating tools that constantly automate the data retrieval and update the model. The creator says of Bitcoin that Bitcoin will stop its supply after 21 million bitcoins. As of now, it has almost reached 90% of its maximum supply. This factor may affect the current model's predictions depending on how the market reacts to this. Therefore, progressively updating the model with new observations and data will lead to up-to-date results close to approximation. In addition to that, the models are built with minimal complexity. In the future, more complex neural networks can be modeled to comprehensively learn the changes in global events and combine an anomaly detection mechanism to assess the stability of Bitcoin and other exogenous factors.

Appendices

Appendix A: Data Retrieval Sources

Bitcoin price index, and Blockchain data were obtained on 18 February 2021 from the Blockchain API (https://blockchain.com/). In addition to that, the other exogenous variables, such as Gold Price and Exchange Rates were extracted from Investing.com: (https://uk.investing.com/commodities/gold,https://uk.investing.com/currencies/usd-cny-historical-data, respectively). Search volume data were retrieved by accessing the Google Trends website (http://www.google.com/trends) on 22 February 2021 and the Wikipedia article traffic statistics site (https://www.wikishark.com/) on 2 February 2021.

All Bitcoin and Blockchain historical data were validated and cross-checked for accuracy by comparing data in popular sources like CoinDesk (https://www.coindesk.com/price/bitcoin/), and Yahoo Finance (https://finance.yahoo.com/).

Appendix B: Model Data Pre-processing

LSTM: When training a network with data with large range of values, as large input values can slow down the learning and sometimes can prevent the network from learning effectively. Therefore, the dataset was scaled using the MinMax Scaler available in the sklearn library. This process scales the dataset values to fit between 0 and 1.

ARIMA: Selecting the order of the model is crucial to build a good model. Firstly, the data was explored as discussed in the Results section, unit root tests were conducted to identify the best order of differencing, and in search of the best model, a grid search with different orders was iteratively fit. Finally, the best model was determined by plotting the residuals and comparing AIC scores.

Appendix C: ADF Test Scores

	Differencing order		
Column	No-diff	diff(0)	diff(1)
Close	0.868240		0.000000e+00
Open	0.858296		0.000000e+00
High	0.873539		0.000000e+00
Low	0.856106		0.000000e+00
Estimated-transaction-volume-usd	0.361389		3.769493e-21
n-transactions	0.077388		1.535323e-19
Hash-rate	0.899546		4.206086e-26
Difficulty	0.948766		1.827273e-21
Cost-per-transaction	0.577504		2.476612e-14
Gold price	0.885700		6.186569e-21
Output-volume	0.023501	0.023501	
Trade-volume	0.080964		6.361866e-22
USD-CNY Price	0.693700		9.939304e-21
SVI	0.006120	0.006120	
Wikiviews	0.016293	0.016293	

Appendix D: Findings

The boxplot shows the closing prices of each month throughout all the years, the price gradually grew during the first 4 months and began to decline over the next 5

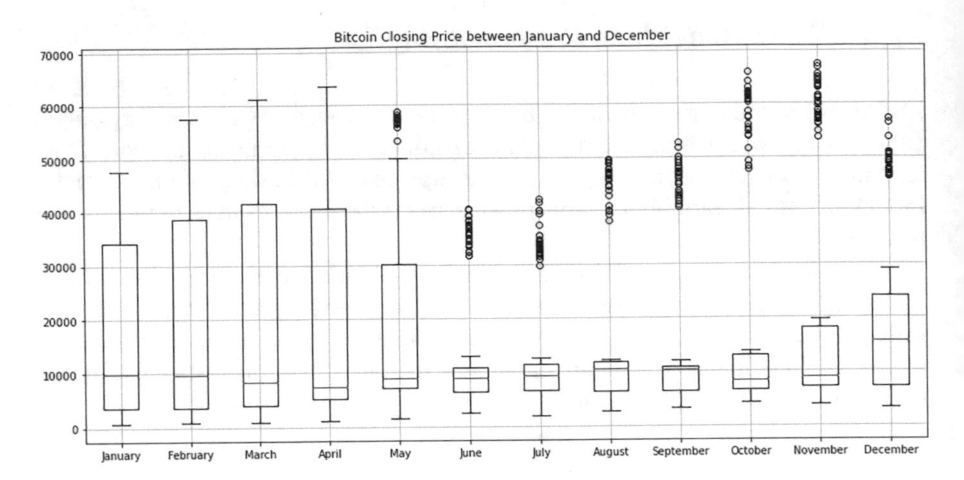

Fig. 17 BTC Closing Price throughout the decade

months, and finally beginning to rise again. It must also be noted there are plenty of outliers following May, this could be due to the high gains for BTC during 2017 and 2019 May, which accumulated for over 50%. This could be what is causing the model to predictions with higher MAE, the smaller size the of observations was not comprehensive for the model to identify the outliers (Fig. 17).

References

1. Teker, D., Teker, S. and Ozyesil, M. (2019). Determinants of Cryptocurrency Price Movements. In: LAHSS-19, MEEIS-19 Nov. 12–14, 2019, Paris (France). 12 November 2019. Higher Education And Innovation Group. [Online]. Available at: doi:10.17758/HEAIG6.H1119510 [Accessed 18 November 2021].
2. Elder, J. and Kennedy, P. E. (2001). Testing for Unit Roots: What Should Students Be Taught? The Journal of Economic Education, 32 (2), pp.137–146. [Online]. Available at: https://doi.org/10.1080/00220480109595179 [Accessed 18 November 2021].
3. Selva, P. (2019). Augmented Dickey Fuller Test (ADF Test) – Must Read Guide. Augmented Dickey Fuller Test (ADF Test) – Must Read Guide. [Online]. Available at: https://www.machinelearningplus.com/time-series/augmented-dickey-fuller-test/ [Accessed 18 November 2021].
4. Lamothe-Fernández, P. et al. (2020). Deep Learning Methods for Modeling Bitcoin Price. Mathematics, 8 (8), p.1245. [Online]. Available at: https://doi.org/10.3390/math8081245 [Accessed 18 November 2021].
5. Balcilar, M. et al. (2017). Can volume predict Bitcoin returns and volatility? A quantiles-based approach. Economic Modelling, 64, pp.74–81. [Online]. Available at: https://doi.org/10.1016/j.econmod.2017.03.019.
6. Kristoufek, L. (2015). What Are the Main Drivers of the Bitcoin Price? Evidence from Wavelet Coherence Analysis. Scalas, E. (Ed). PLOS ONE, 10 (4), p.e0123923. [Online]. Available at: https://doi.org/10.1371/journal.pone.0123923.

7. Choi, H. and Varian, H. (2009). Predicting the present with Google trends. Google Research. Google Inc., Mountain View, CA.
8. Da, Z., Engelberg, J. and Gao, P. (2011). In Search of Attention. The Journal of Finance, 66 (5), pp.1461–1499. [Online]. Available at: https://doi.org/10.1111/j.1540-6261.2011.01679.x.
9. Bouoiyour, J. and Selmi, R. (2015). What Does Bitcoin Look Like? Annals of Economics and Finance, 16, pp.449–492.
10. Hyndman, J. R. and Athanasopoulos, G. (2018). Forecasting: Principles and Practice (2nd ed). [Online]. Available at: https://Otexts.com/fpp2/ [Accessed 9 December 2021].
11. Guyon, I. and Elisseeff, A. (2003). An Introduction to Variable and Feature Selection. p.26.
12. Ariyo, A. A., Adewumi, A. O. and Ayo, C. K. (2014). Stock Price Prediction Using the ARIMA Model. In: 2014 UKSim-AMSS 16th International Conference on Computer Modelling and Simulation. March 2014. pp.106–112. [Online]. Available at: https://doi.org/10.1109/UKSim.2014.67.
13. McNally, S., Roche, J. and Caton, S. (2018). Predicting the Price of Bitcoin Using Machine Learning. In: 2018 26th Euromicro International Conference on Parallel, Distributed and Network-based Processing (PDP). March 2018. pp.339–343. [Online]. Available at: https://doi.org/10.1109/PDP2018.2018.00060
14. Wołk, K. (2020). Advanced social media sentiment analysis for short-term cryptocurrency price prediction. Expert Systems, 37 (2), p.e12493. [Online]. Available at: https://doi.org/10.1111/exsy.12493.
15. Tandon, C. et al. (2021). How can we predict the impact of the social media messages on the value of cryptocurrency? Insights from big data analytics. International Journal of Information Management Data Insights, 1 (2), p.100035. [Online]. Available at: https://doi.org/10.1016/j.jjimei.2021.100035.
16. Brownlee, J. (2016). Feature Selection For Machine Learning in Python. Machine Learning Mastery. [Online]. Available at: https://machinelearningmastery.com/feature-selection-machine-learning-python/ [Accessed 13 December 2021].
17. Bitcoin.org. FAQ – Bitcoin. 2021 [Online]. Available at: https://bitcoin.org/en/faq [Accessed 14 December 2021].
18. Greaves, A. and Au, B. (2015). Using the Bitcoin Transaction Graph to Predict the Price of Bitcoin. p.8.
19. Kristoufek, L. (2013). BitCoin meets Google Trends and Wikipedia: Quantifying the relationship between phenomena of the Internet era. Scientific Reports, 3 (1), p.3415. [Online]. Available at: https://doi.org/10.1038/srep03415.
20. Urquhart, A. (2016). The inefficiency of bitcoin. Economics Letters, 148, 80–82.
21. Tinawi, I. (2019). Machine Learning for Time Series Anomaly Detection. p.55.
22. Rather, A. M., Agarwal, A. and Sastry, V. N. (2015). Recurrent neural network and a hybrid model for prediction of stock returns. Expert Systems with Applications, 42 (6), pp.3234–3241. [Online]. Available at: https://doi.org/10.1016/j.eswa.2014.12.003.
23. Giles, C. L. (2001). Noisy Time Series Prediction using Recurrent Neural Networks and Grammatical Inference. p.23.
24. Valencia, F., Gómez-Espinosa, A. and Valdés-Aguirre, B. (2019). Price Movement Prediction of Cryptocurrencies Using Sentiment Analysis and Machine Learning. Entropy, 21 (6), p.589. [Online]. Available at: doi:https://doi.org/10.3390/e21060589.
25. Guo, Q. et al. (2021). MRC-LSTM: A Hybrid Approach of Multi-scale Residual CNN and LSTM to Predict Bitcoin Price. arXiv:2105.00707 [cs, q-fin]. [Online]. Available at: http://arxiv.org/abs/2105.00707 [Accessed 17 December 2021].
26. Wang, J.-J. et al. (2012). Stock index forecasting based on a hybrid model. Omega, 40 (6), pp.758–766. [Online]. Available at: https://doi.org/10.1016/j.omega.2011.07.008.
27. Cermak, V. (2017). Can Bitcoin Become a Viable Alternative to Fiat Currencies? An Empirical Analysis of Bitcoin's Volatility Based on a GARCH Model. SSRN Scholarly Paper, Rochester, NY: Social Science Research Network. [Online]. Available at: https://doi.org/10.2139/ssrn.2961405 [Accessed 17 December 2021].

28. Xie, Y., Ueda, Y. and Sugiyama, M. (2021). A Two-Stage Short-Term Load Forecasting Method Using Long Short-Term Memory and Multilayer Perceptron. Energies, 14 (18), p.5873. [Online]. Available at: https://doi.org/10.3390/en14185873.
29. Fair, R. C. (1986). Chapter 33 Evaluating the predictive accuracy of models. In: Handbook of Econometrics. 3. Elsevier. pp.1979–1995. [Online]. Available at: https://doi.org/10.1016/S1573-4412(86)03013-1 [Accessed 19 January 2022].
30. Matta, M., Lunesu, I. and Marchesi, M. (2015). Bitcoin Spread Prediction Using Social And Web Search Media. p.10.
31. Nochai, R. and Nochai, T. (2006). ARIMA MODEL FOR FORECASTING OIL PALM PRICE. p.7.
32. Khashei, M., Bijari, M. and Raissi Ardali, G. A. (2012). Hybridization of autoregressive integrated moving average (ARIMA) with probabilistic neural networks (PNNs). Computers & Industrial Engineering, 63 (1), pp.37–45. [Online]. Available at: https://doi.org/10.1016/j.cie.2012.01.017.
33. Pai, P.-F. and Lin, C.-S. (2005). A hybrid ARIMA and support vector machines model in stock price forecasting. Omega, 33 (6), pp.497–505. [Online]. Available at: https://doi.org/10.1016/j.omega.2004.07.024.
34. Zhu, Y., Dickinson, D. and Li, J. (2017). Analysis on the influence factors of Bitcoin's price based on VEC model. Financial Innovation, 3. [Online]. Available at: https://doi.org/10.1186/s40854-017-0054-0.
35. Olah, C. (2015). Understanding LSTM Networks – colah's blog. [Online]. Available at: http://colah.github.io/posts/2015-08-Understanding-LSTMs/ [Accessed 17 February 2022].
36. Barão, S. M. M. (2008). Linear and Non-linear time series analysis: forecasting financial markets. [Online]. Available at: https://www.semanticscholar.org/paper/Linear-and-Non-linear-time-series-analysis%3A-markets-Bar%C3%A3o/c61536dba552c6a874f43c556e529eeb6e5409e3 [Accessed 17 February 2022].
37. Buchholz, M. et al. (2012). Bits and Bets Information, Price Volatility, and Demand for Bitcoin. p.48.
38. Paul, S. (2020). Feature Selection Tutorial in Python Sklearn. [Online]. Available at: https://www.datacamp.com/community/tutorials/feature-selection-python [Accessed 10 March 2022].
39. Edwards, J. (2022). Bitcoin's Price History. [Online]. Available at: https://www.investopedia.com/articles/forex/121815/bitcoins-price-history.asp [Accessed 17 March 2022].
40. Aggarwal, G. et al. (2019). Understanding the Social Factors Affecting the Cryptocurrency Market. arXiv:1901.06245 [cs]. [Online]. Available at: http://arxiv.org/abs/1901.06245 [Accessed 17 March 2022].
41. Mudassir, M. et al. (2020). Time-series forecasting of Bitcoin prices using high-dimensional features: a machine learning approach. Neural Computing and Applications. [Online]. Available at: https://doi.org/10.1007/s00521-020-05129-6 [Accessed 24 March 2022].
42. Raudys, A., & Mockus, J. (1999). Comparison of ARMA and multilayer perceptron based methods for economic time series forecasting. *Informatica, 10*(2), 231–244.
43. Box, G. E. P. and Jenkins, G. M. 1967. *Statistical Models for Prediction and Control*, Madison, Wisconsin: Department of Statistics, University of Wisconsin. Technical Reports #72, 77, 79, 94, 95, 99, 103, 104, 116, 121, and 122
44. Ciaian, P., Rajcaniova, M., & Kancs, D. A. (2016). The economics of BitCoin price formation. *Applied economics, 48*(19), 1799–1815.

Blockchain-Based Novel Solution for Online Fraud Prevention and Detection

Ankit Mundra, Jai Mishra, Harshit Jha, and Chityanj Sharma

1 Introduction

In recent years, online fraud and identity theft have affected many mentally and physically. According to studies and surveys conducted in recent years, approximately one in 15 people are likely to be a victim of online fraud. The most popular type of fraud in an online space is identity or financial data theft, mostly leaked through suspicious online or mobile applications. No system or human is 100% secured from fraud. However, fraud detection algorithms built using artificial intelligence (AI) techniques can help to reduce these risks substantially. Misusing this private information includes opening new accounts, controlling the victim's credit accounts, and obtaining government benefits [1]

Further, as per Facebook security Scam: In 2012, a popular scam got caught involving tech giant and social media website Facebook Hackers attacked and stole various IDs, accounts, passwords, and personal information like credit card information, etc. using Facebook accounts and they use that information to influence and chat with another account by creating fake identity [2]. They used Facebook replicas to get users' information and steal credit cards and various personal information, which led to one of the biggest frauds related to identity theft.

Thus, it is necessary to modify and upgrade the current web system to detect and prevent the identity theft problem, which can help users completely control their personal data and enhance security from online fraud.

In this paper, we are presenting our proposed system of having a plug-and-play identity based on blockchain-based identity verification, whereby it is a system that stores individual personal data on the blockchain. This blockchain can be hosted on users' local storage or stored at any point on another popular blockchain network.

A. Mundra (✉) · J. Mishra · H. Jha · C. Sharma
Department of Information Technology, Manipal University Jaipur, Jaipur, Rajasthan, India

Y. Maleh et al. (eds.), *Blockchain for Cybersecurity in Cyber-Physical Systems*, Advances in Information Security 102, https://doi.org/10.1007/978-3-031-25506-9_12

The system uses security features of blockchain such as transparency, immutability, and non-repudiation. So that you can control your data and know who has access to your identity, this system will have three types of consumers: user, authority, and third-party users. The user can give access to their data to any third-party websites and also, and he can access the list of requestors. The authority will be able to approve and verify the user's data uploaded on the blockchain, and the third-party websites can request data from the user. Blockchain was introduced with the invention of Bitcoin in 2008. Its practical implementation occurred in 2009. For this chapter, it is sufficient to review Bitcoin very briefly. However, it is essential to refer to Bitcoin because without it, the history of blockchain is not complete, as blockchain is a decentralized ledger consisting of blocks connected like a linked list where each block has an address that records ownership and is continuously updated after being verified.

The problems identified in the current traditional system of verifying identities such as identity theft and leakage of data through various websites. Secondly, personal information is kept in a centralized database which may lead to identity theft and data misuse. Our proposed system is assumed to be a storage plug-and-play identity system for an individual's personal information, and others can access it to verify identity during registration processes. This identity which is verified, will be considered authentic, and the user will be able to use it to connect to various websites.

1.1 Why Blockchain?

Blockchain technology provides secure transactions, reduce intermediary costs, is fast, and it also helps in the verification of every transaction. Every node connected to the blockchain network receives a copy of every blockchain transaction between any two users. Blockchain technology works with smart contracts and audits the origin of the product. It ensures non-repudiation i.e. it provides assurance that the sender receives proof of delivery and the recipient will get the proof of identity of the sender. Non-repudiation means no one can deny the authentication of the transaction done between users [3].

1.2 Advantages of Using Blockchain

(a) A decentralized distributed ledger is available to every user on the blockchain, and it is immutable i.e., it cannot be modified a transaction after it is recorded.
(b) All the transactions are fast and transparent, and transaction record is updated automatically on every node on the system.
(c) the Whole process is encrypted, so it is always secure.

(d) The system is decentralized, meaning there is no intermediatory node between any two users on the blockchain, i.e., no central node. This implies that no intermediary fee is required [5].
(e) Authentication and verification can be done on any node connected to the system and confirmed by the participants.

1.3 Applications of Blockchain Banking Application

In the current system, banks charge transfer fees in transactions which is both more costly and time-consuming. For overseas payments, there are other additional charges which are also added to the transaction amount. With the help of blockchain technology, we can directly eliminate the cost of middlemen. It provides a peer-to-peer payment system that is highly secure and least costly [3, 4].

1.3.1 Cyber Security

Cyber-attacks are widespread nowadays and cause a significant threat to users. So we require an effective solution to provide security against unauthorized access to our data. Blockchain provides peer-to-peer connections, which can easily identify any malicious attack by any third-party user.

1.3.2 Supply Chain Management

It is a fast-growing and demanding sector, and it faces various challenges in maintaining communication between multiple departments working under it. Without blockchain, it faces a lack of transparency and coordination, which affect the overall cost of the product it offers. With the help of blockchain, as the system is decentralized, it provides easy traceability across the entire supply chain. Each transaction is recorded in the blockchain and stored in a block. Anyone can verify the authenticity of the product being delivered.

1.3.3 Health Sector

Currently, in the traditional system, the user must collect reports and data from the other hospitals. There is a high chance of data leaks and corruption as this data is stored in the local storage of that particular hospital. Also, this process is a little bit difficult for the patients as it is a slow procedure. Blockchain also removes the central authority that enables patients to access their data anywhere at any time instantly. Also, as it's a transparent system and a copy of every transaction is available to every other person, it is challenging for a hacker to corrupt the data, which is very beneficial for the proper functioning of health sectors.

1.3.4 Governance

As In traditional voting systems, to reduce fraud and rigged voting, we can implement blockchain by which every vote is counted with higher accuracy, and we can count only one vote from a unique digital ID. Voters need not disclose their identity in public. At the same time, voting can also ensure safety and security and might increase the number of voters, which is a good thing in a democratic election. As soon as the vote is added to the public ledger, the information can never be erased or modified in the future.

The next section discusses the surveys done in the past, and we will conclude why we need a new system that can overcome these problems and provide remarks on the future scope of this work.

2 Types of Online Fraud

As we know, online and especially e-commerce websites have now become a common target for attackers in this industry is increasing. This era has presented these fraudulent attackers with a great opportunity to get a considerable amount of personal data. This data is easily available around the website's users frequently visit, making it easier for these fraudsters to fetch data and maintain anonymity. By using this information, attackers get a great opportunity to attack victims. According to 2017 Experian report, there was an increase of about 30% increase of attacks of online fraud as compared to previous years. Additionally, there were around 16.7 million reports in which the attackers fetched victims' identities, eventually leading to fraud attacks and transactions. This shows how vulnerable and important this problem is.

One of the most important and common types of fraud includes transaction and personal details, which attackers generally use. These details usually are fetched by the websites that attackers target.

2.1 Types of Transactional Frauds

Identity Fraud and Payment Theft This is the most vulnerable and famous kind of online fraud, which has grown by around 71% in recent years (year-wise details shown in Fig. 1). In most frauds, attackers use this method to gain illegal access to users' personal information. This fraud sometimes involves card and transaction

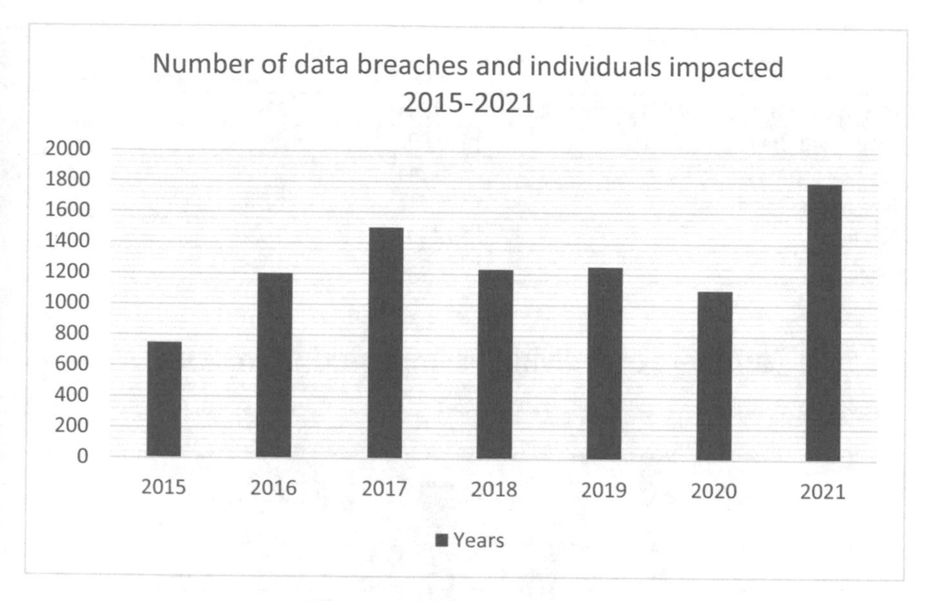

Fig. 1 Number of data breaches and individuals impacted, 2015–2021

details that attackers gain from websites users frequently visit. This paper and the proposed system have a primary target to deal with this fraud [13, 14].

Friendly Fraud This type of fraud happens when the attackers make contact with the owner like they convince them as they are from a credit card company and trying to convince them to give personal information. Once the attackers retrieve the information, they can conduct anonymous activities, eventually leading to transactional fraud.

Triangulation Fraud This is also one of the common frauds in which attackers create a replica of the website. The authentic customer places the order on these false websites and never gets any response or product they ordered. The goods usually don't exist on these websites or are never shipped. In this method, the attacker usually receives the target customer's payment details. Thus, triangulation fraud is getting users' credentials and using them for other online purchases.

Clean Fraud The next type of fraud on the list is Clean Fraud. These are the transactions that usually do not get detected because the transactions appear to be legitimate.

Fraudsters get all the private and personal information of the user's data, including card details, and they use that details to make a fraudulent transaction which is not easily detected by usual transaction systems. Retailers never get information about the transactions' authenticity; attackers primarily use this point

Table 1 Top five types of identity theft

Types of identity theft	No. of reports	Percent of total top five
Government benefits applied for/received	385,264	31.0%
Credit card -fraud new accounts	363,092	29.2%
Miscellaneous identity theft	300,244	24.1%
Business/personal load	105,711	8.5%
Tax fraud	89,649	7.2%
Total	12,43,960	100%

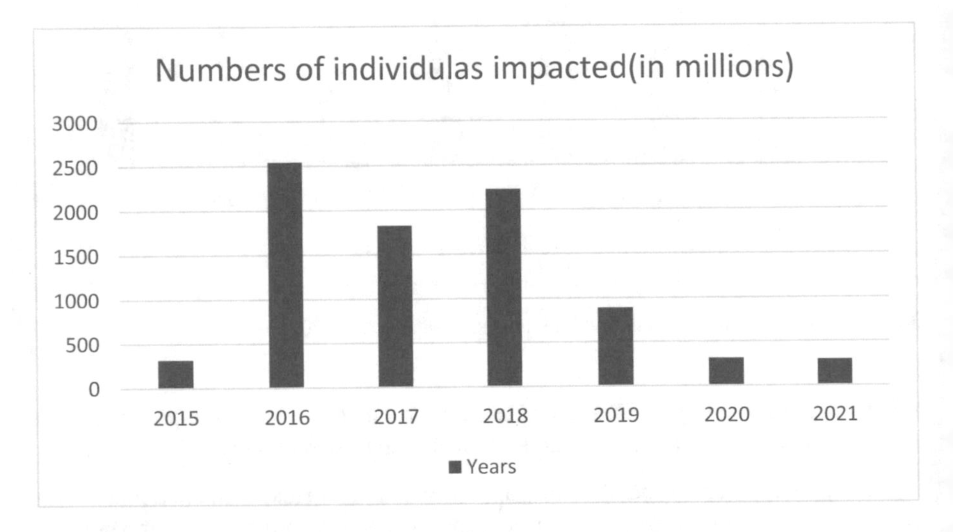

Fig. 2 Number of individuals impacted (in millions)

for this kind of fraud. Our proposed system also attempted to deal with these kinds of fraud.

Affiliate Fraud This kind of fraud usually targets affiliate websites. Many websites run affiliate programs to grab customers' attention. Hackers usually generate false traffic by which merchant thinks they are getting the real and legitimate attention of the users. But in reality, those customers never exist. We can think of refreshing a website multiple times to generate traffic [7].

Further, Table 1 shows the different types of identity thefts, the number of reports filed against each type, and their percentage contribution to overall online fraud incidents. In Fig. 2, we presented the number of individuals (fraud victims) impacted (in millions) through different types of identity theft incidents [6].

3 Literature Review

In this section, we have elaborated on the literature finding on online fraud prevention and detection techniques. Further, attackers, hackers, and unauthorized people are getting smarter daily in this fast-paced digital world. Around 15 billion user credentials from various data breaches are circulating among these cyber-criminals [12].

Direct visibility of these credentials is very risky as these data fraudsters can do illegal activities related to the user's bank account and credentials [9].

This leads to the point that normal username - password data is highly vulnerable and won't be sufficient. To get more information about the present scenario, we must first learn about different authentication systems and their drawbacks.

3.1 Types of Traditional Authentication Systems

Using Username/Email and Password This is one of websites' oldest and most vulnerable authentication factors.

Limitations. Attackers can easily get access to the whole account if they steal the credentials.

One Time Password (OTP) Based Authentication This is very common and used in various websites as a factor of authentication. Users must enter the authentication code received in their emails or registered mobile numbers [10].

Limitations.

(i) Security tokens used in this authentication can fail or break.
(ii) There are various brute force attacks where attackers can steal information, and this security measure can fail.

Biometric Authentication These are also very popular and advanced types of authentication methods. Users must upload their biometrics by face recognition, retina, fingerprint, etc. This information is stored in the database and used as a method to access it.

Limitations.

(i) It is considered a risky method as if they are exposed, to the individual whose personal data is compromised.
(ii) Cost of using this authentication is higher, and also it involves a database, which can be risky.

Normal Blockchain-Based Authentication Blockchain is a decentralized public ledger that records data exchange on its network. So, for data exchange security, this

Table 2 Comparison of different machine learning classification techniques

Methods	Accuracy	Precision	MCC
Local outlier factor	0.8990	0.0038	0.0172
Isolation forest	0.9011	0.0147	0.1047
Support vector machine	0.9987	0.7681	0.5257
Logistic regression	0.9990	0.8750	0.6766
Decision tree	0.9994	0.8854	0.8356
Random forest	0.9994	0.9310	0.8268

authentication method can be used as it verifies the user's identity while protecting the password and user's credentials. The whole blockchain network is used to enhance data integrity [11].

Limitations.

(i) In traditional systems, there is no check of human and identity verification, leading to security vulnerability.

Further, various machine learning-based classification techniques have been mentioned in [8] [9] to make systems and transactions more secure. In [8] author proves that the Random Forest Technique would give the best result for performing classification. Table 1 presents the performance comparison of various machine learning techniques used for fraud detection based on transaction activities [8] (Table 2).

4 Proposed Solution

We have already discussed the type of current and traditional security measures and their limitations in the previous section. This section discusses our proposed solution, which involves creating a plug-and-play digital identity for the user.

Our proposed system comprises two phases:

(A) **Prevention**

This phase comprises our blockchain-based login system, as discussed earlier. As this part deals with the prevention and protection of user data, we call it the prevention part of our system.

Users' data is easily accessible to the third party using traditional and outdated security features. As this data is stored on their servers, the users don't have any knowledge or control over how the data provided to them is used. By using this approach, users will have complete control over their data. They can authorize who can access it or see who they have authorized to use their data. Also, we are using artificial intelligence and machine learning technique for face recognition which will surely optimize security and human verification to prevent fake or impersonated digital identities.

Our main idea revolves around building a system that includes a blockchain-based NFT (Non-Fungible Token) that will have the user's personal data on a decentralized blockchain system. The blockchain can be started on the user's local host, or the data can be stored on popular public chains like Ethereum.

If a user visits the website for the first time, they need to create their NFT. To begin that, the user must provide the details, face recognition, and human verification. Once all the required details have been made, the user will get the confirmation of that created NFT, and all the data the user has provided will be stored in the NFT. Hence, the data is completely safe as it got stored in the blockchain, which cannot be accessed without permission. So, instead of storing the data in the company's or website database, it is stored in the NFT, which any third party does not breach.

Now whenever users visit the website and want to do some tasks, they can access their account by connecting their digital identity NFT and perform the searching and transaction, thus creating a plug-and-play system.

(B) **Detection**

This part deals with fraudulent or anonymous activity on a user account or during a transaction.

We have analyzed various machine learning algorithms proposed in a previous research paper and their performance for fraud detection. Our findings suggested that the random forest technique is best suitable and provides the highest accuracy.

Our proposed solution uses blockchain and a face recognition system. Blockchain provides security to data so that users' data is neither saved on the database nor can be attacked easily. Also, our system uses face recognition, so human verification can be done using this system feature.

Also, our system uses a random forest algorithm to detect any fraudulent transaction or activity on the user's side.

Thus, blockchain is enhancing the network and user's data security, and by using artificial intelligence, we are providing more security for the transaction and human verification part.

Further, Fig. 3 presents the flow for the proposed methodology.

In Fig. 3 we have shown the proposed system's flow chart. Further, the following steps explain the workflow of the proposed approach:

- Step1: The first user will get the login or signup page of the website. This page has our integrated advanced blockchain login/signup system.
- Step:2 There will be two possibilities: either the user is new or an existing user.
- Step:3 Now, if there is a new user, they must create their NFT by filling in the required details and performing checking.
- Step:4 Once the user has created their NFT they will get confirmation of successful login and creation of NFT.
- Step:5 If the user is an existing user, they will just connect the wallet for login to the system.

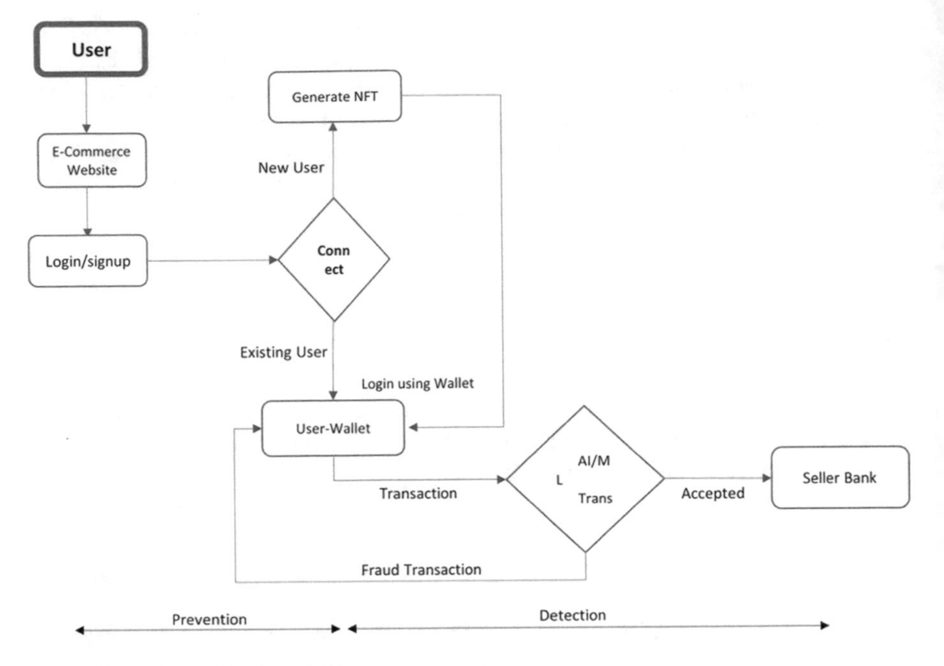

Fig. 3 Flow chart of the proposed system

Table 3 Technical stack of the proposed solution

Front end	Back end
React	Cloud vision
JavaScript	Firebase
Bootstrap	Next.js
Flask	Truffle
Tailwind	Eth

- Step:6 For transaction activity user must fill in the credential and details to perform the transaction.
- Step:7 In the backend, our machine learning algorithm will check whether the transaction is legitimate or not.

Step:8 User will get acknowledgment whenever any anonymous activity is associated with the account.

5 Implementation

This section explains the implementation from the user's perspective and the entire process's data flow. All the technology stacks used in implementing this proposed model are shown in Table 3.

Fig. 4 Login page

Using the above technical stack, we have created a testbed shown in Fig. 4 to Fig. 9. Figures 5 and 6 represents the initial stage of the proposed system. Initially, the data entered by the user will be passed to local storage or IPFS and checked for if there is any previous existence of data. If the data exists, the user will be automatically logged in. Any new user who wants to generate the NFT will initially have to go through a login page represented by Fig. 4 and then have to select 'Don't have an account?' to jump on the sign-up page represented by Fig. 5. Then the user will enter all the required login credentials which will be stored and can further be used to access the minted NFT.

The next step is shown in Fig. 6, which consists of a face recognition system. It is just an alternative for captcha verification for humans. After this segment, your images from the webcam will be passed to google cloud vision, and the AI algorithms will check your emotions and the results will be passed to verify if you are human or not.

The data generated from the image capture will also be passed on to local storage, which will later be used for verification purposes.

The form will capture all your personal data, and clicking submits button will encrypt the data through function. The encrypted data will be passed on to a variable and then the NFT mint function will be called on passing the data through it to make a NFT with your data. Now it will call the function to loadweb3 and start

Sign up

Email Address *

Password *

Confirm Password *

SIGN UP

Already have an account? Login

Fig. 5 Signup page

Crypto Identity

Fig. 6 Face recognition system

metamask to complete the transaction and give all your data on all the blockchain nodes, represented by Fig. 7.

Figure 8 shows the user account in which there will be a connect wallet component. Clicking on it will call a function to loadweb3 and metamask wallet, then you will enter your metamask wallet credentials. The website will read your identity NFT and store it in an array. On the frontend the data from the array will be parsed and shown. Also, this data will be used for other interactions.

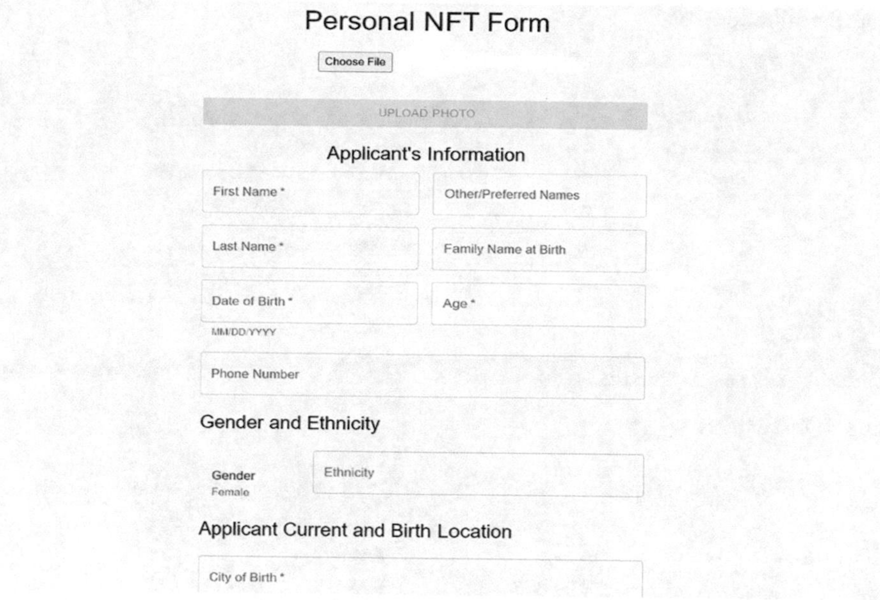

Personal NFT Form
Choose File
UPLOAD PHOTO
Applicant's Information
First Name *
Other/Preferred Names
Last Name *
Family Name at Birth
Date of Birth *
Age *
MM/DD/YYYY
Phone Number
Gender and Ethnicity
Gender
Female
Ethnicity
Applicant Current and Birth Location
City of Birth *

Fig. 7 User details form

Connect Wallet
Ethereum Mainnet
Connected
Account 2
0xC50...967D
0 ETH
$0.00 USD
Buy
Send
Swap
Assets
Activity
0 ETH
$0.00 USD
Don't see your token?
Refresh list or import tokens

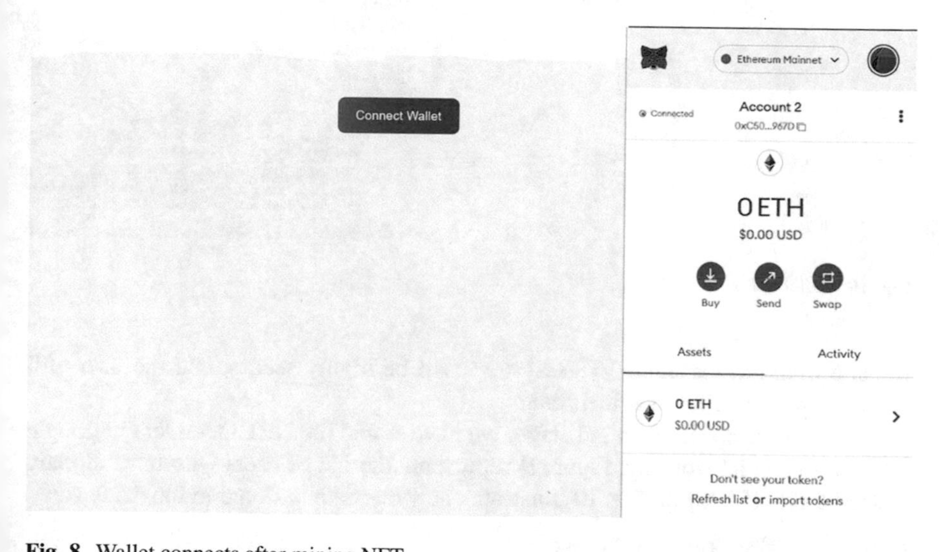

Fig. 8 Wallet connects after mining NFT

As the token is created, it can be used during any transaction on any website. The whole purpose of doing this is to avoid giving all our personal data to every other website, which is vulnerable to hacking. Instead, we are storing it on the blockchain through which we will provide single-point access to the data when required by any

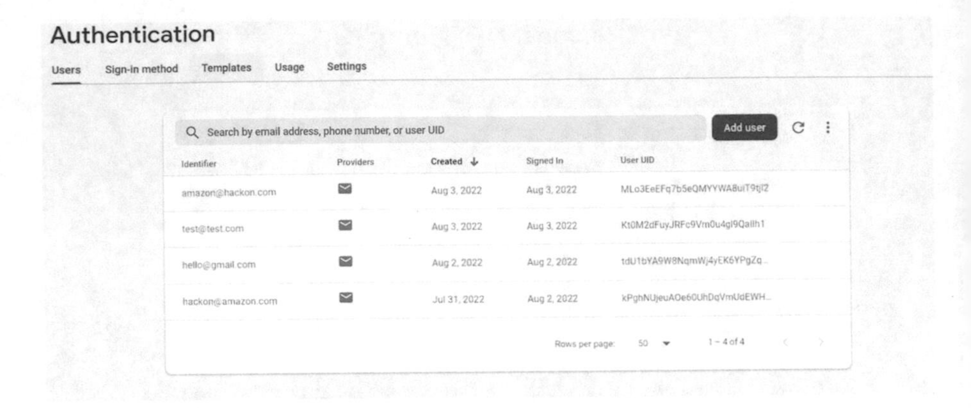

Fig. 9 List of users

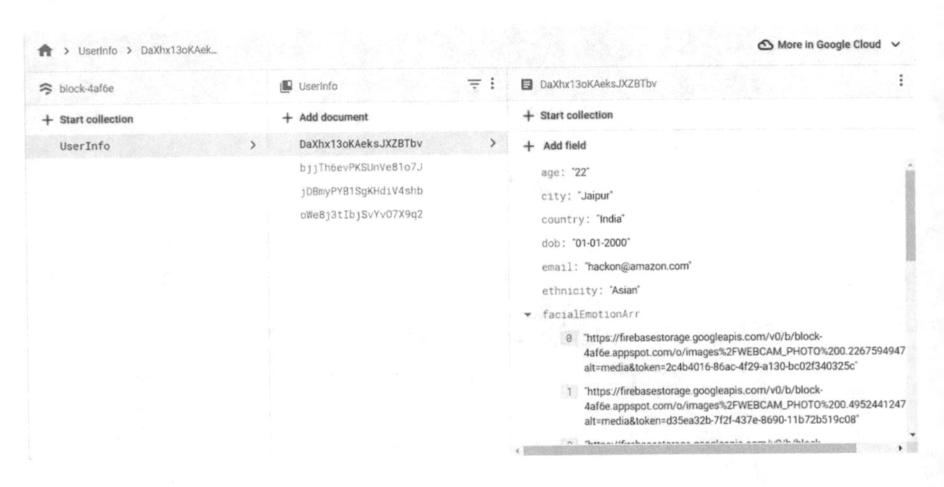

Fig. 10 Database

website during transactions. This system will be highly secure, and the user will have complete control over their data.

As this project is in its initial stage, we have stored the NFT meta-data in google firebase to test its working. Figure 9 represents the list of users who have already created their NFT, while Fig. 10 illustrates how our data is stored in the database.

Detection Using Random Forest

As mentioned earlier, our proposed approach works in two phases: Detection and Prevention. In the detection part, we have used the Random Forest algorithm to detect unusual activities at the user's end. After referring to previous research and implementing AI/ML techniques on financial data and fraud detection, we can conclude that the Random Forest technique can be the most suited for our system (as shown in Table 4).

Table 4 List of users. Analysis of random forest technique

	Precision	Recall	F1-Score	Support
0 (false)	1.00	1.00	1.00	56,962
1 (true)	0.97	0.79	0.87	98
Accuracy				
Macro-Avg.	0.99	0.89	0.93	56,962
Weighted-Avg.	1.00	1.00	1.00	56,962

6 Results and Analysis

In this section, we have discussed the results and analysis of our proposed solution. The current system has problems regarding user data and fraudulent anonymous transactions and activity. In traditional approaches, user data can be breached by a third party. Also, they do not have acknowledgment and control over the data, as a database of the company is involved. Therefore, these approaches can lead to the problem of data breaches, and the website's current transaction system is not highly secure. Our proposed system resolves all these issues.

Our proposed approach solves data sharing issues since instead of sharing users' personal information to multiple websites, we are creating single-point access to the user's data, which is a much better and more secure approach. A single website can jeopardize your real data if there is a security breach or policy lapse.

We are storing the user data on the blockchain through the NFT minted by the user's crypto wallet. Every NFT is assigned a unique digital ID (DID) containing user data. Simultaneously, we are also using AI/ML algorithms to detect any discrepancies in transactions or behavior of the user.

As this system uses features of blockchain, it will allow everyone to know who has access to their data which will help them to prevent identity theft. This system also uses AI/ML algorithms using users' past behavior analytics to detect fraud during interactions with the website.

Therefore, with this system:

(a) Users have control over the data and cannot be breached.
(b) Users have transparency over the data.
(c) Users have an acknowledgment of the anomaly in the transaction.

7 Limitations

Further, to mention the limitations of the proposed solution, if the site is already compromised, then during the creation of NFT and login. The intruder can access the data when the user provides all the required information in the form. As initially, there is no real-world site data to train the model. For any new users, we have to train the model, and for every initial transaction, they will receive an acknowledgment to proceed further, which can affect the user experience.

8 Conclusions

A significant change from simple digital usage to new advanced technologies powered by blockchain and Artificial Intelligence (AI), has forced industries and organizations to think and re-evaluate some key industrial habits that are usually ignored and practiced.

Various systems are directly engaged with users' data and security. We have proposed a login and transaction system that is more secure regarding users' data and transactions. This project aims to modify or upgrade the current approach to detect and prevent the identity theft problem using technology. As we know, user data and trust are the key to the building and growth of the organization.

A login and transaction system has been built to provide security measures designed to prevent unauthorized access to confidential data. Combining it with newly unpenetrated technology involving blockchain NFTs and artificial intelligence can be a more secure improvement process, providing benefits to efficiency and productivity in both productive and service-based organizations. This proposed method can be used for growth, expand that number and promote more significant and faster development across the organization. However, this system must undergo several tests to get into final action. As we know, users' data and trust are the keys to the building and growth of the organization. Therefore, the traditional login and fraudulent transaction combined with artificial intelligence and blockchain-based upgradation can be very helpful in providing security and trust.

References

1. Chen, Z., & Zhu, Y. (2017). Personal Archive Service System using Blockchain Technology: Case Study, Promising and Challenging. In 2017 IEEE International Conference on AI & Mobile Services (AIMS). Honolulu, HI: IEEE. Retrieved from https://ieeexplore.ieee.org/document/8027275/
2. Soliman, A., Bahri, L., Carminati, B., Ferrari, E., & Girdzijauskas, S. (2015). DIVa: Decentralized identity validation for social networks. In 2015 IEEE/ACM International Conference on Advances in Social Networks Analysis and Mining (ASONAM). Paris: IEEE. Retrieved from https://ieeexplore.ieee.org/document/7403568/
3. Abou Jaoude, Joe, and Raafat George Saade. "Blockchain applications–usage in different domains." *IEEE Access* 7 (2019): 45360–45381.
4. Monrat, Ahmed Afif, Olov Schelén, and Karl Andersson. "A survey of blockchain from the perspectives of applications, challenges, and opportunities." *IEEE Access* 7 (2019): 117134–117151.
5. Al-Shuaili, S., Ali, M., Jaharadak, A., & Al-Shekly, M. (2019). An Investigate on the Critical Factors that can Affect the Implementation of E-government in Oman. 2019 IEEE 15Th International Colloquium on Signal Processing & Its Applications (CSPA). https://doi.org/10.1109/cspa.2019.8695988
6. Identity Theft Resource Center, 2021 in review, Data Breach Report LNCS Homepage, http://www.springer.com/lncs, last accessed 2016/11/21.
7. Federal Trade Commission, Consumer Sentinel Network.

8. Dornadula, Vaishnavi Nath, and Sa Geetha. "Credit card fraud detection using machine learning algorithms." *Procedia computer science* 165 (2019): 631–641.
9. Kaggle Dataset: https://www.kaggle.com/mlg-ulb/creditcardfraud/download
10. KM, Anil Kumar, and Jemal Abawajy. "Detection of False Income Level Claims Using Machine Learning." *International Journal of Modern Education & Computer Science* 14, no. 1 (2022).
11. Singh, Yathartha, Kiran Singh, and Vivek Singh Chauhan. "Fraud Detection Techniques for Credit Card Transactions." In *2022 3rd International Conference on Intelligent Engineering and Management (ICIEM)*, pp. 821–824. IEEE, 2022.
12. 5 types of fraud that is used to target e-commerce retailers https://www.ravelin.com/blog/5-types-of-fraud-that-is-used-to-target-e-commerce-retailers
13. Gupta, Priyanka, and Ankit Mundra. "Online in-auction fraud detection using online hybrid model." In International Conference on Computing, Communication & Automation, pp. 901–907. IEEE, 2015.
14. Mundra, Ankit, Nitin Rakesh, and S. P. Ghrera. "Empirical study of online hybrid model for internet frauds prevention and detection." In 2013 International Conference on Human Computer Interactions (ICHCI), pp. 1–7. IEEE, 2013.

Proactive AI Enhanced Consensus Algorithm with Fraud Detection in Blockchain

Vinamra Das, Aswani Kumar Cherukuri, Qin Hu, Firuz Kamalov, and Annapurna Jonnalagadda

1 Introduction

The problem statement is how to create a dynamic consensus algorithm that can predict, prevent and take the system back to stability once any kind of fraud or bias occurs from the leader node's perspective. It has to be emphasized that every user in the system is given a chance to vote for the leader election, and the full control is not shifted to the dynamic algorithm. Hence keeping the concept of 'decentralization' intact even in terms of consensus. The problem statement holds gravity in that blockchains are growing fast, and with the significant number of users, not one user can be fully trusted. Hence the involvement of automated decision-making authority (algorithm) is necessary. While devising the algorithm, it has been kept in mind that parameters such as the node's age, trust of node, and node's stake are not considered while predicting fraud or bias, and only the current 'behavior' as a leader node is taken into consideration. Humans tend to plan and think ahead and not only think 'greedily', meaning their actions need not bring them closer to their target (fraudulent act). In other words, need not maximize their profit function in the immediate step and might have a long last vision before the target is achieved. Those

V. Das · A. K. Cherukuri (✉)
School of Information Technology & Engineering, Vellore Institute of Technology, Vellore, India
e-mail: cherukuri@acm.org

Q. Hu
Department of Computer & Information Science, Indiana University Purdue University, Indianapolis, IN, USA

F. Kamalov
Faculty of Engineering, Canadian University Dubai, Dubai, UAE

A. Jonnalagadda
School of Computer Science and Engineering, Vellore Institute of Technology, Vellore, India

Y. Maleh et al. (eds.), *Blockchain for Cybersecurity in Cyber-Physical Systems*, Advances in Information Security 102, https://doi.org/10.1007/978-3-031-25506-9_13

scenarios are kept out of the scope of this algorithm due to the obvious fuzziness in the logic offered. It also has to be noted that parameters such as 'trust of node' is subjective and computationally difficult to compute.

Proactive artificial intelligence means the introduction of such deep learning methodologies, which can detect fraud or bias well within time before any significant loss to the blockchain system occurs. Reducing the lapse time and conducting elections on the understanding of AI whenever needed reduces human intervention to a much greater extent.

Proof of Work (PoW) has been the oldest consensus algorithm since 2009. It takes up the concept of 'competitive consensus'. The blockchain system has a defined set of 'miners' who possess the capability to add new nodes to the chain, subject to solving a 'mathematical puzzle' which requires some amount of work to determine the answer and is easier to verify. Once this verification takes place, the miner is allowed to add the nodes. Proof of Stake does solve the 'high computational expense' problem by determining the wealth or stake of the miner in the chain, and higher preference in mining is given to one with the higher stake. This works well since the one with the higher stake in the chain itself is less likely to commit fraud, but the bias in the selection of nodes still exists. The miner's activities with the highest stake are unchecked until someone else surpasses him in stake. Another method currently in use is Delegated Proof of Stake (DPoS), wherein the nodes with higher stakes can vote among themselves to choose a validator. This solves the problem previously mentioned in proof of stake, where the decision of a high stake node is unchallenged until someone surpasses its stake, as a validator is present to keep a check on the miners. DPoS is faster than most currently existing consensus algorithms and comes under 'collaborative consensus' category. Though it is fast and efficient, all the power is vested with the miners, and a validator can become ineffective due to any nexus between them. Another collaborative consensus algorithm is Proof of Reputation (PoR) [1]. In this system, the validator should have a reputation that is 'good enough' and ensures a significant loss to the validator if it commits fraud. It has been proven highly secure until now, but the only apparent problem is quantizing something as subjective as 'reputation'. The choice of highest reputation is left on the participating nodes, and one with the clear majority wins but is still prone to failure. Another consensus algorithm devised is RAFT Consensus Algorithm [2] which ensures that the blockchain system is immune to any node or system failure. Herein the leader sends a 'heartbeat' to all the follower nodes, with each follower node having a random time quantum that expires if no heartbeat is received during that period, and the node with an expired time quantum now stands for being elected as the new leader. This works to make the system tolerant to failures, but detecting any bias from leaders is still out of its scope.

Hence the significant problems this chapter deals with are:

- Detecting fraud or bias from the leader node in the selection of blocks.
- Recognizing faulty behavior in case of bias detection.
- Replacing the strategy of the leader node and taking the blockchain system back to stability.
- Reducing the lapse time taken by the follower nodes to observe fraud or bias.

2 Related Work

2.1 *Blockchain Architecture*

A blockchain architecture consists of data structures known as 'blocks' that hold data relevant to the use case of the implementation. These blocks are connected via the hash value of the preceding block. These hash values can be considered pointers and are computed based on the data that a block stores. A hash function is chosen as it is a one-way function; hence the data of the block cannot be retraced from the succeeding block, which holds its hash value. In addition, the hash function satisfies 'strong collision resistance' policy; hence it is highly improbable to have two blocks with different data mapping to the same hash value; therefore, uniqueness is restored. To make this even stronger, a 'nonce' is added to the data to create a hash to increase confusion. The version of the block has to be mentioned so that the whole blockchain is compatible with its blocks. The list of transactions is stored along with the sender and receiver's details and the amount transferred. Hash value generated is also used for the signing of the block. A signed block signifies that it is verified by the central authority and is eligible to be a part of the blockchain system. Any signature pattern reflected in the hash has to be constant for every block in the system. The hash value generated on the raw data that the block holds cannot ensure that the pattern set is being followed. Hence, a nonce is appended with the data to satisfy the pattern. Once that specific number (nonce) is reached, the block is signed. Figure 1 shows the general structure of a block.

Merkle chains are seen as the building blocks of blockchains. They are tree data structures wherein the tree structure grows by adding leaf nodes, which are the hash of their parent leaf. Broadly adding hash values to identify the predecessor of any data structures is done in the Merkle chain.

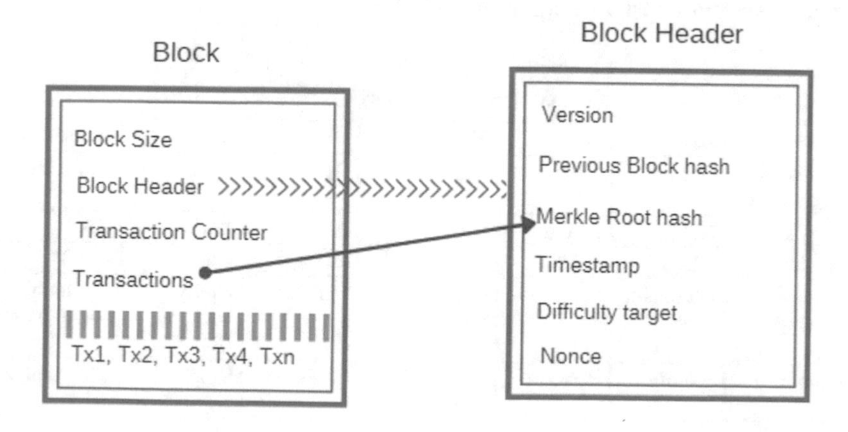

Fig. 1 General structure of a block

2.2 Consensus Algorithm

Consensus in blockchain refers to a protocol following which a mutual agreement is met upon deciding the validity of the stored information in the block. It might not give equal weightage to the vote of confidence of each user, with differences currently being created based on stake, age, and reputation. The nodes with the higher quantified value of these three factors are termed leader nodes. They have a higher level of authority in the blockchain, using their power to add or reject transactions. Although they are selected based on higher reliability, their decision is considered optimum. Still, the presence of human factors leaves some amount of uncertainty. A reward system can be set up to reduce the chances of fraud from the leader nodes, still leaving some chances of bias or fraud depending majorly upon the gain function. Hence an unbiased decision-making element has to be added to the consensus algorithm to make it more efficient and reliable.

Marwala et al. [3] discuss the intermingling of the AI and blockchain technologies defining the various domains wherein these are used individually and can be used together. These two technologies can be useful in various other areas, such as improving data models, smart retail, and improving financial models, as the use of blockchain allows a huge amount of unstructured data to be accumulated in a small space with a level of trust and security. Using AI will help make sense of that data and turn it into desired information. An overview of the amalgamation of deep learning, machine learning, and blockchains is discussed in [4]. Chen et al. [5] discuss another AI-based consensus algorithm focusing on selecting super nodes with the whole system consisting of super nodes, random nodes, and unknown nodes. A convolutional neural network is trained to obtain the nth node's average transaction number (ATN). The one with a higher predicted value of ATN is chosen as it ensures fairness and smooth functioning of the chain. An analysis of using both the fields together with their challenges and scopes has been done in [6]. Table 1 compares various consensus algorithms [7].

Table 1 A comparison of various consensus algorithms

Property	PoW	PoS	PBFT	DPOS	Ripple	Tendermint
Node identity management	Open	Open	Permissioned	Open	Open	Permissioned
Energy saving	No	Partial	Yes	Partial	Yes	Yes
Tolerated	<25%	<51%	<33.3%	<51%	<20%	<33.3%
Power of adversary	Computing power	Stake	Faulty replicas	Validators	Faulty nodes in UNL	Byzantine voting power
Example	Bitcoin	Peercoin	Hyperledger fabric	Bitshares	Ripple	Tendermint

2.3 *Identification of Problem*

- Need for a central authority that can check the selection behavior and patterns of the leader node, preferably a deep learning algorithm.
- The voting should be based on each decision of the leader node. Each selection of the leader node should be under scrutiny, and since it is infeasible for all the nodes to conduct voting for every selection, a deep learning model has to be deployed to detect and mine only the transactions which it finds are biased.
- The deep learning model should learn. Hence once a selection pattern goes under scrutiny, that is for the voting by human interference, and it finds out that the pattern is harmless, hence in the future, any selection pattern close to this is not sent to scrutiny.
- Artificial neural networks are a better choice over other neural systems as there is no need to observe recurrence since the selection of every single block is a single independent decision depending solely on the current parameters.

3 Method

3.1 *Leader Selection*

From the genesis block until the 50th block, a single leader node is chosen randomly from a total of n nodes ($n <= 50$). A pseudorandom number generator generates the output ranging from 1 to n. After the total number of blocks in the system reaches 50, randomly 2% of the nodes are chosen onwards as leader nodes. All the nodes have an equal probability of being selected as the leader node with no priority given based on stake, reputation, or time for which the node has been a part of the system. A total of 50 samples is taken for neural network training using a thumb rule of multivariate statistics of having at least 10 data points for each parameter. Since 5 parameters are taken, we stick to a minimum of 50 samples since increasing the sample size increases neural network training time, thus delaying consensus.

3.2 *Parameters/Inputs*

The following values are taken as input parameters for bias detection by the neural network:

- Sender Node: The sender node refers to the node sending the stipulated units of monetary/non-monetary goods to a single user recorded in that transaction.
- Receiver Node: The receiver node refers to the node which receives the stipulated amount of monetary/non-monetary goods from a single user, which is recorded in that transaction.
- Sender Balance: The sender balance is the amount of money the sender currently holds in his account balance before the transaction being referred to takes place.

For instance, if sender A holds $5000 before sending the amount of $100 to user B, who currently has a balance of $2000. Then the A's balance equals $5000 instead of $4900 (the amount left with him after the transaction).

- Receiver Balance: The receiver balance refers to the standing amount the receiver (B in the above example) has in his account before he receives the total amount from the sender (A). Taking the above example, the receiver balance is taken as $2000 and not $2100 (standing balance after the transaction has taken place).
- Transaction Amount: The amount transferred from one node to another in that single transaction.

These are the parameters over which any kind of bias in the addition of blocks could happen from the leader's point of view since. These values are known to the leader nodes while deciding whether or not the particular block is being added to the chain.

3.3 Neural Network Architecture

Neural network architecture has to be set in such a way so that the best generalization results are obtained, avoiding overfitting the parameters. Hence the number of hidden layers and neurons has to be set in a way that best avoids overfitting [8]. Given constraints to the neural network architecture include:

- 5 input neurons: Each representing the known 5 parameters – sender, receiver, sender balance, receiver balance, and transaction amount. Note that while inputting the values in the neural network, all these values have been scaled on a scale of 0 to 1, by dividing these by the highest values into their respective columns.
- 1 output neuron: One output neuron contains a value ranging from 0 to 1, with 1 signifying that the block is accepted and 0 suggesting that the block gets rejected. Output is continuous between 0 and 1, with values closer to 1 suggesting a higher probability of the block getting selected by the leader node and values closer to 0 suggesting that the block has a higher probability of getting rejected by the leader node.
- The following formula is taken as a thumb rule to determine the number of hidden layers [9]:

$$N_h = N_s/\alpha * (N_i + N_o) \tag{1}$$

N_h = count of hidden neurons.
N_i = count of input neurons.
N_o = output neurons count.
N_s = total number of training samples in the database.
α = scaling factor with values ranging from 2 to 10.

For the calculations here, the number of hidden layers is taken as 2, since this deep neural network can detect almost any pattern. The scaling factor is plugged in as $\alpha = 2$ as it is considered optimal since it can perform the generalization of ambiguous patterns while avoiding overfitting. A total of 100 sample sets are considered for finding the patterns in the leader node's selection habits. Plugging in the values:

$N_i = 5, N_o = 1, N_s = 100, \alpha = 4$, into the equation:

$$N_h = N_s/\alpha * (N_i + N_o)$$

We get,

$$N_h = 100/\left(2^* \left(5 + 1\right)\right)$$
$$= 8.33 \approx 8$$

Hence with a total number of hidden neurons as 8 they have to be divided into 2 hidden layers. This is done to maximize the number of parameters in a neural network. The given structure 5-6-2-1 architecture fits best as it has the highest number of parameters (53 parameters: 9 biases and 44 weights) with these constraints.

- The hyperbolic tangent is taken as the activation function [10]. Figure 2 shows the network architecture.

Following are the steps proposed:

- Step 1: Add the blocks to the system until the total number reaches 50. Till that point node adding the genesis block is the leader and not subject to any supervision.
- Step 2: Select 2% of the total number of nodes (let it be x) randomly as leaders.
- Step 3: Initialize x neural networks with 5-6-2-1- architecture. 5 input neurons and 1 output neuron suggest the selection chances of that block from the respective leader on a scale of 0 to 1. 1 suggests maximum chances of selection and 0 suggests least.
- Step 4: Find the correlation of each of the 5 input parameters over the output using sensitivity testing. This shows the effect of each parameter on the output.
- Step 5: If the weight of any of the parameters equals or exceeds 0.40, then make a list of those parameters. Else continue with the same leader as the leader node without further investigation.
- Step 6: If further investigation happens, keep track of the next 10 blocks in the chain added by that respective leader and let the neural network predict whether or not these blocks will be selected.
- Step 7: If the neural network prediction and leader's prediction match 5 or more times then send the list of these parameters (with weights exceeding or equal to 0.4) to the follower nodes along with the last 10 parameter data.

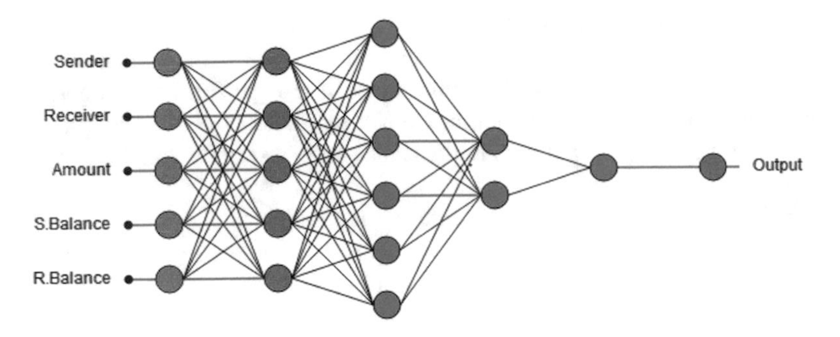

Fig. 2 Neural network architecture deployed

Table 2 The scaled input format for user values

User name	Scaled input format
A	0.17 (1/6)
B	0.33 (2/6)
C	0.50 (3/6)
D	0.67 (4/6)
E	0.83 (5/6)
F	1.00 (6/6)

- Step 8: If more than 50% of the nodes confirm that selection behavior is faulty or biased, then the leader is replaced randomly.

Note: Neural network training is done on the Neural Designer platform.

3.4 *Case Studies on Output and Dominant Feature Extraction*

Consider a blockchain system with 6 members A, B, C, D, E and F represented as 1, 2, 3, 4, 5, and 6 for the neural network input. As mentioned earlier, these values are scaled from 0 to 1. Table 2 shows the corresponding values.

- Each member's transaction values and current standing balances have also been reduced on a scale of 0 to 1 by dividing the highest value into their respective columns.
- Table 3 and Fig. 3 show the sample data set taken, representing the current leader's selection patterns.
- 0 in the output signifies rejected status, whereas 1 signifies selected status.

Note: S.Balance corresponds to the sender balance before sending the stipulated amount to the receiver. R.Balance corresponds to the receiver balance before receiving the stipulated amount from the sender. Table 3 shows the sample values.

It can be easily observed that the leader node is not allowing the transactions to occur whenever the sender value is 0.17 (mapping to user A) and the receiver value

Table 3 Sample data set for selection patterns of current leader

Sender	Receiver	Amount	S.Balance	R.Balance	Output
0.17	0.67	0.08	0.42	0.58	1
0.33	0.67	0.12	0.38	0.7	1
0.17	1	0.035	0.385	0.535	1
0.33	0.17	0.2	0.18	0.585	1
0.83	0.5	0.1	0.49	0.51	1
0.67	0.33	0.055	0.445	0.235	1
0.33	0.5	0.2	0.035	0.69	1
0.17	0.83	0.3	0.19	0.885	0
0.33	0.5	0.01	0.7	0.025	1
0.17	0.83	0.045	0.235	0.2	0
0.83	0.17	0.05	0.05	0.25	0
0.5	0.33	0.08	0.065	0.6	1
0.17	0.83	0.09	0.35	0.45	0
1	0.5	0.16	0.8	0.54	1
1	0.83	0.45	0.6	0.86	1
1	0.67	0.21	0.32	0.4	1
0.17	0.33	0.087	0.06	0.433	1

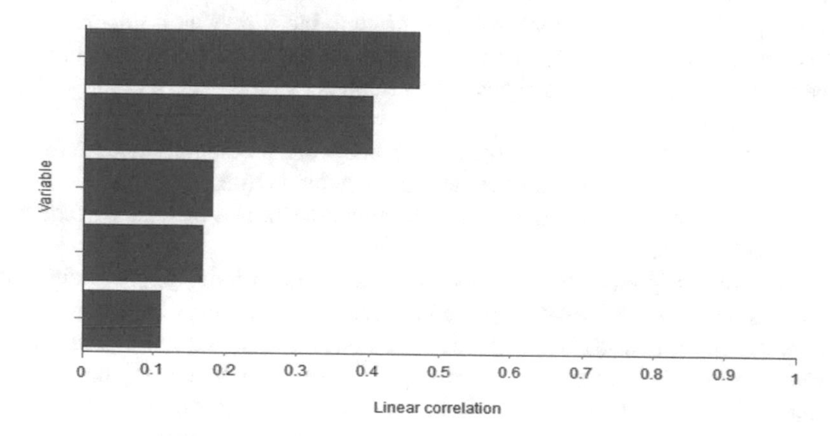

Fig. 3 The weight of each parameter in the output

is 0.83 (mapping to user E). Hence dominant features are Sender and receiver for the leader to choose. Upon training the neural network and finding the effects of each input parameter over the output [11], shown in Fig. 3.

Figure 3 presents the weight of each parameter in the output determination labeled on the Y axis from the top as Sender, Receiver, Amount, Sender balance, and Receiver Balance.

It is observed that the parameters of sender and receiver exceed 0.4. Therefore, these are deterministic factors in the decision of the current leader node.

Note: Ideally, with 5 parameters, all the factors must weight 0.2 in the final output, and hence any parameter exceeding that number by 100% or more (that is, a weight of 0.4 or more) is considered the dominating parameter.

Hence it is evaluated that only the sender and receiver name are deciding factors for the decision of the current leader node. It is also important that till now, the practice of leader nodes adding blocks on these factors might or might not be faulty. It relies on human understanding of whether or not these factors are creating bias, giving rise to any fraud, or are completely harmless to the system.

3.5 *Pattern Validation and Back to Stability*

Once the neural network is trained over the given data set, it has to be validated whether the expected pattern is still being followed or not. It has to be validated whether the pattern that the neural network detected is being followed or not or is just a coincidence [12]. The following flow is followed:

- A trained neural network (over the selection patterns of the current leader node) predicts the chances of selection of the new incoming block.
- If the prediction is that block will get rejected, but it does not, this implies that the pattern might not be binding. In contrast, if the prediction is that block will be rejected, and it does get rejected, this implies that the pattern predicted by the network might be true.
- It is tested for the next 10 blocks. If neural network prediction matches for more than 5 then it is assumed that the pattern predicted is true.
- If this does not hold good, normal functioning is sustained with the current leader as a leader node.
- If the pattern is predicted truly, then the parameters with the highest weights and their values wherein they got rejected are sent to the follower nodes.
- Suppose more than 51% of the follower nodes verify that the parameters and values driving his selection decisions are biased. In that case, the leader is changed and blacklisted (never eligible to become a leader node again), and the new leader is randomly selected from the remaining follower nodes.
- Suppose more than 51% of the follower nodes suggest that his selection practices are not biased. In that case, the parameters are saved, and no further investigation occurs regarding the specified parameter combination with the current leader node acting as the leader.

The effectivity of using a PRNG (pseudo-random number generator) is shown below with 6 users and their frequencies of being selected as leaders over a large sample set (1000 in this case), in Table 4.

Sender node:1, Receiver node:5, Transaction Amount:2500 (scaled to 0.25), Sender Balance:4000 (scaled to 0.4), Receiver Balance: 6000 (scaled to 0.6) as shown in Table 5.

Table 4 Frequency distribution of 6 participants being selected as leader nodes in 1000 iterations

User Number	Frequency
1	163
2	172
3	163
4	163
5	167
6	172

Table 5 Sample output for the chances of selection of transaction with given specifications

	Value
Sender	0.17
Receiver	0.83
Amount	0.25
Sender balance	0.4
Receiver balance	0.6
Output	0.3885531676

Table 6 User names against their corresponding scaled inputs

User Name	Scaled Input Format
A	0.2
B	0.4
C	0.6
D	0.8
E	1.0

It was specified in the above example, f that the leader does not allow transactions between the sender as 1 and the receiver as 5. Our neural network verifies this as it predicts that only a chance of 0.385531676 exists for selecting this block by the current leader node. Hence our assumption is confirmed. In another case study, the leader node does not allow transactions with an amount less than $3000, represented as 0.3 in the data set, since the maximum amount is $10,000. The system consists of 5 members A, B, C, D, and E, with cardinal numbers 1, 2, 3, 4, and 5, respectively. They are hence scaled from 0 to 1 with values as shown in Table 6.

The neural network output possesses the following dependencies on the inputs (scaled from 0 to 1), as shown in Fig. 4.

It is hence accurately observed that the 'amount' is a deterministic factor in the selection pattern of the leader node. As this is detected, the neural network code predicts his future decisions based upon the five mentioned factors of the incoming new blocks. For the next 10 blocks, if more than 5 are predicted correctly, the pattern is truly predicted and reported to the follower nodes. It is left to the follower node's discretion whether or not this practice is faulty, and general voting is held to decide whether or not the follower node should be allowed to continue his position or should be overthrown.

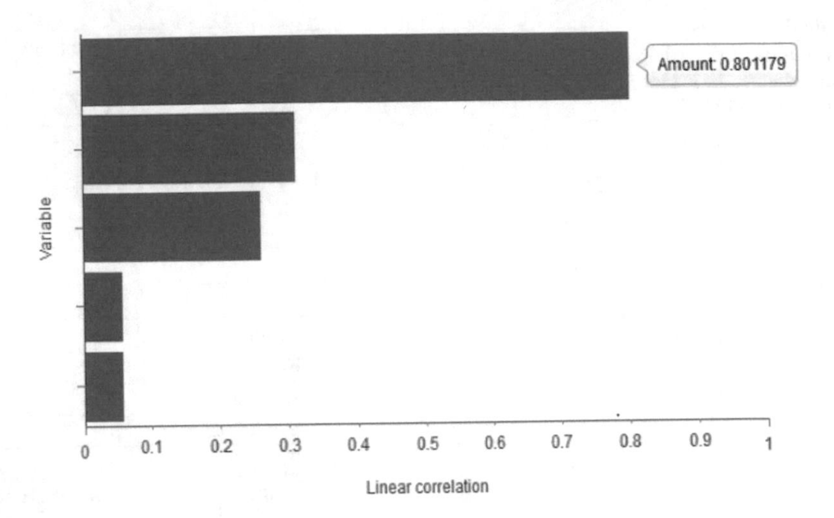

Fig. 4 The weight of the 'Amount' factor is the highest with 0.801179

Table 7 Selection chances of transaction with the given specifics from the above leader node

	Value
Sender	0.6
Receiver	0.4
Amount	0.24
Sender balance	0.55
Receiver balance	0.65
Output	0.000468558636

Table 7 shows the chances of selecting a leader node as predicted by the neural network once trained upon the above-given data set. Sender: 3 (scaled as 0.6), Receiver: 2 (scaled as 0.4), Amount: $2400 (scaled as 0.24), Sender Balance: $5500 (scaled as 0.55), Receiver Balance: $6500 (scaled as 0.65). It was taken as a constraint from our side that amount less than $3000 will not be selected by this leader node and is hence truly detected upon a random data set.

4 Learning Methodology for Bias Pattern Validation

The use of deep learning methods is irrelevant if the central authority designed using ANN does not get better over time. Once it finds that bias is happening, the neural network sends the details to the users to decide whether it is the kind of bias that can lead to some harm or is just a use case constraint. For instance: Suppose a bank cannot allow transactions of greater than $10,000, then it is a bias that is bound to get detected by the neural network every time, but is a system constraint, and hence nothing is wrong with this pattern. Therefore, once this pattern is sent to the users,

Table 8 Values of the biased pattern detected earlier versus current biased values of the same parameter

Unbiased values stored in stack (x_i)	Values of the same parameter detected presently (y_i)
0.60	0.55
0.62	0.58
0.68	0.62
0.70	0.75
0.72	0.78
0.74	0.84
0.79	0.85
0.80	0.87
0.82	0.88
0.86	0.91

they will confirm that the pattern is unbiased. The neural network sends this data to a stack with the parameters detected as deciding factors for this selection pattern. A sample of 10 values of this specific pattern is also stored in this stack. If this bias is detected in the future, the neural network checks whether or not the deciding patterns are present in the stack. If it is present, then it finds the error of the values of the parameters of the past with the present. If the deviation is less than or equal to 20%, that is 0.2 (since all the parameters are on a scale of 0 to 1); hence the pattern is similar to that observed in the past and is unbiased. Suppose the transaction values noted to be unbiased are as shown in Table 8.

Root mean squared value calculated for the above values:

$$Error = \sqrt{\left(\Sigma\left(x_i - y_i\right)^2 / n\right)} \tag{2}$$

$$Error\ calculated = 0.18947$$

Hence it is less than 0.20; therefore, the pattern observed is similar to the past and is not biased, therefore not sent to the user nodes for further voting.

Note:

- Error is calculated using the root mean squared method.
- If more than one parameter is detected, then every parameter's deviation should not exceed 0.20. If the error value exceeds, then only the selected pattern is sent to the nodes for further scrutiny.
- Values x_i and y_i are used for error calculation in a sorted format (of highest correlation parameter detected) and then taken pairwise. This enables the calculation of errors based upon comparable values in the two data sets.

5 Results

The training gets completed in 102 iterations, with elapsed time being less than 0.5 seconds (rounded off to 0). This saves the time to analyze every block through a public consensus algorithm and only brings them into play if any bias is prevalent from the leader node's perspective. Experimentally a correlation value of 0.4 for each parameter is the bias threshold. A variation of 0.2 or greater represents two different selection patterns. It has to be noted that these two values may differ if the number of parameters is increased or more constraints exist in the data. Table 9 indicates the training outcomes of ANN.

Commenting on time costs for neural network training, it takes less than 1 s and around 102 iterations to train on 50 sample data points. An addition of one second on the higher end will occur on a per-block basis. The final contributions of the paper are summarized in the points below:

- The hierarchy of trust in human-driven systems had to be broken somewhere. Hence replacing the leader nodes with an AI system removes any bias in node selection.
- Time consumption in consensus is reduced, since the leader node only does the selection and the power of overruling his verdict goes to the nodes only when the system detects any anomaly. Every node need not go through the voting process.
- A neural network detects fraud and learns whether or not the perceived pattern was fraud/biased over time since the memory of these patterns is matched by a similarity measure in future selections.
- The need to reward the leader nodes is now removed since he is under constant supervision and the appointment of the next leader node is random.

Cyber-physical systems (CPS) is an integrated technology that combines the devices (sensors) that monitor the physical environments with the computing systems with the help of suitable communication mechanisms to provide abstraction and analysis capabilities. Several blockchain applications are reported in the literature [14], including Cyber-physical systems [15, 16, 17, 18, 19, 20]. Securing the blockchain in the application area is also a major concern. Guo and Yu. [21] have discussed the security risks associated with blockchain technology and provided measures for securing the blockchain. Narula et al. [22] have discussed a trust model for

Table 9 Training outcomes of the ANN

	Value
Final parameters norm	3.55
Final loss	0.00381
Final selection loss	0.241
Final gradient norm	0.00958
Iterations number	102
Elapsed time	0
Stopping criterion	Gradient norm goal

blockchain in supply chain management networks. Baniata et al. [23] have discussed the optimized neighbor node selection for smart blockchains. Bashar et al. [24] have discussed the multileader consensus protocol for healthcare blockchain. The proposed model is aimed at addressing these concerns.

6 Conclusion and Future Scope

The work carried out in this paper shows how artificial neural networks in amalgamation with majority voting consensus schemes can be used to monitor the selection behavior of the leader node and detect if any fraud or bias is committed from the leader's side [13]. The system turns back to stability as the leader is overthrown, and a new leader is elected at random. A random choice of leader is chosen so that no manipulation can be done from the side of the nodes who want to be the leader, as there is no factor upon which their selection is based. In summary, the proposed algorithm simplifies the data upon which bias can occur and presents it to the follower nodes, who decide whether or not his actions are faulty. In brief, this method reduces decision time for the removal of a fraudulent leader as it mines the dominant selection features and makes only a small sample of his fraud or bias activities visible to the follower nodes. Hence easing the process of analyzing his selection patterns for the users. Another track of further research can be more smart data mining and making smart systems using Internet of Things, enabling bigger data sets and a higher number of systems.

References

1. Gai, Fangyu & Wang, Baosheng & Deng, Wenping & Peng, Wei. (2018). Proof of Reputation: A Reputation-Based Consensus Protocol for Peer-to-Peer Network. https://doi.org/10.1007/978-3-319-91458-9_41.
2. Nakagawa, Takuro & Hayashibara, Naohiro. (2018). Energy Efficient Raft Consensus Algorithm. 719–727. https://doi.org/10.1007/978-3-319-65521-5_64.
3. Marwala, Tshilidzi & Xing, Bo. (2018). Blockchain and Artificial Intelligence. SSRN Electronic Journal. https://doi.org/10.2139/ssrn.3225357.
4. Adoma, Francisca. (2018). Big Data, Machine Learning and the BlockChain Technology: An Overview. International Journal of Computer Applications. 180. 1–4. https://doi.org/10.5120/ijca2018916674.
5. Chen, Jianwen & Duan, Kai & Zhang, Rumin & Zeng, Liaoyuan & Wang, Wenyi. (2018). An AI Based Super Nodes Selection Algorithm in BlockChain Networks.
6. Salah, Khaled & H Rehman, M & Nizamuddin, Nishara & Al-Fuqaha, Ala. (2018). Blockchain for AI: Review and Open Research Challenges. IEEE Access.
7. Wang, Huaimin & Zheng, Zibin & Xie, Shaoan & Dai, Hong-Ning & Chen, Xiangping. (2018). Blockchain challenges and opportunities: a survey. International Journal of Web and Grid Services. 14. 352. https://doi.org/10.1504/IJWGS.2018.10016848.
8. Caruana, Rich & Lawrence, Steve & Lee Giles, C. (2000). Overfitting in Neural Nets: Backpropagation, Conjugate Gradient, and Early Stopping. Advances in Neural Information Processing Systems. 13. 402–408.

9. http://hagan.okstate.edu/NNDesign.pdf#page=469
10. B. M. Wilamowski, "Neural network architectures and learning algorithms," in *IEEE Industrial Electronics Magazine*, vol. 3, no. 4, pp. 56–63, Dec. 2009. https://doi.org/10.1109/MIE.2009.934790
11. Novak, Roman & Bahri, Yasaman & A. Abolafia, Daniel & Pennington, Jeffrey & Sohl-Dickstein, Jascha. (2018). Sensitivity and Generalization in Neural Networks: an Empirical Study.
12. Kim, Tai-Hoon. (2010). Pattern Recognition Using Artificial Neural Network: A Review. 76. 138–148. https://doi.org/10.1007/978-3-642-13365-7_14.
13. Sarda, Paul & Chowdhury, Mohammad & Colman, Alan & Kabir, Ashad & Han, Jun. (2018). Blockchain for Fraud Prevention: A Work-History Fraud Prevention System. 1858–1863. https://doi.org/10.1109/TrustCom/BigDataSE.2018.00281.
14. Wahi, V., Cherukuri, A. K., Srinivasan, K., & Jonnalagadda, A. (2021). CryptoCert: A Blockchain-Based Academic Credential System. In *Recent Trends in Blockchain for Information Systems Security and Privacy* (pp. 293–313). CRC Press.
15. Zhao, W., Jiang, C., Gao, H., Yang, S., & Luo, X. (2020). Blockchain-enabled cyber–physical systems: A review. *IEEE Internet of Things Journal*, *8*(6), 4023–4034.
16. Maleh, Y., Lakkineni, S., Tawalbeh, L. A., & AbdEl-Latif, A. A. (2022). Blockchain for Cyber-Physical Systems: Challenges and Applications. *Advances in Blockchain Technology for Cyber Physical Systems*, 11–59.
17. Khalil, A. A., Franco, J., Parvez, I., Uluagac, S., Shahriar, H., & Rahman, M. A. (2022, June). A literature review on blockchain-enabled security and operation of cyber-physical systems. In *2022 IEEE 46th Annual Computers, Software, and Applications Conference (COMPSAC)* (pp. 1774–1779). IEEE.
18. Latif, S. A., Wen, F. B. X., Iwendi, C., Li-li, F. W., Mohsin, S. M., Han, Z., & Band, S. S. (2022). AI-empowered, blockchain and SDN integrated security architecture for IoT network of cyber physical systems. *Computer Communications*, *181*, 274–283.
19. Pal, S., Dorri, A., & Jurdak, R. (2022). Blockchain for IoT access control: Recent trends and future research directions. *Journal of Network and Computer Applications*, 103,371.
20. Sinha, U., Hadi, A. A., Faika, T., & Kim, T. (2019, April). Blockchain-based communication and data security framework for IoT-enabled micro solar inverters. In *2019 IEEE CyberPELS (CyberPELS)* (pp. 1–5). IEEE.
21. Guo, H., & Yu, X. (2022). A Survey on Blockchain Technology and its security. *Blockchain: Research and Applications*, *3*(2), 100,067.
22. Narula, S., Jonnalagadda, A., & Cherukuri, A. K. (2021). A Dynamic Trust Model for Blockchain-Based Supply Chain Management Networks. In *Recent Trends in Blockchain for Information Systems Security and Privacy* (pp. 59–73). CRC Press.
23. Baniata, H., Anaqreh, A., & Kertesz, A. (2022). DONS: Dynamic Optimized Neighbor Selection for smart blockchain networks. *Future Generation Computer Systems*, *130*, 75–90.
24. Bashar, G. D., Holmes, J., & Dagher, G. G. (2022). ACCORD: A Scalable Multileader Consensus Protocol for Healthcare Blockchain. *IEEE Transactions on Information Forensics and Security*, *17*, 2990–3005.

Printed in the United States
by Baker & Taylor Publisher Services